THE SPATIAL DISTRIBUTION OF MICROBES IN THE ENVIRONMENT

The Spatial Distribution of Microbes in the Environment

Edited by

Rima B. Franklin
*Department of Biology, Virginia Commonwealth University,
Richmond, VA, U.S.A.*

and

Aaron L. Mills
*Laboratory of Microbial Ecology, University of Virginia,
Charlottesville, VA, U.S.A.*

A C.I.P. Catalogue record for this book is available from the Library of Congress.

ISBN 978-1-4020-6215-5 (HB)
ISBN 978-1-4020-6216-2 (e-book)

Published by Springer,
P.O. Box 17, 3300 AA Dordrecht, The Netherlands.

www.springer.com

Printed on acid-free paper

All Rights Reserved
© 2007 Springer
No part of this work may be reproduced, stored in a retrieval system, or transmitted
in any form or by any means, electronic, mechanical, photocopying, microfilming, recording
or otherwise, without written permission from the Publisher, with the exception
of any material supplied specifically for the purpose of being entered
and executed on a computer system, for exclusive use by the purchaser of the work.

Contents

Contributing Authors — vii

Preface — ix

Acknowledgements — xi

INTRODUCTION
RIMA B. FRANKLIN & AARON L. MILLS — 1

STATISTICAL ANALYSIS OF SPATIAL STRUCTURE
  IN MICROBIAL COMMUNITIES
RIMA B. FRANKLIN & AARON L. MILLS — 31

BACTERIAL INTERACTIONS AT THE MICROSCALE – LINKING
  HABITAT TO FUNCTION IN SOIL
NAOISE NUNAN, IAIN M. YOUNG, JOHN W. CRAWFORD,
  & KARL RITZ — 61

SPATIAL DISTRIBUTION OF BACTERIA AT THE MICROSCALE
  IN SOIL
ARNAUD DECHESNE, CÉLINE PALLUD,
  & GENEVIÈVE L. GRUNDMANN — 87

ANALYSIS OF SPATIAL PATTERNS OF RHIZOPLANE
  COLONIZATION
GUY R. KNUDSEN & LOUISE-MARIE DANDURAND — 109

MICROBIAL DISTRIBUTIONS AND THEIR POTENTIAL
   CONTROLLING FACTORS IN TERRESTRIAL SUBSURFACE
   ENVIRONMENTS
   R. MICHAEL LEHMAN   135

SPATIAL ORGANISATION OF SOIL FUNGI
   KARL RITZ   179

SPATIAL HETEROGENEITY OF PLANKTONIC
   MICROORGANISMS IN AQUATIC SYSTEMS
   BERNADETTE PINEL-ALLOUL & ANAS GHADOUANI   203

THE INTERRELATIONSHIP BETWEEN THE SPATIAL
   DISTRIBUTION OF MICROORGANISMS
   AND VEGETATION IN FOREST SOILS
   SHERRY J. MORRIS & WILLIAM J. DRESS   311

INDEX   331

Contributing Authors

John W. Crawford
SIMBIOS Center, University of Abertay Dundee, Bell Street, Dundee, DD1 1HG, UK

Louise-Marie Dandurand
Department of Plant, Soil, and Entomological Sciences, University of Idaho, Moscow, ID 83844, USA

Arnaud Deschesne
Institute of Environment and Resources, Technical University of Denmark, Kgs. Lyngby, Denmark

William J. Dress
Science Department, Robert Morris University, Moon Township, PA 15108, USA

Rima B. Franklin
Department of Biology, Virginia Commonwealth University, Richmond, VA 23284 USA

Anas Ghadouani
Aquatic Ecology and Ecosystem Studies, School of Environmental Systems Engineering, The University of Western Australia, Perth, Western Australia, Australia

Genevieve L. Grundmann
Laboratoire d'Ecologie Microbienne, Université de Lyon1, Villeurbanne cedex, France

Guy R. Knudsen
Department of Plant, Soil, and Entomological Sciences, University of Idaho, Moscow, ID 83844, USA

R. Michael Lehman
US Department of Agriculture, North Central Agricultural Research Laboratory, Brookings, SD 57006, USA

Aaron L. Mills
Laboratory of Microbial Ecology, University of Virginia, Charlottesville, VA 22904, USA

Sherry J. Morris
Biology Department, Bradley University, Peoria, IL 61625, USA

Naoise Nunan
CNRS, BioEMCo, Bât. EGER, Centre INRA-INAPG, 78850 Thiverval-Grignon, France

Céline Pallud
Soil and Environmental Biogeochemistry, Department of Geological and Environmental Sciences, Stanford University, Stanford, CA 94305, USA

Bernadette Pinel-Alloul
Groupe de Recherche Interuniversitaire en Limnologie (GRIL), Université de Montréal, Montréal, Québec, Canada

Karl Ritz
National Soil Resources Institute, Cranfield University, Silsoe, MK45 4DT, UK

Iain M. Young
SIMBIOS Center, University of Abertay Dundee, Bell Street, Dundee, DD1 1HG, UK

Preface

In my first microbiology class in 1968, Richard Wodzinki opened his first lecture with "Wodzinski's Laws of Bacteriology." Those laws were (1) Bacteria are very very small, (2) Bacteria are our friends, and (3) Bacteria always have the last word. These simple statements motivated a career of curiosity, and started me on a wild ride of discovery with my miniscule colleagues. The realization that an entity so tiny could mediate critical ecological processes observed across scales of kilometers begs for an explanation of how populations and communities are distributed within those large spaces. How big is a microbial community? Where does one stop and another start? Are there rules of organization of the communities into spatially discrete patches, and can those patches be correlated with observed processes and process rates?

Over the years I have added what I tell my classes are "Mills' Corrolaries to Wodzinski's Laws." With respect to the topic of this volume, the corollaries to the first law are: (1a) But there are a whole lot of them, and (1b) They can grow very very fast. Again, distribution in space and time is a central theme, and it has motivated much of my effort over the last 30 years.

Whether they were inspired by catchy (but meaningful and correct) phrases, or not, the theme has also captured the attention of some excellent microbiologists, and they have agreed to share their observations with us in this volume. There is much activity in this area at present, and this volume will no doubt be incomplete when it becomes available. Nevertheless, we all hope that it helps our current and future colleagues to appreciate the importance of spatial scale and spatial distribution in the understanding of just how the "very small" microbes manage to do so very much.

Aaron L. Mills, January 2007

Acknowledgements

The authors gratefully acknowledge the help of Jeff Chanat who worked out the derivation of the interaction distance for bacterial cells in Chapter 1. Additionally, Jeff's thoughtful consideration of spatial dynamics for individual microbial cells helped define the framework within which this volume is set.

Chapter 1

INTRODUCTION
The importance of microbial distribution in space and spatial scale to microbial ecology

Rima B. Franklin[1] and Aaron L. Mills[2]
[1]*Department of Biology, Virginia Commonwealth University, Richmond, VA 23284 USA;*
[2]*Laboratory of Microbial Ecology, University of Virginia, Charlottesville, VA 22904-4123 USA*

Abstract: Microorganisms are very small, and their individual effects are equally miniscule. Their effects on ecosystems, however, are felt at the landscape scale. To understand how their aggregate activities are arranged on these landscapes, microbes must be studied at a variety of scales, from the microscopic to the regional, and those scales must eventually be reconciled.

Keywords: bacteria, spatial distribution, community analysis, multiscale, interaction scale

1. INTRODUCTION

Because individuals can react only to their local environment, all ecological interactions are intrinsically spatial. It is the local environment that affects nutrient or food uptake, competition, or predation risk, and therefore indirectly controls growth, movement, reproduction, and survival. Conversely, individuals can only influence conditions in the area immediately adjacent to themselves. For microorganisms, the "local" environment is quite small because the microbes themselves are very small, e.g., individual bacteria are usually <2 μm long. While some aspects of the environment (e.g., temperature and pressure) may be the same at both the macro- and microscales, bulk measurements of other environmental variables (e.g., nutrient concentration and moisture content) may not accurately reflect the local conditions affecting an individual microorganism or a microbial assemblage. Nevertheless, most studies in microbial ecology are performed at larger spatial scales, using sample sizes that are determined by the observer's perception of environmental variability or by the particular analytical technique to be employed, and rarely consider the small spatial scale at which individuals may actually interact with one another and the environment. Furthermore, because the importance of microorganisms to ecosystems is in terms of their mediation of larger-scale ecological processes, investigations generally reflect the spatial extents thought to be appropriate for those processes.

Brock (1987) proposed that in order to appropriately study the ecology of microorganisms, i.e., bacteria, efforts must focus on scales important to individual bacterial cells. He maintained that this is the "only way we can really see organisms in their actual environments," and "without knowing where these organisms lived, (how) can we make any sense" out of ecological analyses. Given that a single milliliter (1 cm^3) of unpolluted surface water typically contains around 10^6 bacteria, and soils can contain up to 10^{10} microorganisms in a single gram, it is a daunting task to study microbial ecology at the level of the individual. Moreover, the small size of microorganisms and the hyperdiversity of microbial communities mean that Brock's challenge is still currently insurmountable for all but a few narrowly defined questions.

Not only are ecological approaches that rely on the identification and classification of individual members of a microbial community impractical, once completed, such a study would only be able to describe a very small geographic area within an ecosystem. The primary interest of most environmental scientists is in how microbial activity is manifested at larger spatial scales and how that activity controls nutrient cycling, decomposition, primary productivity, and other microbially mediated ecosystem functions at scales that are relevant to humans. The total capacity of microbial communities at these larger scales can be thought of as the sum of the activity of several "unit communities" of microorganisms (Swift, 1984), in separate microhabitats, whose individual activities are pooled into what scientists observe at the field or landscape scale. In order to understand well how these units fit together and how their combined activity contributes to overall ecosystem function, the small-scale spatial distribution of microorganisms and microbial communities must be understood better. In particular, the size and distribution of these unit communities (patches) must be known along with the biological implications of the interactions among neighboring patches, and how variations in the macro-environment may alter these relationships and influence the activity of the patches. Furthermore, we need to know how the distribution of those patches controls the distribution of observed activities on the landscape. Indeed, one might consider individual microbes to be elements of a mosaic that becomes a coherent

Introduction 3

image in terms of the distribution of functions within the landscape only when viewed from a much larger scale (Fig. 1-1).

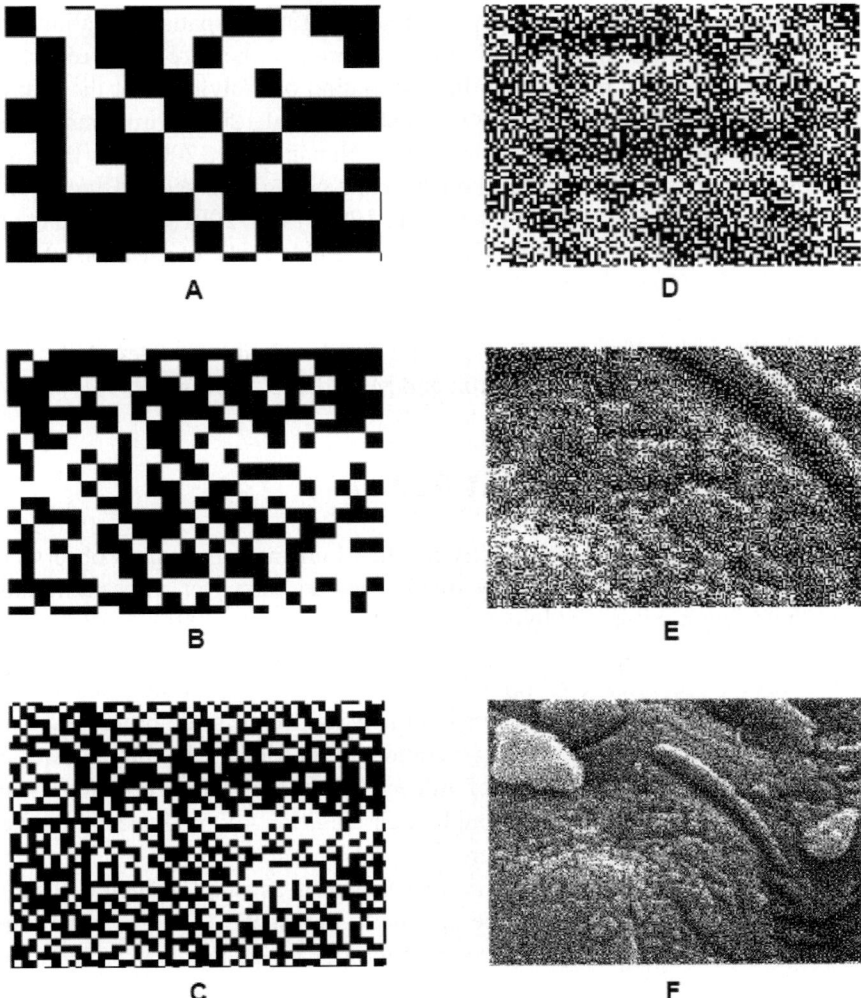

Figure 1-1. Illustrative example of scale resolution for microbial communities. At the scale of the organism (**A**), patchiness and inability to sample adequately make correlation of biological properties with physical and chemical observations impossible. When the microorganisms are "remotely sensed" as communities, the correlation between community and system properties becomes more coherent, until at some point (usually at the scale at which physical and chemical measurements can be extrapolated across reasonable characteristic lengths) the picture of community ecosystem interactions becomes coherent (**E–F**). (Photograph from NASA.)

Despite the importance of spatial variability in environmental microbiology, studies that specifically consider spatial scale when examining the distribution patterns of microorganisms are still rare. Most often, when microbial ecologists publish papers describing the "spatial variation" or "spatial distribution" of bacteria in the environment, they are either reporting the results of studies monitoring the distribution of individuals at the microscale (Dandurand et al., 1995, 1997; Dechesne et al., 2003; Grundmann and Normand, 2000; Jordan and Maier, 1999; Nunan et al., 2001), or they are discussing patterns observed at the landscape or regional scale (Blum et al., 2004; Cho and Tiedje, 2000; Finlay et al., 1996, 1999, 2001; Fulthorpe et al., 1998; GarciaPichel et al., 1996; Green et al., 2004; Teske et al., 2000). Much less research effort has been directed toward understanding spatial scale and variation in microbial communities at distances intermediate to those discussed above, i.e., from centimeters to a few hundreds of meters (see, e.g., Franklin et al., 2000, 2002; Franklin and Mills, 2003; Parkin, 1993).

2. THE INDIVIDUAL SCALE

Compared with the patches of activities that human investigators observe at the field scales, microbes are very small. They are, however, very numerous, and their communities are hyperdiverse (Roberts et al., 2004). Furthermore, turnover of microbes occurs very rapidly as compared with other organisms. As a result, the impact of microbes on the landscape is significant. Our miniscule friends have the last word in many landscape scale processes, e.g., the oxidation of carbon, the regeneration of nutrients, the breakdown of contaminants, the acceleration of mineral weathering, and microbes may even affect climate change through their addition of methane and other greenhouse gases to the atmosphere.

As individuals, the effect of microbes is nearly as small as their cell sizes and their impact on the landscape is equally miniscule. However, when the ecology of the microbes as opposed to microbes in ecology is examined, the small scale becomes the relevant scale. Competition, or cooperation, between individual cells and cell–cell communication occur at scales similar to that of the microbial cell itself. At scales of individual bacteria, diffusion processes dominate transfers of dissolved materials to and from the cells.

An "interaction distance" can be defined as the distance over which a cell can effect a change in the concentration of some solute. For example, one might ask at what distance from a cell will dissolved oxygen be noticeably reduced from the bulk solution concentration, if the cell is using oxygen as

Introduction

fast as it can be supplied to the cell by diffusion? In essence, this describes a steady-state situation for a perfectly absorbing sphere. (Note that cell growth and division are ignored for purposes of this illustration.) This situation would define the maximum interaction distance, as the concentration of the diffusing material should be zero at the cell, thus yielding the greatest concentration gradient possible for that compound. The problem has been addressed by several investigators with concerns about fluxes of materials to and from microbial cells (Berg, 1983; Schulz and Jørgensen, 2001) as opposed to the effect of the cells on the surrounding environment. The illustration presented here focuses on the distance over which a single cell would alter chemical concentrations in its local environment.

The flux of material to a spherical microbe that is using a limiting (nutrient) material as quickly as it can be delivered to the cell from the medium can be used to determine the concentration of the nutrient at a point some distance from the cell. That flux is represented by the advection–dispersion equation:

$$\frac{\partial C}{\partial t} = D \frac{\partial^2 C}{\partial x^2} - u \frac{\partial C}{\partial x} \qquad (1)$$

where C is the concentration of the dissolved substance (e.g., dissolved oxygen) ($M \cdot L^{-3}$), D is the diffusion coefficient ($L^2 \cdot T^{-1}$) (because dispersion in the static case is due entirely to molecular diffusion), u is the advection coefficient ($L \cdot T^{-1}$), and x is distance (L). In the static case, there is no water movement and therefore no advection (i.e., $u = 0$), and Eq. (1) simplifies to

$$\frac{\partial C}{\partial t} = D \frac{\partial^2 C}{\partial x^2} \qquad (2)$$

If steady-state conditions are assumed, then $\partial C/\partial t = 0$, and the equation further simplifies to

$$D \frac{\partial^2 C}{\partial x^2} = 0 \qquad (3)$$

which is a form of the Laplace equation. In turn, the Laplace equation can be rewritten for spherical coordinates as

$$D\left[\frac{\partial^2 C}{\partial r^2} + \frac{2}{r} \cdot \frac{\partial C}{\partial r}\right] = 0 \qquad (4)$$

where r = the radius of the sphere as defined earlier (O'Neil, 1991). The solution to this equation is lengthy but it has been published in a number of places (O'Neil, 1991). The solution does not require inclusion of D, as the steady-state assumption means that the perfect-sorption property of the microbial cell exactly counterbalances the delivery of the nutrient material to the cell. Specifically, the solution reduces to

$$C_r = C_\infty \left(1 - \frac{r_0}{r}\right) \qquad (5)$$

where C_r is the solution concentration at distance, r, from the center of the cell, r_0 is the radius of the cell. The specific boundary conditions are $C_{(r_0)} = 0$ (the concentration within the cell, and at its surface, is zero) and C_∞ = concentration in the bulk solution. Because $C_r \to C_\infty$ asymptotically as $r \to \infty$, the solution must be considered over some finite distance. Note that this solution is a rearrangement of the equation (derived differently) presented by Berg (1983) in his discussion of diffusion processes in biological systems. Using Eq. (5), the distance at which the solution concentration, C_r, has been reduced by 5% can be determined as:

$$C_r = 0.95 \cdot C_\infty = C_\infty \left(1 - \frac{r_0}{r}\right)$$

$$\frac{r_0}{r} = 1 - 0.95 \qquad (6)$$

$$r = \frac{r_0}{0.05}$$

Thus, at a distance from the center of the cell equal to 20 times the cell radius, the concentration of the diffusing nutrient will be 95% of the bulk solution concentration. Thus, for a 1 µm diameter cell, that distance is only 10 µm (Fig. 1-2). For smaller changes in concentration, the (interaction)

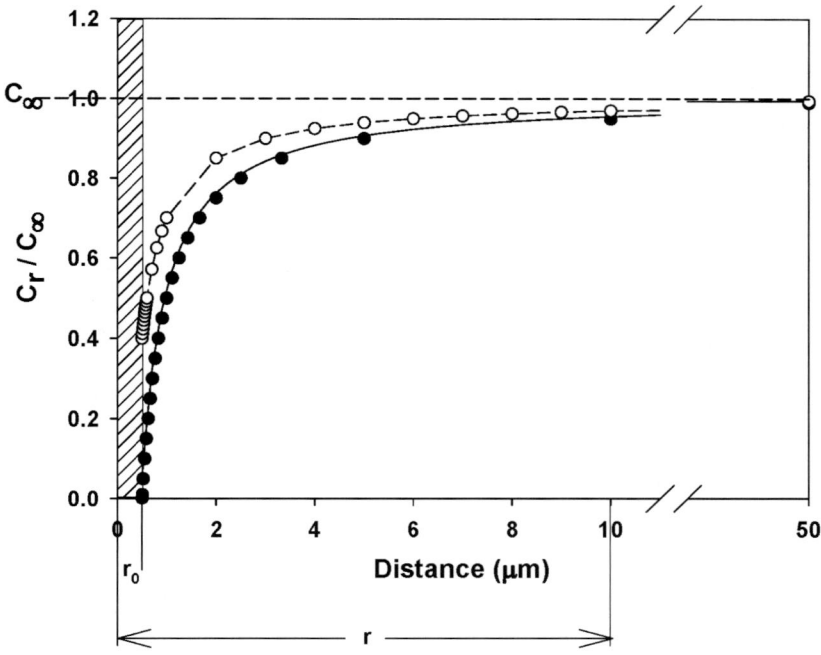

Figure 1-2. Change in concentration with distance from the center of a 1 μm diameter perfectly absorbing cell (filled circles and solid line). Hasured area represents the radius of the cell so that the concentration profile begins at the cell surface. Equation (5) can also be used to generate a profile for cells that are not perfectly absorbing. In such cases the steady-state assumption still holds, but applies to the uptake of the diffusing solute. The open circles and dotted line show the profile for a cell that absorbs the solute to maintain a steady-state concentration at the surface of 40% of the bulk solution concentration.

distance increases, such that for a 1% reduction in the bulk-solution concentration, the distance is 50 μm (Table 1-1).

Although this example considers the case of consumption, an analog for production of a compound could easily be constructed, as long as the concentration of the product produced by the cell and diffusing away from the cell is known at the cell surface (in these simple calculation, it has been assumed that the concentration on the surface equals the internal cellular concentration). Obviously, the interaction distance calculated here is affected by the presence of other cells separated by distances less than twice the interaction distance, and it should be obvious that more cells will substantially alter the concentration of solute in the vicinity of the group. This calculation does, however, demonstrate the scale at which the effects of

Table 1-1. Distance, under steady-state conditions, at which solute concentration is some specific fraction of the bulk solution concentration. The second and third columns refer to the distance from the center of the cell and the distance from the cell surface, respectively.

C_r/C_∞	R (μm)	$r - r_0$ (μm)
0.999	500	499.5
0.99	50	49.5
0.95	10.0	10.5
0.90	5.0	4.5
0.75	2.0	1.5
0.50	1.0	0.5
0.25	0.67	0.17
0.10	0.56	0.06

single cells may be felt, and demonstrates how important are the large numbers of organisms (Whitman et al., 1998) in effecting processes at the landscape scale.

3. METHODOLOGICAL LIMITATIONS ASSOCIATED WITH COMMUNITY ANALYSIS

When macroorganismal ecologists set out to investigate community organization and spatial variability, the studies usually involve identifying the individuals in an area and recording their locations, relative to one another. However, there are a number of attributes of microbial communities that limit the use of such an approach, and many methodological constraints have thus far hampered the ability to study microbial diversity. In particular, the small size of microorganisms means that they are difficult to visualize. Even with the aid of a microscope, the lack of morphological distinctiveness among types makes the visual classification of individuals into different taxonomic groups impossible. Moreover, the tremendous abundance of organisms in microbial communities means that the task of sorting them is overwhelming. Another difficulty is in developing and implementing sampling methods that preserve the spatial distribution of microorganisms within the native environmental matrix during sample collection and processing. In addition, the hyperdiversity of microbial communities means that the use of such an individual-based approach is impractical in many cases; for example, microbial communities in soil have been shown to contain up to 10,000 types (species) in a single 30 g sample (Torsvik et al., 1996, 1998), and it has been proposed that the oceans may contain 2×10^6 types (Curtis et al., 2002).

Culture-based studies provide the framework from which microbial ecologists derive much of their current understanding of microbial interactions and community dynamics. However, it is well documented that cultural techniques are both selective and unrepresentative of the total microbial community. Some studies propose that <1% of microorganisms in the environment can be cultured in the laboratory (Holben, 1994), and that between only 1% and 5% of the microorganisms on earth have even been identified and named (Kennedy and Gewin, 1997), though there is little specific evidence to support the accuracy of these estimates. Since the application of molecular biological methods to microbial ecology in the mid-1980s, many new, previously uncultivated, microorganisms have been identified. Whole groups of organisms, known only from molecular sequences, are now believed to be quantitatively significant in many environments. In particular, the use of 16S-rRNA gene sequences has brought about a new era of microbial systematics, and it has become quite popular to survey microbial community diversity using polymerase chain reaction (PCR) and 16S-rRNA/DNA-based methods. The 16S-rRNA genes contain highly conserved sequence domains interspersed with more variable regions, and comparative analysis of rRNA sequences can identify "signature sequence motifs" that are targets for evolutionary-based identification (Theron and Cloete, 2000). The use of PCR amplification of 16S-rRNA genes and subsequent cloning has allowed "identification" of a number of new "species"; however, a tremendous portion of the microbiota remains unexplored.

4. "WHOLE-COMMUNITY" APPROACHES TO MICROBIAL COMMUNITY ANALYSIS

In order to comprehend the full extent of the relationships within a microbial community, and between a community and its surroundings, researchers must be able to evaluate attributes for the assemblage without relying on microbial growth and culture-based techniques for detection or identification of individuals. This need has led to the development of several approaches that use "whole-community samples" for analysis. The basic premise behind the approach is that all of the organisms in a sample are analyzed as a unit, and relative comparisons are made between communities based on overall characteristics manifested by the different mixtures of organisms. In order to monitor structural differences in microbial communities, most of the research has focused on the analysis of whole-community DNA samples, and several new molecular genetic approaches have recently emerged (to be discussed). Similarly, the lipid content of microbial cells (e.g., phospholipid ester-linked fatty acids, PLFA (Tunlid and White, 1990) and fatty acid methyl esters, FAME (Kennedy, 1994)) may be used to monitor community composition (Laczko et al., 1997; Zogg et al., 1997). However, many subsets of the microbial community respond to stressful conditions in their microenvironment by shifting lipid composition (Kieft et al., 1997; White et al., 1997), confounding the interpretation of the phospholipid patterns and signatures. Another commonly used whole-community approach is community-level physiological profiling (CLPP), where patterns in carbon substrate utilization are compared for different communities (Garland and Mills, 1991; Garland et al., 2003).

5. MOLECULAR GENETIC TECHNIQUES FOR COMPARING COMMUNITY STRUCTURE

Modern molecular approaches have been extended from examination of populations to the community level as well. For more detail on the use and development of these methods (see Dahllof, 2002; Johnsen et al., 2001; Kozdroj and van Elsas, 2001; Theron and Cloete, 2000; Torsvik et al., 2002). As discussed above, most of the recent research using the whole-community approach in microbial ecology has focused on the analysis of the combined genetic material (either DNA or RNA) from a community sample. However, many of these techniques (e.g., DNA hybridization (Griffiths et al., 1996; Lee and Fuhrman, 1990, 1991), percent $G + C$ content (Holben and Harris, 1995), or DNA reassociation kinetics (C_0t curves) (Torsvik et al., 1990, 1994)) require a fairly large environmental sample in order to obtain enough genetic material for analysis. The need for large quantities of DNA often means that sample collection can be very time-consuming (e.g., filtering large volumes of water), and that samples may need to be gathered over a relatively large area (e.g., several grams of soil), making it impossible to examine small-scale spatial differences in community structure. Moreover, the analyses themselves are very time-consuming, which further limits the feasibility of large and comprehensive studies of microbial community dynamics. Technological development over the last several years has helped reduce this problem, and the introduction of PCR-based methods now permits more rapid analysis using smaller sample sizes.

Recently, the use of PCR-based "DNA fingerprinting" for the analysis of microbial communities has become very popular. Commonly used PCR-based DNA fingerprinting techniques include: denaturing gradient gel electrophoresis (DGGE; Muyzer, 1999; Muyzer et al., 1993), amplified ribosomal DNA restriction analysis (ARDRA; Massol-Deya et al., 1995), terminal restriction fragment length polymorphism (T-RFLP; Liu et al., 1997; Marsh, 1999), randomly amplified polymorphic DNA (RAPD; Franklin et al., 1999; Wikstrom et al., 1999, 2000; Williams et al., 1993), and amplified fragment length polymorphism (AFLP; Franklin et al., 2001; Franklin and Mills, 2003; Zabeau and Vos, 1993). These methods can be broadly categorized into two groups: (i) approaches where specific primers, designed to amplify certain known genes or sections of a genome, are used to direct the PCR (e.g., DGGE, ARDRA, and T-RFLP), or (ii) approaches where the PCR amplification is based on the distribution of random sequences throughout the DNA sample (e.g., RAPD and AFLP). When specific primers are used to study microbial communities, the 16S rRNA gene is most often considered.

There are several additional molecular biological techniques that have recently emerged for the study of microbial communities, and should be briefly mentioned. Specifically, the novel application of nucleic acid array technology to microbial community analysis may provide an efficient means to assess the presence of organisms or the expression of genes in communities. However, the performance of microarray hybridization in environmental studies has yet to be carefully evaluated, and a number of technological challenges need to be solved before this method can reliably inventory complex samples (Zhou, 2003; Zhou and Thompson, 2002). Another important technique that is being refined is fluorescence *in situ* hybridization (FISH) with rRNA-targeted probes in combination with microscopy or flow cytometry (Handelsman and Smalla, 2003). FISH has the unique potential to study the composition of bacterial communities *in situ* and may also be used to provide new ways to link structure and function in microbial ecology studies (Wagner et al., 2003). While these techniques present a tremendous opportunity to examine microbial community dynamics in a wide variety of systems, they are nevertheless confined to "accessible" and previously encountered organisms. In order to apply either FISH or DNA microarrays, some portion of the genetic sequence of the individuals of interest must be available.

6. SPATIAL HETEROGENEITY IN MICROBIAL SYSTEMS

In natural systems, environmental heterogeneity arises as a result of the interaction of a hierarchical series of interrelated variables that fluctuate at many different spatial and temporal scales. These physical, chemical, and biological variables may combine to influence the abundance, diversity, and activity of microorganisms at many different spatial scales. These properties do not vary independently; rather, the general perception is that any such variable measured at a certain point in space and time is the outcome of several processes, all of which are spatially variable. It is thought that the relative role of different environmental forces may vary across scales and among ecological systems, and one of the major challenges for the discipline of ecology is to measure the relative strengths of these factors in natural ecosystems, examine the interactions among them, and combine this information in an effort to explain the patterns of organism distribution, abundance, and function.

Studies of spatial organization in microbial systems may be broadly categorized into four scales of interest: microscale, plot scale, field or landscape scale, and regional scale (Parkin, 1993). Within each of these scales/categories, multiple levels of organization may exist. Often, the hierarchical levels are nested so that high-level units consist of aggregations of lower-level units, though the boundaries between levels are not usually visible. Many of the studies that have considered spatial variability in microbial ecology focus on a single scale, though it has been suggested that, because of the hierarchical nature of spatial variability, multiscale analyses of spatial variability are needed in order to fully represent the complexity of natural systems (Benedetti-Cecchi, 2001).

Because spatial variability can manifest at many different scales, the patterns one observes depend greatly on the scale of observation (Avois, et al., 2000; Levin, 1992). In sampling theory, spatial scale is defined by several characteristic properties: grain size, sampling interval, and extent (Legendre and Legendre, 1998). *Grain size* is the size of the elementary sampling units (e.g., the volume of sample), and defines the resolution of the study (Schneider, 1994). *Sampling interval* is the average distance between sampling units, and the *extent* is the total area included in the study. Depending on the ecological question being addressed, and what is already known about the scale of the process of interest, the dimensions of these components vary. For a given sampling design, no structure can be detected that is smaller than the grain size or larger than the extent of the study. In this way, the sampling design defines the observational window for spatial pattern analysis (Legendre and Legendre, 1998). These concepts will be further discussed in Chapter 2.

6.1. "Local" controls on the spatial distribution of microorganisms

To study adequately the spatial distribution of microbial populations, guilds, and communities at larger spatial scales (centimeters to plot scale), it

Introduction

is important to understand the factors that control the distribution of individuals and populations at the microscale. Microorganisms are generally said to inhabit "microhabitats," but this term is poorly defined, and the meaning differs for different types of organisms (e.g., fungi versus bacteria) and in different systems (e.g., soil versus aquatic). For example, bacterial development in soils is probably influenced by conditions within only a few microns (see earlier analysis), while a fungus has the advantage of being able to extend beyond its immediate surroundings, using its hyphae in much the same way as a plant uses its root system (Harris, 1994). For this reason, a fungus may experience a degree of averaging of soil conditions, and is not restricted to as small of a microhabitat as a bacterium (Parkin, 1993). In aquatic systems, the more diffuse nature of the environmental matrix may mean that microbes are impacted by environmental variability existing at a broader spatial scale, compared to a more highly structured soil matrix. Like the interaction distance mentioned earlier, the size of a microhabitat may be defined by the physical and chemical environment directly adjacent to the microbial cell or colony (Parkin, 1993), and, in this regard, is not a fixed unit. Its size is operationally dependent upon the specific process or microorganism under study, and the nature of the environmental matrix within which the organism resides.

The distance between, and "reachability" of, different microhabitats is an important issue that may help control the spatial distribution of microorganisms and microbial community composition. For an organism to be present in a system, it must either evolve there or be transported from another site, so the spatial continuity of microhabitats may help control the distribution patterns of microorganisms at many different scales. Moreover, spatial continuity and transport of microorganisms may influence the response of a microbial community to a disturbance. In particular, the frequency or extent to which a disturbed system is inoculated with new organisms from "nearby" or "connected" communities could have a strong influence on system recovery. Similarly, this type of information could be useful for predicting the distribution and persistence of a nonnative or invasive microorganism in an ecosystem. Practical applications include determining the distribution of plant pathogens in an agricultural system or judging the success of an intentionally introduced organism placed in a contaminated environment for the purposes of bioremediation. At this point, it is unclear what the relationship is between spatial heterogeneity and colonization success for these types of organisms. Colonization success may be greater in a heterogeneous system, because a spatially heterogeneous environment is more likely to include a microenvironment that is hospitable to the new organism. However, in a diverse and spatially heterogeneous habitat, the number of occurrences of this ideal microenvironment may be small, in which case spatial heterogeneity may make it more difficult for an invasive/introduced organism to achieve dominance and thus have a major influence on the ecosystem.

Bacterial colonization can occur due to active movement of an organism to a new site, or through passive transport by other agents (e.g., water or animals) (Harris, 1994). Though little is known about the importance of bacterial motility on colonization, it is generally assumed that active movement is relatively small compared to other dispersive processes. More research is necessary to investigate the relative importance of these two transport pathways, and the spatial extend over which each may be important. A central question that follows from an investigation of transport of microorganisms is to what extent can an isolated cell survive and successfully colonize a given location. After a cell arrives at a colonization site, it could exist in a resting stage for some period of time, it could die, or it could grow and reproduce, potentially providing a seed for further colonization of another location. Which of these scenarios takes place is likely determined by resource availability and by interactions with other community members. The ability to predict colonization efficiency then requires increased research into several questions of fundamental ecological importance, including habitat suitability, invasibility of existing communities, and interactions among community members (e.g., competition, predation, and synergistic or mutualistic relationships). As scientists learn more about these phenomena, especially at scales relevant to individual microbes, they will become better able to predict the persistence of unique organisms in a new habitat. This type of knowledge about the microscale variation in microbial communities is necessary for understanding the mechanisms behind microbial community formation and maintenance, and for evaluating the stability and resilience of these communities.

In addition to the topics discussed above, spatial heterogeneity may help control community composition and diversity by altering biological interactions among organisms and through habitat partitioning. In particular, it is thought that spatial heterogeneity plays an important role in determining diversity, as spatial structure in microenvironments can increase niche complexity. This increased niche complexity may create favorable habitat space for many types of organisms, with very different physiological requirements, within a rather small area. Similarly, if the habitat is subdivided into many separate pockets of resources, populations may avoid competition by physical isolation, and this is thought to contribute to the tremendous microbial diversity seen in soils (Zhou et al., 2002). Habitat partitioning can also influence predation, and thus exert a strong indirect control on community composition.

Introduction

The plausibility of spatial structure (e.g., patchiness) at small scales in microbial systems has been intensively debated in the past (Azam and Ammerman, 1984; Fenchel, 1984; Lehman and Scavia, 1982; Levin and Segal, 1976; Sieburth, 1984); however, a great deal of evidence is now available to demonstrate that this type of microscale patchiness is widespread (Blackburn and Fenchel, 1999; Blackburn et al., 1998; Duarte and Vaqué, 1992; Grundmann and Debouzie, 2000; Krembs et al., 1998; Long and Azam, 2001; Nunan et al., 2002). A prerequisite for such an analysis is the conservation of the native state of the environmental sample such that the *in situ* distribution of the inhabitants and the environmental components are preserved. One strategy for investigating microorganisms within their natural spatial distribution is by embedding the samples in a material such as agarose (Macnaughton et al., 1996), paraffin wax (Licht et al., 1996; Poulsen et al., 1994; Rothemund et al., 1996), and hard setting resins (Kawaguchi and Decho, 2002; Manz et al., 2000; Nunan et al., 2001) prior to analysis. In aquatic systems, a spatial information preservation (SIP) method has been applied, which is based on rapidly freezing small samples of water as a means of maintaining the three-dimensional particle distribution for microscopic analysis (Krembs et al., 1998).

In soils, one-dimensional microscale data has been collected along soil transects (Grundmann and Debouzie, 2000) and along plant roots (Dandurand et al., 1995), and nonrandom spatial patterns of bacteria have been identified. More recently, efforts have focused on analyzing the two- (Dandurand et al., 1995; Nunan et al., 2001) and three-dimensional (Dechesne et al., 2003; Grundmann et al., 2001) distribution of microorganisms by integrating the analysis of multiple "microsamples." The results indicate that the microhabitat distributions in soil probably involve an array of colonized patch sizes, and the location of different nutrient sources is thought to be one of the major factors determining the distribution of bacteria in soil (Dechesne et al., 2003). For example, the distribution of particulate carbon may have a strong influence on the small-scale variations in bacteria abundance (Parkin, 1987; Wachinger et al., 2000). However, the situation is more complex, and less well understood, for soluble substrates (Dechesne et al., 2003). In aquatic systems, most of the previous work on very small-scale patchiness has been based on "cluster" hypotheses, including the proposal that bacteria actively congregate around phytoplankton cells (i.e., the phycosphere concept (Azam and Ammerman, 1984; Bell and Mitchell, 1972)) or particulate organic matter (Long and Azam, 2001) to enhance their exposure to growth substrates. For example, direct manipulation of water samples via the addition of algal detritus has been shown to stimulate the formation of nanoscale patches of lake bacterioplankton (Krembs et al., 1998). In order to determine the biological and environmental significance of this type of patchiness, it will be necessary to determine how common the phenomenon is in space and time and in different environments. If patchiness at these scales is widespread, as many investigators now believe, it may mean that rate processes that are concentration dependent are being miscalculated (Krembs et al., 1998).

6.2. Variability at larger spatial scales

In general, the grain size used for collecting environmental samples of microbial communities is too large to permit analysis of the location or activity of individual organisms, and most of the work looking at microbial community spatial variability examines larger scales. Studies in agricultural soils have demonstrated that significant spatial heterogeneity may exist for microbiological processes (Bending et al., 2001; Grundmann and Debouzie, 2000), community structure (Balser and Firestone, 1996; Cavigelli et al., 1995; Franklin and Mills, 2003), and abundance (Nunan et al., 2001; Wollum and Cassel, 1984); patch size estimates range widely from as small as 2 mm (Grundmann and Debouzie, 2000) to nearly 10 m (Franklin and Mills, 2003). Similar studies have been conducted in grassland and forest soils (Both et al., 1992; Kuperman et al., 1998; Morris and Boerner, 1999; Ritz et al., 2004; Robertson and Tiedje, 1988; Saetre and Bååth, 2000), in a shallow coastal aquifer (Franklin et al., 2000), and in the open ocean (Duarte and Vaqué, 1992; Mackas, 1984). For salt marsh and marine sediments, variation has been examined at small scales (i.e., <1 m^2; Berardesco et al., 1998; Danovaro et al., 2001; Franklin et al., 2002; Scala and Kerkhof, 2000), at intermediate (<150 m; Moran et al., 1987; Scala and Kerkhof, 2000), and at larger distances (kilometer; Green et al., 2004; Scala and Kerkhof, 2000). In general, all of these studies concluded that microbial communities are organized at a variety of spatial scales, which likely reflect the scales of heterogeneity in the distribution of physical and chemical properties for the environment under investigation. Most of this work has considered more general community properties (e.g., total abundance, biomass, or activity (Duarte and Vaqué, 1992; Moran et al., 1987; Morris and Boerner, 1999)), while relatively few studies have examined the distribution of microbial community structure (Balser and Firestone, 1996; Both et al., 1992; Mackas, 1984; Saetre and Bååth, 2000).

6.3. Hierarchical scales of organization

Given that environmental factors do not necessarily operate independently, or at distinct spatial scales, studying microbial systems using a single analytical scale cannot provide a complete understanding of community dynamics. Multiscale comparisons, in which patterns are analyzed at several different spatial scales, may be more useful when trying to identify the factors that control community development. Conclusions about the organization of microbial communities, the effect of disturbance, or the roles of various limiting factors are likely to differ at different spatial scales (Wiens et al., 1986). Moreover, the characterization of microbial communities at several different scales may help explain paradoxes that arise when different investigators, studying similar communities but at different scales, arrive at different conclusions about the factors that structure those communities. These disagreements may reflect viewpoints of different scales, and not differences in the way communities are organized (Rahel, 1990).

Recently, scientists have begun to focus on multiscale comparisons, and have found evidence for nested scales of spatial structure in microbial communities (Ettema and Wardle, 2002; Robertson and Gross, 1994; Saetre and Bååth, 2000; Stenger et al., 2002). For example, Nunan et al. (2002) studied the spatial distribution of soil bacteria at three different scales, ranging from micrometers to meters, and found that the distribution of individual bacterial cells was organized at two scales in the subsoil, and at a single scale in the topsoil. Franklin and Mills (2003), examined spatial structure in a detailed examination of an agricultural (top) soil and found definite multiscale structure (Fig. 1-3). Arrangements of communities that could produce such a structure have not been documented, but one possibility that could explain the nested structure would be "clusters of clusters" (Fig. 1-4). Such an arrangement suggests that patch distribution might be described by some fractal geometry, although no specific studies on such a description have yet been reported.

Studies conducted in agricultural and shrub-steppe ecosystems suggest that microbial biomass and activity may be spatially dependent at scales <1 m, nested within a larger scale related to variations at the landscape level (Robertson et al., 1997; Ronimus et al., 1997; Smith et al., 1994). The presence of nested scales of variation suggests that the various factors regulating the development of microbial communities in the soil ecosystems may operate at different scales (Robertson and Gross, 1994), and a simultaneous analysis of the multiscale spatial variability of microbial community structure and the associated microenvironment could help identify these factors and determine their relative influence.

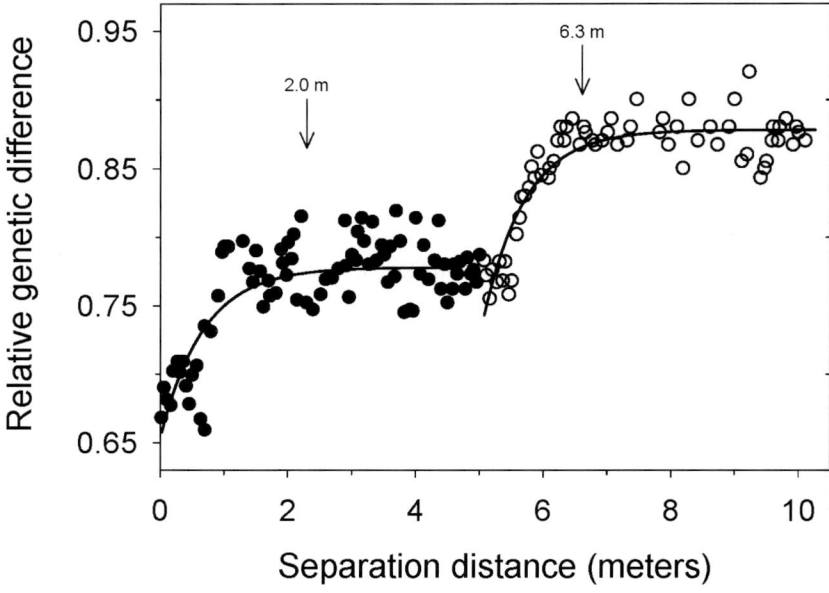

Figure 1-3. Geostatistical analysis of similarity data for soil microbial communities in an eastern Virginia wheat field showing nested spatial structure. All the points are from the same data set, but the models were developed by splitting the set into two parts. The soild circles represent the data used to make the shorter range "variogram" and the open circles represent the longer range (range values are presented with the appropriate data).

7.   ECOLOGICAL CONTEXT AND MOTIVATION

Increased research into the spatial distribution of microorganisms and microbial communities has many ecological and environmental applications. For example, scientists are often interested in understanding issues of scale (spatial and temporal), in part, because of a desire to make predictions about ecosystem processes using information gathered at a smaller scale, or vice versa, i.e., upscaling and downscaling (Stein et al., 2001). This is a pressing issue because calculations of the effects of human activities on ecosystems often need to be made at spatial scales that far exceed the scale at which

Introduction

Figure 1-4. Arrangement of objects into "clusters of clusters" that could produce a semivariogram showing nested spatial structure.

measurements are made (Schneider, 1994). For example, rates of nutrient processing through an ecosystem are generally calculated based on information measured for a few sampling locations. However, a direct scale-up of these rates is not appropriate unless one assumes that the factors that influence nutrient cycling are distributed homogenously across the landscape and over time. The reliability of such an estimate may be greatly increased by incorporating some information about the spatial and temporal distribution of the process of interest into any models or calculations (Schneider, 1998). Moreover, the increase in spatial scale may result in new interactions and relationships, and a change in system organization, so that a change in the level at which one wishes to understand or quantify a process cannot necessarily be addressed by simply changing scale (O'Neill and King, 1998).

Another important issue for scientists designing and planning field experiments is resolution, and the need to make decisions about the appropriate scale for collecting data. In some cases, information collected at finer scales may be too noisy, and may obscure the detection of large-scale relationships. Alternately, the ability to detect relationships between large-scale processes may be inhibited by the loss of fine-scale information; for example, Hewitt et al. (1998) detected fewer relationships between environmental variables and communities using coarser resolution in a comparative study. Ultimately, the scale used for an analysis must be determined based on the processes of interest, and different scales may be appropriate for different ecological questions. This information is necessary for scientists to design effective sampling schemes for the environment, and changing the number, location, and size of samples collected may influence one's results. For instance, Parkin et al. (1987) studied the effect of sample size on determination of soil denitrification rates, and found that smaller samples provided significantly lower estimates of the mean denitrification rate than did larger samples. Similarly, Kang and Mills (2006) demonstrated that the community structure recovered using molecular approaches (DGGE) varied in the same sample depending on the size of the sample collected from grassland soil.

While many ecological theories and models acknowledge that elements that are close to one another in space or time are more likely to be influenced by the same generating processes, the same energy inputs, or a similar physical environment (Legendre and Fortin, 1989), the classical statistical procedures employed to analyze these phenomena assume independence of observations. Statisticians generally count one degree of freedom for each independent observation, which allows them to choose an appropriate statistical distribution for testing; the lack of independence that arises from the presence of spatial autocorrelation makes it difficult (in many cases, impossible) to accurately determine the number of degrees of freedom and correctly perform tests such as correlation, regression, or analysis of variance. Positive autocorrelation reduces within-group variability, artificially increasing the amount of among-group variance, and often leads to the determination that differences among groups are significant, when in fact they are not (Legendre et al., 1990). Violations of the assumption of independence and inappropriate application of these statistical procedures to spatially autocorrelated data may lead to incorrect conclusions. Therefore, understanding the type and extent of spatial variation in microbial systems is necessary in order to perform appropriate statistical analyses and to design reliable sampling schemes for the environment.

Increased knowledge of the spatial distribution of microorganisms and microbial communities in the environment also has many environmental applications, e.g., determining the impact of various land management practices on microbial communities or estimating biodegradation rates.

In particular, agricultural land management practices have been shown to reduce heterogeneity in soil characteristics, which may influence the microbial community and nutrient cycling. Webster et al. (2002) found a decrease in the diversity of certain microbial populations in response to a fertilizer application, and Parry et al. (1999) correlated differences in denitrification rates between pasture and cropped soil with differences in pore space structure in soil clods. Biodegradation rates may also be strongly influenced strongly by the spatial heterogeneity of environmental conditions and microbial distributions at many different scales. At the microscale, the placement of certain organisms, relative to transport pathways through the soil matrix or substrate availability, could be particularly important. For example, the spatial distribution of bacteria in soil, relative to pore networks or organic matter deposits, is thought to influence the degradation of groundwater pollutants (Nunan et al., 2001). At larger scales, high spatial variability is a key problem when quantifying methane emissions from soils at both the meter scale (Adrian et al., 1994; Wachinger et al., 2000) and the landscape scale (Valentine et al., 1994), and it is thought that different processes are responsible for this variation at different scales. Wachinger et al. (2000) demonstrated that CH_4 production was strongly correlated to the presence of fresh organic carbon at the centimeter- to meter-scale, while hydrologic regime was important at larger scales.

Given the necessary scale at which scientists are forced to view the interactions of microbes with their surroundings, microbial ecology has some analogy to remote sensing: The forest can be viewed reasonably well, but the trees and understory constitute detail that is below the limit of useful resolution. While it may be possible to determine important properties of the relationships of the individual microbial cells with their surroundings (perhaps including their relationships with other cells), the scale is not relevant to that which is of interest to an individual attempting to understand the bulk behavior of the entire system. As one moves to larger scales, detailed information on the individual cells is lost, and the patterns displayed by the community begin to resemble the bulk patterns (Fig. 1-1). While it is necessary to examine the populations within communities to determine the role of these organisms in providing requisite functions to the assemblage, but the target scale of integration must be the community.

When considered from the perspective of theoretical ecology – with reference to species assembly "rules" and hypotheses that predict function and stability of ecosystems – the issue of scale in the analysis of microbial communities is particularly crucial. One point that is so fundamental, but often overlooked, is the question of how to define the limits of a microbial community. Most often, the boundaries used to define a "community" are utilitarian and dictated by the required sample size and the researcher's perception of environmental variability. The size of the sample then determines the number of bacteria that will be analyzed and evaluated. Generally, the "mere fact that organisms coexist at the moment of sampling is taken as evidence that they are part of a community… (while) the 'community' may be no more than a disparate collection of organisms that happen to found within a sample of a particular size" (Harris, 1994). How one defines a community and the limits of these interactions may determine the extent to which ecological theories (initially developed for application in communities of macroorganisms) may be adapted to apply to microbial systems. Continuing the forest analogy – the question becomes whether these theories would be most appropriately applied to a group of individuals of a certain abundance or mass (e.g., 10,000 trees), a unit of area with particular dimensions (e.g., 100 km^2 plot), or a seemingly discrete functioning assemblage (e.g., an island or forest).

In the environment, it is difficult to delimit microbial communities and separate the cells and species that are members of a community from those that are not. Functionally defined microbial communities exist in continuity with one another, and the distinctions between them blur. These "communities" are connected by fluxes of organisms, materials, energy, and information. The localized activity combines to mediate processes that are important at the field and landscape scale, while activity at the ecosystem level results from the further combined interaction of these larger-scale communities. One's perspective within this hierarchy can be crucial to understanding the pattern or process of interest, and studies that consider multiple spatial scales when studying environment–community interactions are particularly valuable. It is important to point out that most systems cannot be neatly subdivided into hierarchical scales of organization, and there is no single natural scale on which ecological phenomena should be studied (Levin, 1992). The description of the system will vary with the choice of scales, and, rather than trying to determine the "correct" scale, ecologists must try to understand how the system description changes across scales (Levin, 1992). Moreover, learning to scale up from individual measures of environmental samples to processes at the field and landscape scale requires an understanding of how information is transferred from fine scales to broadscales – this requires that scientists learn to aggregate and simplify, retaining essential information without getting bogged down in unnecessary detail (Levin, 1992).

8. INTRODUCTION TO THIS VOLUME

The purpose of this volume is to review the newly emerging study of spatial arrangement of microbial populations and communities. The collected work examines the distribution of microbes (bacteria, fungi, and viruses) at a variety of spatial scales and in several different environments. Given that microorganisms play such a fundamental role in establishing the biogeochemical cycles necessary for the long-term functioning of ecosystems, one major goal of this project is to summarize and relate patterns of microbial community structure and composition with patterns of microbial activity. In addition, specific attention is paid to the issue of how to integrate information across spatial scales, which is necessary in order to understand and predict how these tiny organisms can have such a profound effect on landscape and ecosystem-level processes. Progress in this area requires the quantitative analysis of large data sets and rigorous description of complex patterns. As the mathematics required for such analyses are generally not a part of the training of most microbiologists, a chapter is included presenting an overview of the more common techniques used in analysis of spatial patterns. Readers should be able to use this chapter as a reference for the statistical methods that they encounter during the reading of any remaining chapters. In addition to presentation, discussion, and integration of results of scale investigations, methods of study will be presented to the readers that will allow for explicit consideration of spatial scale as an essential and integral part of any consideration of miniscule organisms that, in aggregate, mediate landscape-scale processes.

REFERENCES

Adrian, N. R., J. A. Robinson, and J. M. Suflita, 1994, Spatial variability in biodegradation rates as evidenced by methane production from an aquifer, *Appl. Environ. Microbiol.* **60**:3632–3639.

Avois, C., P. Legendre, S. Masson, and B. Pinel-Alloul, 2000, Is the sampling strategy interfering with the study of spatial variability of zooplankton communities, *Can. J. Fish. Aquat. Sci.* **57**:1940–1956.

Azam, F., and J. W. Ammerman, 1984, Cycling of organic matter by bacterioplankton in pelagic marine ecosystems: microenvironmental considerations, in: *Flows of Energy and Materials in Marine Ecosystems*, M. J. R. Fasham, ed., Plenum Press, New York, pp. 345–360.

Balser, T. C., and M. K. Firestone, 1996, Sources of variability in BIOLOG assays of soil microbial communities: spatial and analytical. Presented at the Conference on Substrate Use for Characterization of Microbial Communities in Terrestrial Ecosystems, Innsbruck, Austria.

Bell, W., and R. Mitchell, 1972, Chemotactic and growth responses of marine bacteria to algal extracellular products, *Biol. Bull.* **143**:265–277.

Bending, G. D., E. Shaw, and A. Walker, 2001, Spatial heterogeneity in the metabolism and dynamics of isoproturon degrading microbial communities in soil, *Biol. Fert. Soil.* **33**:484–489.

Benedetti-Cecchi, L., 2001, Variability in abundance of algae and invertebrates at different spatial scales on rocky sea shores, *Mar. Ecol. Prog. Ser.* **215**:79–92.

Berardesco, G., S. Dyhrman, E. Gallagher, and M. P. Shiaris, 1998, Spatial and temporal variation of phenanthrene-degrading bacteria in intertidal sediments, *Appl. Environ. Microbiol.* **64**:2560–2565.

Berg, H. C., 1983, *Random Walks in Biology*. Princeton University Press, Princeton, NJ.

Blackburn, N., and T. Fenchel, 1999, Influence of bacteria, diffusion and shear on micro-scale nutrient patches, and implications for bacterial chemotaxis, *Mar. Ecol. Prog. Ser.* **189**:1–7.

Blackburn, N., T. Fenchel, and J. G. Mitchell, 1998, Microscale nutrient patches in planktonic habitats shown by chemotactic bacteria, *Science* **282**:2254–2256.

Blum, L. K., M. S. Roberts, J. L. Garland, and A. L. Mills, 2004, Microbial communities among the dominant high marsh plants and associated sediments of the United States east coast, *Microb. Ecol.* **48**:375–383.

Both, G. J., S. Gerards, and H. J. Laanbroek, 1992, Temporal and spatial variation in the nitrite-oxidizing bacterial community of a grassland soil, *FEMS Microbiol. Ecol.* **101**: 99–112.

Brock, T. D., 1987, The study of microorganisms *in situ*: progress and problems, in: *Ecology of Microbial Communities*, M. Fletcher, T. R. G. Gray, and J. G. Jones, eds., Cambridge University Press, Cambridge, UK, pp. 1–17.

Cavigelli, M. A., G. P. Robertson, and M. J. Klug, 1995, Fatty-acid methyl-ester (FAME) profiles as measures of soil microbial community structure, *Plant Soil.* **170**:99–113.

Cho, J. C., and J. M. Tiedje, 2000, Biogeography and degree of endenicity of fluorescent *Pseudomonas* strains in soil, *Appl. Environ. Microbiol.* **66**:5448–5456.

Curtis, T. P., W. T. Sloan, and J. W. Scannel, 2002, Estimating prokaryotic diversity and its limits, *Proc. Natl. Acad. Sci. USA* **99**:10491–10499.

Dahllof, I., 2002, Molecular community analysis of microbial diversity, *Curr. Opin. Biotechnol.* **13**:213–217.

Dandurand, L. M., G. R. Knudsen, and D. J. Schotzko, 1995, Quantification of *Pythium-ultimum var sporangiiferum* zoospore encystment patterns using geostatistics, *Phytopathology* **85**:186–190.

Dandurand, L. M., D. J. Schotzko, and G. R. Knudsen, 1997, Spatial patterns of rhizoplane populations of *Pseudomonas fluorescens*, *Appl. Environ. Microbiol.* **63**:3211–3217.

Danovaro, R., M. Armeni, A. Dell'Anno, M. Fabiano, E. Manini, D. Marrale, A. Pusceddu, and S. Vanucci, 2001, Small-scale distribution of bacteria, enzymatic activities, and organic matter in coastal sediments, *Microb. Ecol.* **42**:177–185.

Dechesne, A., C. Pallud, D. Debouzie, J. P. Flandrois, T. M. Vogel, J. P. Gaudet, and G. L. Grundmann, 2003, A novel method for characterizing the microscale 3D spatial distribution of bacteria in soil, *Soil Biol. Biochem.* **35**:1537–1546.

Duarte, C. M., and D. Vaqué, 1992, Scale dependence of bacterioplankton patchiness, *Mar. Ecol. Prog. Ser.* **84**:95–100.

Ettema, C. H., and D. A. Wardle, 2002, Spatial soil ecology, *Trends Ecol. Evol.* **17**:177–183.

Fenchel, T., 1984, Suspended marine bacteria as a food source, in: *Flows of Energy and Materials in Marine Ecosystems*, M. J. R. Fasham, ed., Plenum Press, New York, pp. 301–316.

Finlay, B. J., J. O. Corliss, G. F. Esteban, and T. Fenchel, 1996, Biodiversity at the microbial level: the number of free-living ciliates in the biosphere, *Quart. Rev. Biol.* **71**:221–237.

Finlay, B. J., G. F. Esteban, K. J. Clarke, and J. L. Olmo, 2001, Biodiversity of terrestrial protozoa appears homogeneous across local and global spatial scales, *Protist* **152**: 355–366.

Finlay, B. J., G. F. Esteban, J. L. Olmo, and P. A. Tyler, 1999, Global distribution of free-living microbial species, *Ecography* **22**:138–144.

Franklin, R. B., and A. L. Mills, 2003, Multi-scale variation in spatial heterogeneity for microbial community structure in an eastern Virginia agricultural field, *FEMS Microbiol. Ecol.* **44**:335–346.

Franklin, R. B., D. R. Taylor, and A. L. Mills, 1999, Characterization of microbial communities using randomly amplified polymorphic DNA (RAPD), *J. Microbiol. Methods* **35**:225–235.

Franklin, R. B., D. R. Taylor, and A. L. Mills, 2000, The distribution of microbial communities in anaerobic and aerobic zones of a shallow coastal plain aquifer, *Microb. Ecol.* **38**:377–386.

Franklin, R. B., J. L. Garland, C. H. Bolster, and A. L. Mills, 2001, The impact of dilution on microbial community structure and functional potential: a comparison of numerical simulations and batch culture experiments, *Appl. Environ. Microbiol.* **67**:702–712.

Franklin, R. B., L. K. Blum, A. McComb, and A. L. Mills, 2002, A geostatistical analysis of small-scale spatial variability in bacterial abundance and community structure in salt-marsh creek bank sediments, *FEMS Microbiol. Ecol.* **42**:71–80.

Fulthorpe, R. R., A. N. Rhodes, and J. M. Tiedje, 1998, High levels of endemicity of 3-chlorobenzoate-degrading soil bacteria, *Appl. Environ. Microbiol.* **64**:1620–1627.

GarciaPichel, F., L. PrufertBebout, and G. Muyzer, 1996, Phenotypic and phylogenetic analyses show *Microcoleus chthonoplastes* to be a cosmopolitan cyanobacterium, *Appl. Environ. Microbiol.* **62**:3284–3291.

Garland, J. L., and A. L. Mills, 1991, Classification and characterization of heterotrophic microbial communities on the basis of patterns of community-level-sole-carbon-source utilization, *Appl. Environ. Microbiol.* **57**:2351–2359.

Garland, J. L., M. S. Roberts, L. F. Levine, and A. L. Mills, 2003, Community-level physiological profiling using an oxygen-sensitive fluorophore in a microtiter plate, *Appl. Environ. Microbiol.* **69**:2994–2998.

Green, J. L., A. J. Holmes, M. Westoby, I. Oliver, D. Briscoe, M. Dangerfield, M. Gillings, and A. J. Beattie, 2004, Spatial scaling of microbial eukaryote diversity, *Nature* **432**: 747–750.

Griffiths, B. S., K. Ritz, and L. A. Glover, 1996, Broad-scale approaches to the determination of soil microbial community structure: application of the community DNA hybridization technique, *Microb. Ecol.* **31**:269–280.

Grundmann, G. L., and D. Debouzie, 2000, Geostatistical analysis of the distribution of NH_4^+ and NO_2^--oxidizing bacteria and serotypes at the millimeter scale along a soil transect, *FEMS Microbiol. Ecol.* **34**:57–62.

Grundmann, G. L., and P. Normand, 2000, Microscale diversity of the genus Nitrobacter in soil on the basis of analysis of genes encoding rRNA, *Appl. Environ. Microbiol.* **66**:4543–4546.

Grundmann, G. L., A. Dechesne, F. Bartoli, J. P. Flandrois, J. L. Chasse, and R. Kizungu, 2001, Spatial modeling of nitrifier microhabitats in soil, *Soil Sci. Soc. Am. J.* **65**:1709–1716.

Handelsman, J., and K. Smalla, 2003, Techniques: conversations with the silent majority, *Curr. Opin. Microbiol.* **6**:271–273.

Harris, P. J., 1994, Consequences of the spatial distribution of microbial communities in soil, in: *Beyond the Biomass: Compositional and Functional Analysis of Soil Microbial Communities*, K. Ritz, J. Dighton, and K. E. Giller, eds., Wiley, Chichester, UK, pp. 239–247.

Hewitt, J. E., S. F. Thrush, V. J. Cummings, and S. J. Turner, 1998, The effect of changing sampling scales on our ability to detect effects of large-scale processes on communities, *J. Exp. Mar. Biol. Ecol.* **227**:251–264.

Holben, W. E., 1994, Isolation and purification of bacterial DNA from soil, in: *Methods of Soil Analysis. Part 2: Microbiological and Biochemical Properties*, R. W. Weaver, S. Angle, P. Bottomley, D. Bezdicek, S. Smith, A. Tabatabai, and A. Wollum, eds., Soil Science Society of America, Madison, WI, pp. 727–751.

Holben, W. E., and D. Harris, 1995, DNA-based monitoring of total bacterial community structure in environmental samples, *Molec. Ecol.* **4**:627–631.

Johnsen, K., C. S. Jacobsen, V. Torsvik, and J. Sorensen, 2001, Pesticide effects on bacterial diversity in agricultural soils – a review, *Biol. Fert. Soil.* **33**:443–453.

Jordan, F. L., and R. M. Maier, 1999, Development of an agar lift-DNA/DNA hybridization technique for use in visualization of the spatial distribution of Eubacteria on soil surfaces, *J. Microbiol. Methods* **38**:107–117.

Kang, S., and A. L. Mills, 2006, The effect of sample size in studies of soil microbial community structure, *J. Microbial Meth.* **66**:242–250.

Kawaguchi, T., and A. W. Decho, 2002, *In situ* microspatial imaging using two-photon and confocal laser scanning microscopy of bacteria and extracellular polymeric secretions (EPS) within marine stromatolites, *Mar. Biotechnol.* **4**:127–131.

Kennedy, A. C., 1994, Carbon utilization and fatty acid profiles for characterization of bacteria, in: *Methods of Soil Analysis. Part 2, Microbiological and Biochemical Properties*, Vol. 5, R. W. Weaver and J. S. Angle, eds., Soil Science Society of America, Madison, WI, pp. 543–556.

Kennedy, A. C., and V. L. Gewin, 1997, Soil microbial diversity: present and future considerations, *Soil Sci.* **162**:607–617.

Kieft, T. L., E. Wilch, K. O'Connor, D. B. Ringelberg, and D. C. White, 1997, Survival and phopholipid fatty acid profiles of surface and subsurface bacteria in natural sediment microcosms, *Appl. Environ. Microbiol.* **63**:1531–1542.

Kozdroj, J., and J. D. van Elsas, 2001, Structural diversity of microorganisms in chemically perturbed soil assessed by molecular and cytochemical approaches, *J. Microbiol. Methods* **43**:197–212.

Krembs, C., A. R. Juhl, and J. R. Strickler, 1998, The spatial information preservation method: sampling the nanoscale spatial distribution of microorganisms, *Limnol. Oceanogr.* **43**: 298–306.

Krembs, C., A. R. Juhl, R. A. Long, and F. Azam, 1998, Nanoscale patchiness of bacteria in lake water studied with the spatial information preservation method, *Limnol. Oceanogr.* **43**:307–314.

Kuperman, R. G., G. P. Williams, and R. W. Parmelee, 1998, Spatial variability in the soil foodwebs in a contaminated grassland ecosystem, *Appl. Soil Ecol.* **9**:509–514.

Laczko, E., A. Rudaz, and M. Aragno, 1997, Diversity of anthropogenically influenced or disturbed microbial communities, in: *Microbial Communities: Functional Versus Structural Approaches*, H. Insam and A. Rangger, eds., Springer, Berlin, pp. 57–67.

Lee, S., and J. A. Fuhrman, 1990, DNA hybridization to compare species composition of natural bacterioplankton assemblages, *Appl. Environ. Microbiol.* **56:**739–746.

Lee, S., and J. A. Fuhrman, 1991, Spatial and temporal variation of natural bacterioplankton assemblages studied by total genomic DNA cross-hybridization, *Limnol. Oceanogr.* **36:**1277–1287.

Legendre, P., and L. Legendre, 1998, *Numerical Ecology*, 2nd Edition. Elsevier, Amsterdam.

Legendre, P., and M. -J. Fortin, 1989, Spatial pattern and ecological analysis, *Vegetatio* **80:**107–138.

Legendre, P., N. L. Oden, R. R. Sokal, A. Vaudor, and J. Kim, 1990, Approximate analysis of variance of spatially autocorrelated regional data, *J. Classif.* **7:**53–75.

Lehman, J. T., and D. Scavia, 1982, Microscale patchiness of nutrients in plankton communities, *Science* **216:**729–730.

Levin, S. A., 1992, The problem of pattern and scale in ecology, *Ecology* **73:**1943–1967.

Levin, S. A., and L. A. Segal, 1976, Hypothesis for origin of planktonic patchiness, *Nature* **259:**659.

Licht, T. R., K. A. Krogfelt, P. S. Cohen, L. K. Poulsen, J. Urbance, and S. Molin, 1996, Role of lipopolysaccharide in colonization of the mouse intestine by *Salmonella typhimurium* studied by *in situ* hybridization, *Infect. Immun.* **64:**3811–3817.

Liu, W. -T., T. L. Marsh, H. Cheng, and L. J. Forney, 1997, Characterization of microbial diversity by determining terminal restriction fragment length polymorphisms of genes encoding 16S rRNA, *Appl. Environ. Microbiol.* **63:**4516–4522.

Long, R. A., and F. Azam, 2001, Microscale patchiness of bacterioplankton assemblage richness in seawater, *Aquat. Microb. Ecol.* **26:**103–113.

Mackas, D. L., 1984, Spatial autocorrelation of plankton community composition in a continental shelf ecosystem, *Limnol. Oceanogr.* **29:**451–471.

Macnaughton, S. J., T. Booth, T. M. Embley, and A. G. O'Donnell, 1996, Physical stabilization and confocal microscopy of bacteria on roots using 16S rRNA targeted, fluorescent-labelled oligonucleotide probes, *J. Microbiol. Methods* **26:**279–285.

Manz, W., G. Arp, G. Schumann-Kindel, U. Szewzyk, and J. Reitner, 2000, Widefield deconvolution epifluorescence microscopy combined with fluorescence *in situ* hybridization reveals the spatial arrangement of bacteria in sponge tissue, *J. Microbiol. Methods* **40:**125–134.

Marsh, T. L., 1999, Terminal restriction fragment length polymorphism (T-RFLP): an emerging method for characterizing diversity among homologous populations of amplification products, *Curr. Opin. Microbiol.* **2:**323–327.

Massol-Deya, A. A., D. A. Odelson, R. F. Hickey, and J. M. Tiedje, 1995, Bacterial community fingerprinting of amplified 16S and 16-23S ribosomal DNA gene sequences and restriction endonuclease analysis (ARDRA), in: *Molecular Microbial Ecology Manual*, A. D. L. Akkermans, J. D. Van Elsas, and F. J. De Bruijn, eds., Kluwer Academic, Dordrecht, The Netherlands, Section 3.3.2, pp. 1–8.

Moran, M. A., A. E. Maccubbin, R. Benner, and R. E. Hodson, 1987, Dynamics of microbial biomass and activity in five habitats of the Okefenokee Swamp ecosystem (Georgia, USA), *Microb. Ecol.* **14:**203–218.

Morris, S. J., and R. E. J. Boerner, 1999, Spatial distribution of fungal and bacterial biomass in southern Ohio hardwood forest soils: scale dependency and landscape patterns, *Soil Biol. Biochem.* **31**:887–902.

Muyzer, G., 1999, DGGE/TGGE a method for identifying genes from natural ecosystems, *Curr. Opin. Microbiol.* **2**:317–322.

Muyzer, G., E. C. DeWaal, and G. Uitterlinden, 1993, Profiling of complex microbial populations by denaturing gradient gel electrophoresis analysis of polymerase chain reaction-amplified genes coding for 16S rRNA, *Appl. Environ. Microbiol.* **59**:695–700.

Nunan, N., K. Ritz, D. Crabb, K. Harris, K. J. Wu, J. W. Crawford, and I. M. Young, 2001, Quantification of the *in situ* distribution of soil bacteria by large-scale imaging of thin sections of undisturbed soil, *FEMS Microbiol. Ecol.* **37**:67–77.

Nunan, N., K. Wu, I. M. Young, J. W. Crawford, and K. Ritz, 2002, *In situ* spatial patterns of soil bacterial populations, mapped at multiple scales, in an arable soil, *Microb. Ecol.* **44**:296–305.

O'Neil, P. V., 1991, *Advanced Engineering Mathematics*, 4th Edition. Brooks/Cole, Pacific Grove, CA.

O'Neill, R. V., and A. W. King, 1998, Homage to St. Michael; or, why are there so many books on scale? in: *Ecological Scale: Theory and Applications*, D. L. Peterson and V. T. Parker, eds., Columbia University Press, New York, pp. 3–15.

Parkin, T. B., 1987, Soil microsites as a source of denitrification variability, *Soil Sci. Soc. Am. J.* **51**:1194–1199.

Parkin, T. B., 1993, Spatial variability of microbial processes in soil – a review, *J. Environ. Qual.* **22**:409–417.

Parkin, T. B., J. L. Starr, and J. J. Meisinger, 1987, Influence of sample size on measurement of soil denitrification, *Soil Sci. Soc. Am. J.* **51**:1492–1501.

Parry, S., P. Renault, C. Chenu, and R. Lensi, 1999, Denitrification in pasture and cropped soil clods as affected by pore space structure, *Soil Biol. Biochem.* **31**:493–501.

Poulsen, L. K., F. S. Lan, C. S. Kristensen, P. Hobolth, S. Molin, and K. A. Krogfelt, 1994, Spatial distribution of *Escherichia coli* in the mouse large intestine inferred from ribosomal-RNA in-situ hybridization, *Infect. Immun.* **62**:5191–5194.

Rahel, F. J., 1990, The hierarchical nature of community persistence – a problem of scale, *Am. Nat.* **136**:328–344.

Ritz, K., W. McNicol, N. Nunan, S. Grayston, P. Millard, D. Atkinson, A. Gollotte, D. Habeshaw, B. Boag, C. D. Clegg, B. S. Griffiths, R. E. Wheatley, L. A. Glover, A. E. McCaig, and J. I. Prosser, 2004, Spatial structure in soil chemical and microbiological properties in an upland grassland, *FEMS Microbiol. Ecol.* **49**:191–205.

Roberts, M. S., J. L. Garland, and A. L. Mills, 2004, Microbial astronauts: assembling microbial communities for advanced life support systems, *Microb. Ecol.* **47**:137–149.

Robertson, G. P., and K. L. Gross, 1994, Assessing the heterogeneity of belowground resources: quantifying pattern and scale, in: *Exploitation of Environmental Heterogeneity by Plants: Ecophysiological Processes Above- and Belowground*, M. M. Caldwell and R. W. Pearcy, eds., Academic Press, San Diego, CA, pp. 237–253.

Robertson, G. P., and J. M. Tiedje, 1988, Deforestation alters denitrification in a lowland tropical rain-forest, *Nature* **336**:756–759.

Robertson, G. P., K. M. Klingensmith, M. J. Klug, E. P. Paul, J. R. Crum, and B. G. Ellis, 1997, Soil resources, microbial activity, and primary production across an agricultural ecosystem, *Ecol. App.* **7**:158–170.

Ronimus, R. S., L. E. Parker, and H. W. Morgan, 1997, The utilization of RAPD-PCR for identifying thermophilic and mesophilic Bacillus species, *FEMS Microbiol. Lett.* **147**: 75–79.

Rothemund, C., R. Amann, S. Klugbauer, W. Manz, C. Bieber, K. H. Schleifer, and P. Wilderer, 1996, Microflora of 2,4-dichlorophenoxyacetic acid degrading biofilms on gas permeable membranes, *Sys. Appl. Microbiol.* **19**:608–615.

Saetre, P., and E. Bååth, 2000, Spatial variation and patterns of soil microbial community structure in a mixed spruce-birch stand, *Soil Biol. Biochem.* **32**:909–917.

Scala, D. J., and L. J. Kerkhof, 2000, Horizontal heterogeneity of denitrifying bacterial communities in marine sediments by terminal restriction fragment length polymorphism analysis, *Appl. Environ. Microbiol.* **66**:1980–1986.

Schneider, D. C., 1994, *Quantitative ecology – spatial and temporal scaling.* Academic Press, San Diego, CA.

Schneider, D. C., 1998, Applied scaling theory, in: *Ecological scale: theory and applications*, D. L. Peterson and V. T. Parker ed., Columbia University Press, New York, pp. 253–269.

Schulz, H. N., and B. B. Jørgensen, 2001, Big bacteria, *Ann. Rev. Microbiol.* **55**:105–137.

Sieburth, J. M., 1984, Protozoan bacteriovory in pelagic marine waters, in: *Heterotrophic Activity in the Sea*, J. E. Hobbie and P. J. L. Williams, eds., Plenum Press, New York, pp. 405–444.

Smith, J. L., J. J. Halvorson, and H. Bolton, 1994, Spatial relationships of soil microbial biomass and C and N mineralization in a semiarid shrub-steppe ecosystem, *Soil Biol. Biochem.* **26**:1151–1159.

Stein, A., J. Riley, and N. Halberg, 2001, Issues of scale for environmental indicators, *Agric. Ecosys. Environ.* **87**:215–232.

Stenger, R., E. Priesack, and F. Beese, 2002, Spatial variation of nitrate-N and related soil properties at the plot-scale, *Geoderma* **105**:259–275.

Swift, M. J., 1984, Microbial diversity and decomposer niches, in: *Current Perspectives in Microbial Ecology*, M. J. Klug and C. A. Reddy, eds., American Society of Microbiology Press, Washington, DC, pp. 8–16.

Teske, A., T. Brinkhoff, G. Muyzer, D. P. Moser, J. Rethmeier, and H. W. Jannasch, 2000, Diversity of thiosulfate-oxidizing bacteria from marine sediments and hydrothermal vents, *Appl. Environ. Microbiol.* **66**:3125–3133.

Theron, J., and T. E. Cloete, 2000, Molecular techniques for determining microbial diversity and community structure in natural environments, *Crit. Rev. Microbiol.* **26**:37–57.

Torsvik, V., J. Goksøyr, and F. L. Daae, 1990, High diversity of DNA of soil bacteria, *Appl. Environ. Microbiol.* **56**:782–787.

Torsvik, V., J. Goksøyr, F. L. Daae, R. Sørheim, J. Michalsen, and K. Salte, 1994, Use of DNA analysis to determine the diversity of microbial communities, in: *Beyond the Biomass*, K. Ritz, J. Dighton, and K. E. Giller, eds., Wiley, Chichester, UK, pp. 39–48.

Torsvik, V., R. Sorheim, and J. Goksøyr, 1996, Total bacterial diversity in soil and sediment communities – a review, *J. Indust. Microbiol.* **17**:170–178.

Torsvik, V., F. L. Daae, R. A. Sandaa, and L. Øvreås, 1998, Novel techniques for analysing microbial diversity in natural and perturbed environments, *J. Biotechnol.* **64**:53–62.

Torsvik, V., L. Øvreås, and T. F. Thingstad, 2002, Prokaryotic diversity – magnitude, dynamics, and controlling factors, *Science* **296**:1064–1066.

Tunlid, A., and D. C. White, 1990, Use of lipid biomarkers in environmental samples, in: *Analytical Microbial Methods*, A. Fox, S. L. Morgan, L. Lennart, and G. Odham, eds., Plenum Press, New York, pp. 259–274.

Valentine, D. W., E. A. Holland, and D. S. Schimel, 1994, Ecosystem and physiological controls over methane production in northern wetlands, *J. Geophys. Res.* **99**:1563–1571.

Wachinger, G., S. Fiedler, K. Zepp, A. Gattinger, M. Sommer, and K. Roth, 2000, Variability of soil methane production on the micro-scale: spatial association with hot spots of organic material and Archaeal populations, *Soil Biol. Biochem.* **32**:1121–1130.

Wagner, M., M. Horn, and H. Daims, 2003, Fluorescence *in situ* hybridization for the identification and characterization of prokaryotes, *Curr. Opin. Microbiol.* **6**:302–309.

Webster, G., T. M. Embley, and J. I. Prosser, 2002, Grassland management regimens reduce small-scale heterogeneity and species diversity of beta-proteobacterial ammonia oxidizer populations, *Appl. Environ. Microbiol.* **68**:20–30.

White, D. C., H. C. Pinkart, and D. B. Ringelberg, 1997, Biomass measurements: biochemical approaches, in: *Manual of Environmental Microbiology*, C. J. Hurst, G. R. Knudsen, M. J. McInerney, L. D. Stetzenback, and M. V. Walter, eds., American Society for Microbiology Press, Washington, DC, pp. 91–101.

Whitman, W. B., D. C. Coleman, and W. J. Wiebe, 1998, Prokaryotes: the unseen majority, *Proc. Natl. Acad. Sci. USA* **95**:6578–6583.

Wiens, J. A., J. F. Addicott, T. J. Case, and J. Diamond, 1986, The importance of spatial and temporal scale in ecological investigations, in: *Community Ecology*, J. Diamond and T. J. Case, eds., Harper & Row, New York, pp. 145–172.

Wikstrom, P., A. -C. Andersson, and M. Forsman, 1999, Biomonitoring complex microbial communities using random amplified polymorphic DNA and principal component analysis, *FEMS Microbiol. Ecol.* **28**:131.

Wikstrom, P., L. Hagglund, and M. Forsman, 2000, Structure of a natural microbial community in a nitroaromatic contaminated groundwater is altered during biodegradation of extrinsic, but not intrinsic substrates, *Microb. Ecol.* **39**:203–210.

Williams, J. K. G., M. K. J. Hanafey, A. Rafalski, and S. V. Tingey, 1993, Genetic analysis using random amplified polymorphic DNA markers, *Meth. Enzymol.* **218**:704–740.

Wollum, A. G., II, and D. K. Cassel, 1984, Spatial variability of *Rhizobium japonicum* in 2 North Carolina (USA) soils, *Soil Sci. Soc. Am. J.* **48**:1082–1086.

Zabeau, M., and P. Vos., 1993, Selective restriction fragment amplification: a general method for DNA fingerprinting, European Patent Application 92402629.7, Publication Number EP 0534858 A1.

Zhou, J., 2003, Microarrays for bacterial detection and microbial community analysis, *Curr. Opin. Biotechnol.* **6**:288–294.

Zhou, J., and D. K. Thompson, 2002, Challenges in applying microarrays to environmental studies, *Curr. Opin. Biotechnol.* **13**:204–207.

Zhou, J., B. Xia, D. S. Treves, L. -Y. Wu, T. L. Marsh, R. V. O'Neill, A. V. Palumbo, and J. M. Tiedje, 2002, Spatial and resource factors influencing high microbial diversity in soil, *Appl. Environ. Microbiol.* **68**:326–334.

Zogg, G. P., D. R. Zak, D. B. Ringelberg, N. W. MacDonald, K. S. Pregitzer, and D. C. White, 1997, Compositional and functional shifts in microbial communities due to soil warming, *Soil Sci. Soc. Am. J.* **61**:475–481.

Chapter 2

STATISTICAL ANALYSIS OF SPATIAL STRUCTURE IN MICROBIAL COMMUNITIES
Overview of methods and approaches

Rima B. Franklin[1] and Aaron L. Mills[2]
[1]*Department of Biology, Virginia Commonwealth University, Richmond, VA 23284 USA;*
[2]*Laboratory of Microbial Ecology, University of Virginia, Charlottesville, VA 22904-4123 USA*

Abstract: This chapter provides a review of the basic statistical techniques used to detect and quantify spatial structure in ecological data as they can be applied to the analysis of microbial communities. It also discusses the general implications of spatial structure in data analysis, including the inappropriate use of parametric statistical tests with spatially autocorrelated data, and suggests possible alternative procedures. Methods discussed include geostatistics and variogram analysis, kriging, correlograms, Mantel and partial Mantel tests, and time-series analysis.

Keywords: spatial structure, microbial communities, statistical analysis, autocorrelation, geostatistics, kriging, scale, spatial autocorrelation

1. INTRODUCTION

Most ecological theories and models acknowledge that elements close to one another in space (or time) are likely to be more similar, as they are influenced by the same generating processes, the same energy inputs, or a connected physical environment. However, the classical statistical procedures employed to analyze these phenomena usually consider the biological organisms, and their controlling variables, to be distributed in a random or uniform way, neglecting the natural spatial structure. While the importance of spatial structure in experimental design and statistical analyses is generally accepted, most ecologists do not fully consider it when designing a sampling scheme or evaluating data. As scientists have become more aware of the importance of the spatial components of the phenomena they study, and as the number of statistical and computational tools available for quantifying these processes has increased, explicit considerations of spatial structure in microbial ecology studies has become more common.

In general, studies of spatial structure begin with exploratory analyses, followed by the application of techniques aimed at detecting and characterizing spatial patterns. Statistically, this entails testing data against the null hypothesis that there is no pattern in the data. In situations where it is determined that the data *are* patterned, analyses that allow one to distinguish competing agents of pattern can be applied. This latter stage involves posing alternative models for the pattern, and comparing these models against each other to find the most likely explanation for the observed structure. Depending on the nature of the data and the patterning agents involved, this stage of analysis can take several directions. The purpose of this chapter is to provide an overview of some of the more commonly available methods for detecting and characterizing spatial structure in ecological data, and to discuss their application in the analysis of microbial communities. Sampling strategies for the environment must also be considered, along with procedures for hypothesis testing that incorporate spatial structure. Most of the chapter focuses on the application of geostatistical approaches, including variograms and kriging, to the analysis of spatial autocorrelation. A brief overview of other methods is included, as are references for those seeking more information. In addition, the reader is advised to consult Chapter 8 for more detailed discussion of advanced multivariate methods and spectral analysis.

2. MOTIVATION FOR STUDYING SPATIAL SCALE

Research into the spatial distribution of microorganisms and microbial communities has many ecological and environmental applications, which have been discussed in detail as part of Chapter 1. In addition, the existence of spatial structure has important consequences for scientists considering other aspects of microbial ecology, even those who are not interested in spatial structure *per se*. For example, a better understanding of spatial variation and scale-dependent patterns is important for the design of field experiments and for the correct application of statistical hypothesis tests. These issues have been extensively reviewed in both the statistical and ecological literature (Bonham and Reich, 1999; Dale and Fortin, 2002; Hoosbeek et al., 1998; Legendre et al., 2004; Sokal et al., 1993; van Es and van Es, 1993), and will be briefly discussed here.

Statistical Analysis of Spatial Structure

2.1. Importance in sampling design

For an ecological hypothesis to be rigorously tested, the experiment must be thoughtfully designed. In so far as possible, the design of a field sampling approach needs to consider spatial scale in order to develop a sampling scheme that allows detection of the pattern or process of interest. For most questions in microbial ecology, there are actually several spatial scales that may be pertinent to understanding a particular process, and the patterns that one can observe are controlled by the sampling design selected. There is no single appropriate scale for studying the relationship of microorganisms to the environment, and multiple scales of study may be necessary in order to understand fully a system.

The term *scale effects* refers to changes in the results of a study due to a change in the scale at which the study is conducted. The effect of changing scale on sampling and experimental design, statistical analyses, and modeling have been well documented in ecology (Dungan et al., 2002; Levin, 1992; Miller et al., 2004; Turner, 1989; Turner and Carpenter, 1999; Wu and Levin, 1994). As scale changes, new patterns and processes may emerge, and controlling factors may shift, even for the same phenomena. For example, observations made at fine scales may miss important processes operating on broader scales; conversely, broadscale observations may not have enough detail necessary to understand fine-scale dynamics and the factors important at the level of individuals and populations.

In addition to potential scale effects, a consideration of spatial structure in sampling design is important because of *spatial autocorrelation*. Spatial autocorrelation refers to the extent to which the similarity of spatial locations (or samples from those locations) is dependent on their separation (Mackas, 1984). The autocorrelation may be either positive (neighboring sites tend to be similar) or negative (neighboring sites less alike than expected by chance). In ecological field surveys, observations are typically gathered at different spatial locations, and thus most data exhibit some degree of spatial autocorrelation, depending on the scale at which they were recorded.

Autocorrelation among samples is important because it means that the observations are not actually independent; instead, knowledge of the value of a parameter at one location provides some information on the value of that parameter at other nearby sites. This lack of independence means that the application of parametric statistical tests for data analysis may be compromised (discussed below). In many cases, the impact of spatial autocorrelation can be reduced at the stage of designing experiments (Legendre et al., 2002, 2004), if consideration of spatial patterns and structure is not one of the goals of the research. One solution to the problem of spatial autocorrelation and lack of independence among samples is to design a sample collection scheme so that there is little spatial structure present in the data, and then use parametric statistical hypothesis tests. In this case, samples must be collected close enough together that they represent replicates of the system under investigation, but they must be placed far enough apart to avoid autocorrelation. The application of this approach is often greatly enhanced by the use of an exploratory pilot study to help in determining the sample size unit, the spacing, and extent of a experiment. When this is not possible, detailed information on historical sampling schemes and hypotheses about the scale of the processes of interest may be useful. Recently, Legendre et al. (2004) used simulations to explicitly address the question of how prior knowledge of the spatial organization, obtained from either previous surveys or a pilot study, can be used to optimize sampling design. The results demonstrate that, for constant effort, experimental designs with a larger number of smaller blocks, broadly spread across the experimental area, lead to tests that have more power in the presence of spatial autocorrelation. Moreover, Legendre et al. demonstrated that randomly positioned experimental units (i.e., a completely randomized design) should only be used when the experimental area is homogeneous at broadscale. It should not be used when spatial autocorrelation, or repetitive deterministic structures such as waves, are present.

In the absence of prior knowledge, ecologists have been taught to rely on systematic or random sampling designs to avoid any influence spatial structure may have in biasing their data analysis. This approach may be adequate when one is trying to estimate the parameters of a local population, since each point has, *a priori*, the same probability of being included in the sample; however, observations may retain some degree of spatial dependence if the average distance between samples is smaller than the zone of influence of the underlying ecological phenomenon (Legendre and Fortin, 1989). In addition to the commonly applied techniques (e.g., uniform random sampling, stratified random sampling, and systematic sampling), several more advanced procedures (e.g., nested designs and cluster sampling) have recently been described and may be applicable to microbial systems (for more information, see Thompson, 2002; Gilbert, 1987). In situations where the detection and characterization of spatial structure is part of the researcher's goals, sampling design issues may become even more complicated. Numerous texts are available to aid in the selection of an appropriate sampling design to maximize statistical power and enhance pattern detection (Brockman and Murray, 1997; Dutilleul, 1993, 1998; Fortin et al., 1989).

2.2. Classical statistics and hypothesis testing

As mentioned above, the effects of spatial autocorrelation must also be considered during data analysis, as well as during sample collection and experimental design. In general, the methods of classical statistics are not appropriate for the analysis of spatially structured data, as one of the most important fundamental assumptions in hypothesis testing is the independence of observations. Statisticians generally count one degree of freedom for each independent observation, which allows them to choose an appropriate statistical distribution for testing. The lack of independence that arises from the presence of autocorrelation makes it difficult (in many cases, impossible) to accurately determine the number of degrees of freedom and correctly perform tests such as correlation, regression, or analysis of variance. Positive autocorrelation reduces within-group variability, artificially increaseing the amount of among-group variance, and often leads to the determination that differences among groups are significant, when in fact they are not (Legendre et al., 1990). Negative autocorrelation may have the opposite effect.

Some procedures exist that allow researchers to make corrections and perform statistical analyses in the presence of spatial autocorrelation. These include randomization tests and "corrected" parametric procedures (for an overview, see Cliff and Ord, 1981; Legendre, 1993; Legendre and Legendre, 1998; Legendre et al., 1990; Oberrath and Bohning-Gaese, 2001). Corrected tests rely on modified estimates of the variance of the statistic, and on adjusted estimates of the effective sample size and the number of degrees of freedom (Legendre and Legendre, 1998). Procedures have been developed to compensate for the presence of spatial autocorrelation in regressions (Cliff and Ord, 1981), t-tests (Cliff and Ord, 1981), and ANOVA (Griffith, 1978; Legendre et al., 1990); however, the application of these techniques is often limited by constraints such as sample size or the physical distribution of sampling locations (e.g., a procedure may require sampling locations to be along a regularly spaced grid (Legendre et al., 1990)). Moreover, the successful application of these approaches depends partly on the spatial structure present in the system of interest; for example, short-ranged spatial autocorrelation has been demonstrated to affect the power of ANOVA tests more than large-ranged spatial autocorrelation (Legendre et al., 2004).

2.3. Application of spatial statistics

There are several procedures available to test for the presence of spatial structure in ecological data (for reviews, see Goovaerts, 1998; Legendre, 1993; Legendre and Fortin, 1989; Robertson, 1987; Rossi et al., 1992).

A summary of the more commonly used methods, classified by the type of ecological question they address, is presented in Table 2-1. Ecologists who are interested in describing spatial structures in quantitative ways usually have one of two primary objectives: (1) to establish that there is no significant spatial autocorrelation present in the data set, so that standard parametric statistical procedures may be used (as discussed above); or (2) to demonstrate the existence of significant spatial pattern in order to use it in conceptual or statistical models. When spatial structure is being explicitly considered as a factor in an experiment, there are a variety of different techniques useful for data analysis. The selection of an appropriate method depends on the type of data and the objectives of the experiment. Tests are available to determine the significance of the spatial structure, to describe the spatial patterns, and to test the influence this structure may have on the ecology of the system. The remainder of this chapter will summarize some of the methods commonly used to examine and quantify these spatial patterns and their potential application in microbial ecology.

3. COMPONENTS OF SCALE

"Scale" is a key concept in both sampling design and the analysis of spatial patterns. However, the term has a huge variety of disparate meanings, depending upon the context of usage and field of the researcher, and it is useful at this time to define some terminology. *Scale* may be used in reference to the physical dimensions of an object or phenomena, in which case the term implies the measurement of some type. The term may also be used to refer to a scale of observation – the spatial dimensions at which and over which phenomena are observed (O'Neill and King, 1998). Discussions may refer to specific units of measurement (e.g., kilometer scale), or, more often, relative comparisons are made (e.g., "broadscale" versus "fine scale").

In sampling theory, spatial scale is defined by several characteristic properties: grain size, sampling interval, and extent (Fig. 2-1). *Grain size* is the size of the elementary sampling units used in an experiment (e.g., the volume of sample), and thus defines the *resolution* of a study (Schneider, 1994). Similarly, grain may refer to the fineness of distinctions recorded in the data, and is important because if the samples are coarse grain relative to the spatial structure of interest (e.g., large or too far apart), the pattern will be invisible statistically. The grain of the data defines the minimum length scale that can be used in data analysis.

Table 2-1. Methods for spatial pattern analysis, classified by ecological questions and objectives. (Adapted from Legendre and Fortin, 1989.)

Objectives	Methods
1) Testing for significant spatial autocorrelation, address one of two goals:	Option 1: Correlograms for a single variable, using Moran's I or Geary's c. Two-dimensional spectral analysis
(1.1) Establish that there is no significant spatial autocorrelation in the data, so that parametric statistical tests may be used	Option 2: Mantel test between the variable (or multidimensional matrix) and space (geographical distance matrix). Mantel test between a variable and a model.
(1.2) Establish that there is significant autocorrelation and determine the kind of pattern	Option 3: Mantel correlogram for multivariate data.
2) Description of spatial structure	Option 1: Correlograms or variograms
	Option 2: Clustering and ordination with spatial constraint
3) Test causal models that use space (location) as a predictor	Partial Mantel test, using three dissimilarity matrices (one matrix contains spatial separation distances)
4) Estimation (via interpolation) and mapping	Option 1: Interpolated map for a single variable (e.g., trend surface analysis)
	Option 2: Interpolation while taking into account the spatial autocorrelation structure function (e.g., variogram). For a single variable, produce a kriging map
	Option 3: Multidimensional mapping: clustering and ordination with spatial constraint

Many of the variables of interest to environmental microbiologists only have values at "points" (in an idealized sense), though the measurement of these variables is associated with a physical sample of a particular size or volume. For example, variables such as the concentration of a chemical compound are generally measured for a sample of a particular size (called the *support size*), though they are reported as if they represented point values. Consideration of support size may have an effect on spatial analyses and modeling, as there may be a significant difference in estimating the average value over a large volume and in estimating the average value over a small volume.

Sampling interval is the average distance between sampling units, and is sometimes referred to as *lag*. Lag may represent the distance between the centroids of the sampling units, or to the distance between the closest boundaries (Dungan et al., 2002). As a geostatistical concept, the term "lag" may have a slightly different meaning, which is discussed later in this chapter. *Extent* is the total area included in a study, and defines the maximum size of the spatial structure that can be detected in an analysis.

Depending on the ecological question of interest, the dimensions of these various scale properties will vary. For a given sampling design, no spatial structure can be detected that is smaller than the grain size or larger than the extent of the study. In this way, the sampling design defines the observational window for spatial pattern analysis (Legendre and Legendre, 1998). Grain size, sampling interval, and extent are important issues in sample design because they are often correlated for logistical reasons. For example, if sampling intensity (number of samples) is governed by issues such as time or cost, the interval (spacing) among samples may be dictated by the extent of the study area rather than ecological or statistical considerations. These factors must be balanced, along with a consideration of theoretical models of spatial variation and what is already known about the pattern or process of interest, in order to determine the optimal sampling scheme for a particular study.

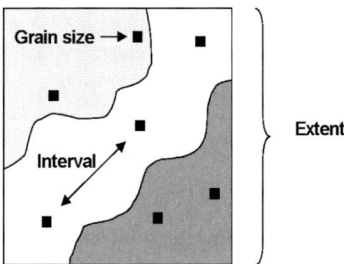

Figure 2-1. Components of scale important in spatial analysis and experimental design. The black squares represent sampling units.

4. APPROACHES TO PATTERN ANALYSIS

4.1. Point pattern analysis

The analysis of spatial patterns in ecology is generally approached using two sets of methods. The first is *point pattern analysis*, which is concerned with the distribution through space of individual objects; the data are the *locations* of some measurement or event of interest. The main purpose of such analyses is to determine whether the geographic distribution of data points is random and, if not, to describe the type of pattern in order to infer what type of process may have generated it. Point pattern analyses are limited to the description of the kind of distribution encountered, and, to a certain extent, to the quantification of the degree of clustering (Rossi, 1996). This family of methods is commonly used in plant science, and includes indices of dispersion, quandrat-density, and nearest-neighbor approaches (Ludwig and Reynolds, 1988; Pielou, 1977; Ripley, 1981); examples include the classic Clark and Evans Index (Clark and Evans, 1954) and many variations of Ripley's K statistic (Ripley, 1976, 1981).

In general, point patterns are analyzed in terms of the average or distribution of point-to-point distances. Methods often include the calculation of simple metrics concerned with nearest-neighbor distances, and evaluation of patterns of distribution (e.g., clumped, random, or regular), with some indication of the scale at which the pattern is expressed. Point pattern analysis is very sensitive to the definition of the study area (since a regularly distributed pattern can be made to seem clustered by including large margins within the study area) and to boundary corrections. These and other issues related to point pattern analysis are covered in some depth by Diggle (1983) and Upton and Fingleton (1985). Reviews by Wiegand and Moloney (2004) and Walter et al. (2005) highlight recent advances in these methodologies, especially as associated with the increased availability of sophisticated computer programs for data analysis.

In microbial ecology, the approach of point pattern analysis is especially useful for studying spatial variability associated with single cells or individual colonies, and could provide information about the relative importance of biological versus environmental heterogeneity in determining the distribution of organisms at the microscale. In general, the small size of bacteria means that the type of data necessary for this approach – micrometer scale measurements of the location of individual cells in an *undisturbed* environmental sample – is difficult to obtain and rarely available. However, some recent studies have reported advances that make the visualization of bacteria *in situ* easier; for example, the preparation of thin sections from soil samples has made it possible to observe the distribution of bacterial cells over spatial ranges from micrometers to centimeters (Nunan et al., 2001). Application of fluorescent *in situ* hybridization (FISH) is another technique that can advance this type of approach, allowing for the identification of individual microbial cells. This method was recently applied by Manz et al. (2000) to look at the spatial distribution of *Desulfovibrionaceae* and *Desulfovibrio desulfuricans* inside marine sponges. One major difficulty that limits the use of these techniques is the problem of preserving the distribution of cells within an environmental matrix, as well as image acquisition and data analysis. For more information on recent work in considering the microscale distribution of bacteria analyzed using both point pattern analysis and geostatistical approaches, the reader is directed toward the chapters by Nunan and Ritz (Chapter 3), Knudsen and Dandurand (Chapter 5), and Grundman and Dechesne (Chapter 4).

4.2. Surface pattern analysis

The second family of methods for analyzing spatial structure is *surface pattern analysis*, which deals with the study of spatially continuous phenomena (Legendre and Legendre, 1998). In this case, the data are actually *measurements* or observations made at specific locations, in contrast to point pattern analysis where the data are the location of a given attribute. Surface pattern analysis includes a large number of methods to answer a variety of questions (Table 2-1), and has been extensively reviewed in the literature (Cliff and Ord, 1981; Legendre, 1993; Legendre and Fortin, 1989; Legendre and Legendre, 1998; Rossi, 1996; Rossi et al., 1992). These procedures are more widely applicable in microbial ecology at the spatial scales commonly studied, compared to point pattern analyses, and provide a means of comparing the spatial distribution of microorganisms and environmental variables. There are a number of different techniques used for the assessment of spatial patterns including geostatistics and variogram analysis, univariate and multivariate correlograms, and methods from time-series analysis, which will be discussed individually later in this chapter. Application of each of these methods at different spatial scales and in different types of environments can be found in later chapters.

4.3. Limitations and assumptions

Many of the statistical methods commonly applied to the analysis of spatial structure (particularly time-series analyses, Mantel correlograms, and univariate correlation coefficients) require the condition of *second-order stationarity* to be satisfied. This condition states that the expected value (mean) and spatial covariance of the variable is the same over the study area, and that the variance is finite (Legendre and Legendre, 1998). This is a rather strong condition, rarely (if ever) met in natural systems. A relaxed form of the stationarity hypothesis is the *intrinsic assumption*, which states that the difference in a variable between any two locations separated by a distance *d* must have zero mean and finite variance. More simply, the difference between the values of two individuals is only a function of the distance between the sampling points, rather than due to the existence of a dominant spatial structure. Under the intrinsic assumption, the increments of the values of the variable are being considered rather than the variable itself, and it is the variance in these increments that is analyzed using *variogram analysis* and geostatistics. These relaxed assumptions, the predictive ability one has from using geostatistical variograms (e.g., ability to generate maps of parameters at unknown locations), and the relative ease with which such analyses can be performed (numerous geostatistical software packages are available) are the main reason to focus on the use of geostatistical tools for the analysis of microbial communities. Another advantage of this technique is that theoretical variograms can be fitted to experimental ones, allowing for the comparison of observed structures with structures derived from hypothesized generating processes (Escudero et al., 2003).

5. GEOSTATISTICAL TECHNIQUES

Geostatistics concentrates on the modeling of spatial dependence, and was originally developed by researchers working in the fields of geology and mining. It is a set of statistical tools for incorporating the spatial coordinates of sampling observations into data processing, and can provide a powerful means of quantitatively describing spatial variation by expressing a measure of association between two samples as a function of the distance separating them. Scientists have recently begun to use geostatistics to study and interpret spatial dependence in ecology (Legendre, 1993; Legendre and Fortin, 1989; Robertson, 1987; Rossi et al., 1992), and the application of these tools in microbial systems is increasing in popularity (Castrignanò et al., 2000; Dobermann et al., 1995; Franklin et al., 2002; Franklin and Mills, 2003; Grundmann and Debouzie, 2000; Mackas, 1984; Morris, 1999; Parkin, 1987; Schlesinger et al., 1996). When applied to microbiological data, most of the analyses have focused on either the micrometer to centimeter-scale distribution of individuals (Dandurand et al., 1995, 1997; Grundmann and Debouzie, 2000; Nunan et al., 2001, 2003), or the larger-scale distribution of total biomass and abundance (Mottonen et al., 1999; Robertson et al., 1997; Saetre, 1999; Smith et al., 1994; Troussellier et al., 1993). The techniques have recently begun to be applied to the analysis of microbial community composition and diversity (Cavigelli et al., 1995; Franklin and Mills, 2003; Mackas, 1984; Saetre and Bååth, 2000).

5.1. Variogram analysis

In geostatistics, spatial patterns are described in terms of dissimilarity (or variance) between observations as a function of distance between each pair of sampling locations. The data are analyzed by plotting these two parameters to create a *variogram* (sometimes called a *semi-variogram*). These plots, known as *experimental variograms*, are then modeled as *theoretical variograms* using a variety of structure functions. Goodness of fit can be determined, and parameter estimates can be derived from the theoretical variograms that are interpretable from an ecological perspective. The models may also be used as a predictive tool for kriging contour maps, which visualize the distribution of the spatial patterns, and to make predictions of the value of the variable of interest at unknown locations. Spatial patterns and autocorrelation structure of different variables may also be quantitatively compared.

5.1.1. Constructing experimental variograms

To construct an experimental variogram, data are analyzed by plotting some measure of sample variability versus separation distance between samples. These plots often take the shape presented in Fig. 2-2, where variability increases with increasing spatial separation (lag distance) over a given range. At distances less than this range, samples are considered to be spatially autocorrelated; points separated by distances greater than the range are uncorrelated. Most often, the measure of sample variability used to construct an experimental variogram is the *semi-variance* (γ) which is computed as half the average squared difference between the components of every data pair (Goovaerts, 1999):

$$\gamma(d) = \frac{1}{2n_d} \sum [y_{(i+d)} - y_{(i)}]^2$$

where y are the observed values and n_d is the number of pairs of points located at distance d from one another. The summation is for i varying from 1 to n_d. Sometimes variograms are standardized by dividing each variogram value by the overall sample variance; this allows for the variograms from different data sets to be compared. Other types of difference/distance plots are also generated in geostatistics (for review, see Goovaerts, 1997; Englund

Statistical Analysis of Spatial Structure

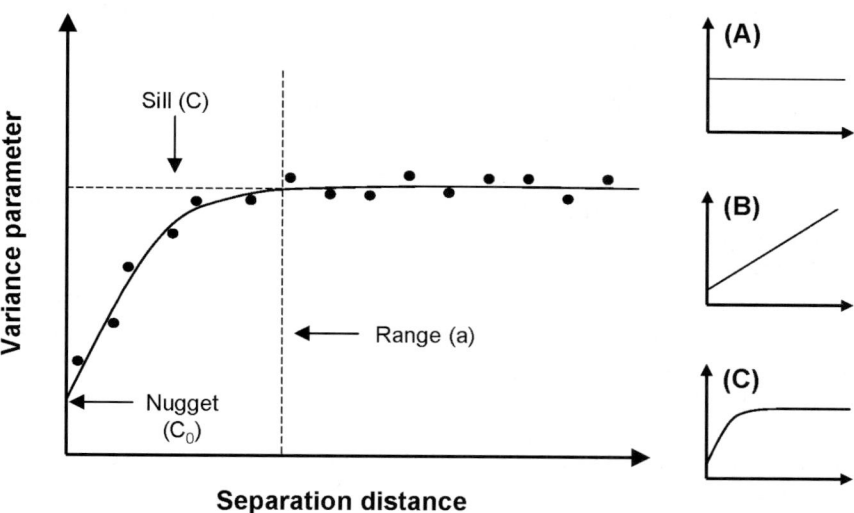

Figure 2-2. Experimental variogram showing the relationship between sample variability (usually semi-variance) and spatial separation distance. The most common patterns obtained in a variogram analysis are nugget (**A**), linear (**B**), and linear-sill (**C**).

and Sparks, 1991). For example, a *madogram* is very similar to a variogram, except that the absolute different between each pair is computed to calculate $\gamma(d)$, rather than the square of the difference. The madogram is less sensitive to extreme values in the data set, compared to the variogram, however it cannot be used to infer nugget variance.

An experimental variogram may take one of three basic forms: nugget/ flat (Fig. 2-2A), linear (Fig. 2-2B) or linear-sill (Fig. 2-2C). Hypothetical variograms and example maps for each pattern are presented in Fig. 2-3. These basic variogram forms may be modeled using a variety of equations including linear, Gaussian, exponential, and spherical functions (Table 2-3, Fig. 2-4). A flat variogram (Fig. 2-2A), also called "pure nugget effect," indicates the absence of a spatial structure to the data (no lag-to-lag spatial continuity) at the scales the observations were made (Rossi et al., 1992). The use of the term "nugget" refers to the mining and geology origins of geostatistical methods. For patterned data with positive spatial autocorrelation, semi-variance increases with the separation distance, reflecting the idea that samples that are nearby spatially (small separation distance) will be more similar than samples that are farther apart. If spatial autocorrelation is present throughout the entire extent of the area sampled, a linear pattern may be obtained (Fig. 2-2B). This would represent a situation where the samples are spatially autocorrelated at all distances measured (samples become more different as the separation distance becomes greater), but they have not yet reached their maximum difference. In situations where the area sampled is larger than the scale of the autocorrelation structure (Fig. 2-2C), the semivariogram stops increasing at a given distance called the *range* (a). The range is the distance over which samples show spatial autocorrelation and is sometimes referred to as the *correlation length scale* (Table 2-2). Beyond the range, the variogram will fluctuate around the *sill* (C), which is roughly equal to the total sample variance.

By definition, the value of a variogram at a lag separation of zero should be zero variance. However, when an experimental variogram is plotted and the line projected back, it often does not point to the origin. The interpolated value of variance when spatial separation is zero is known as the *nugget* (C_0). Nugget arises from two main sources: (1) spatial variability that is present at very small distances (smaller than the shortest sampling interval), and (2) variability that is due to measurement error or sampling error. The ratio of the nugget to the sill provides an estimate of the proportion of the total variance that cannot be accounted for by spatial variation (Murray, 2001).

The proportion of model sample variance (the sill, C) that is explained by the structural variance (C_1) can be used as a normalized measure of *spatial dependence*, and represents the percent of variability in the data that can be accounted for by considering the spatial separation of the sampling locations (Robertson and Freckman, 1995). The ratio of structural variance to sample variance (C_1/C) may vary between 0 and 1; higher values indicate that a larger portion of the total variability among samples is spatially dependent over the range of separation distances modeled. A low level of spatial dependence indicates either that sampling/analytical error is high or that dependence occurs at scales smaller than the average separation distance in the first lag class.

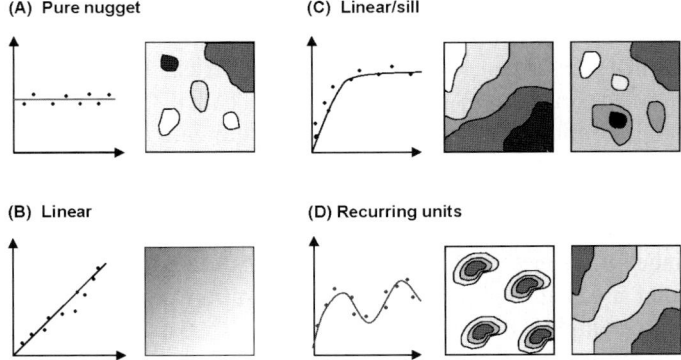

Figure 2-3. Hypothetical variograms with example surface maps: (**A**) Pure nugget – At the scale sampled, no spatial autocorrelation is evident. (**B**) Linear – Autocorrelation present over entire extent sampled. (**C**) Linear/sill – May display either large-scale or smaller scale heterogeneity, range within spatial extent. (**D**) Recurring units – Example of complex pattern, use combinations of theoretical variogram models. (Modified from Ettema and Wardle, 2002.)

Statistical Analysis of Spatial Structure

Table 2-2. Summary of geostatistical model parameters.

Parameter	Term	Description and interpretation
Nugget	C_0	The y intercept of the semi-variogram model. It is the variability that arises from measurement errors or spatial sources of variation at distances smaller than the shortest sampling interval
Sill	C	The y value at which the semi-variogram levels off ($C = C_0 + C_1$)
Range	a	The distance at which the semi-variogram stops increasing. It represents the "range" of spatial dependence for the samples; beyond this distance samples are no longer spatially autocorrelated. Sometimes referred as the "correlation length scale"
Spatial dependence	C_1/C	The ratio of the spatially structured component of the model (C_1) to the total variance (C). It represents the proportion of the total variability in the data set that can be accounted for by considering the spatial separation of samples

5.1.2. Distance classes

Prior to constructing a variogram, it is necessary to segregate the data into distance classes (*bins*). The purpose of binning the data is to obtain the best resolution (maximum detail) at small distances without being misled by structural artifacts due to whatever particular size class is chosen. There are two main ways of dividing distances (lags) into classes, either by forming equal distance classes or by creating classes with equal frequencies (Legendre and Legendre, 1998). When forming equal distance classes, it is necessary to determine the appropriate number of bins for each analysis. Sturge's rule is often used to determine the number of classes in histograms and is also useful for variograms and correlograms; it states that the number of classes = $1 + 3.3 \log_{10} m$, where m is the number of points in either the upper or lower triangle of the distance matrix (Legendre and Legendre, 1998). Variograms are generally not valid beyond 1/2 of the maximum distance between samples (Englund and Sparks, 1991; Rossi et al., 1992), and so the appropriate lag distance may be calculated as the maximum pair distance divided by two, and then subdivided into an appropriate number of size classes.

One problem with using equal distance classes is that the number of points in some bins may be quite small (especially bins at the far right of the variogram), so that the average variance associated with one of these classes may not be a valid estimate of the mean variance at that distance. Two common ways to compensate for this are: (1) to only consider the first two-thirds of the variogram when describing the spatial structure (Englund and Sparks, 1991), or (2) to remove any points in the variogram that are the result of averaging too few pairs (e.g., <30). In general, it is recommended that spatial autocorrelation analysis should not be performed with <30 sampling locations, because the number of pairs of samples in each distance class becomes too small to produce significant results with fewer locations (Legendre and Fortin, 1989). Variograms can be computed for subregions of the sampling grid so long as there are a sufficient number of data (Rossi et al., 1992).

An alternative to distance classes with equal widths is to create distance classes containing the same number of pairs (equal frequency). The advantage to this method is that, by controlling the number of pairs in each distance class, one has a greater assurance of the validity of the variogram at larger distances classes. However, this type of classification makes it much more difficult to calculate and plot the variogram. In general, researchers compute variograms for equal width distance classes, which means that the number of pairs of points used in the computation decreases as distance increases.

5.1.3. Modeling theoretical variograms

Once an experimental semi-variogram has been plotted, the data are fit with a continuous function; this is used for prediction algorithms, and to smooth out sample fluctuations and estimate useful model parameters. The functions most commonly used to model theoretical variograms are presented in Table 2-3 and Fig. 2-4. For complex spatial patterns, models may be combined or applied individually to different portions of the experimental variogram (Legendre and Legendre, 1998).

In general, the function used to model the theoretical variogram is selected by the researcher after visual examination of the experimental variogram. In some cases, where it is difficult to make this determination, several different functions may be applied and the final selection may be made by considering statistical goodness-of-fit. This is particularly relevant in situations where the experimental variogram displays a linear-sill pattern, which can be modeled using either the Gaussian, exponential, or spherical functions. Differences between these nonlinear functions lie mostly in the shape of the left-hand portion of the curves, near the origin. In practice, the spherical and

Statistical Analysis of Spatial Structure

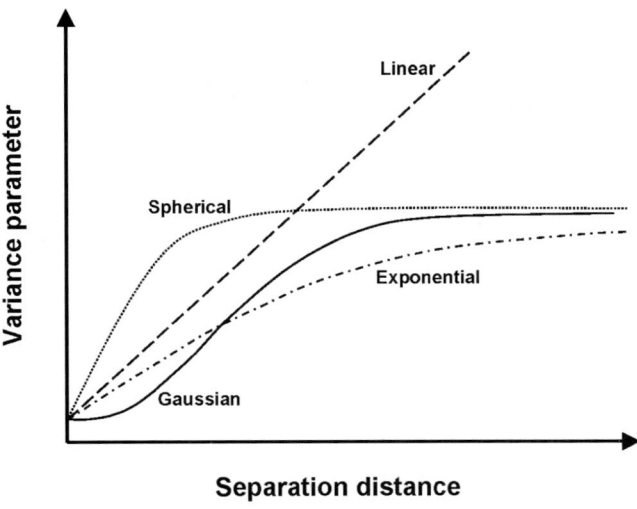

Figure 2-4. Four common theoretical variogram forms, which are fitted to experimental variograms by regression. The form is often selected based on visual inspection of the data; forms may be combined to analyze more complex patterns.

exponential models do not differ much (Legendre and Legendre, 1998). Several authors have warned users of the risk of numerical instability when using the Gaussian model, and it is rarely applied at this point (Goovaerts, 1998). Once the experimental variogram has been fit with a theoretical one, it is possible to determine a number of parameters useful for describing the structure of the spatial phenomenon (Fig. 2-2, Table 2-2) as discussed in section 5.1.1.

A number of researchers have used geostatistical semi-variogram models as a means of studying the spatial variability of microbial properties in the environment (see Chapter 1). In these studies, the data generally follow the linear-sill pattern (increasing variance and then leveling-off), and have been successfully modeled using linear, spherical, exponential models. Using these models, the range and spatial dependence are calculated and then compared for different variables (e.g., microbial community properties and environmental properties) or for the same parameters at different sites (e.g., under different land management regimes). This approach has been successfully used to analyze several types of data (e.g., abundance, biomass, and community structure), collected at a variety of scales. These studies have demonstrated that significant spatial autocorrelation exists in microbial communities, and the amount of variability that can be explained by considering the spatial separation of sampling locations is often quite high.

Table 2-3. Summary of model types, plots of *y* (variance parameter) versus *d* (distance).

Model type	Equation
Nugget	$y = C_0$
Linear	$y = C_0 + bd$ where *b* is the slope and C_0 is the intercept
Exponential	$y = C_0 + C_1 [1 - \exp(-3\,d/a)]$
Gaussian	$y = C_0 + C_1 [1 - \exp(-3\,d^2/a^2)]$
Spherical	$y = C_0 + C_1 [1.5\,d/a - 0.5\,(d/a)^3]$ if $d \leq a$; $y = C$ if $d > a$

5.1.4. Anisotropy and directional variograms

When the autocorrelation function is the same for all geographic directions considered, the phenomenon is said to be *isotropic*. When a variable is isotropic, a single variogram may be computed over all directions of the study area (*omnidirectional variogram*). *Anisotropy* is present in data when the autocorrelation function is not the same for all geographic directions considered, and may be analyzed by computing separate variograms for each direction/dimension of interest (e.g., north/south, east/west). Knowledge of the major directions of anisotropy can be important for estimating the value of a variable at an unsampled location, and can also provide clues to the underlying processes that are controlling the spatial distribution (Murray, 2001). For example, Franklin et al. (2002) analyzed the distribution of bacterial abundance and microbial community structure in salt marsh creek bank sediments, considering isotropy associated with vertical versus horizontal spatial separation. Overall, variability due to horizontal position (distance from the creek bank) was much smaller than that due to vertical position (elevation) for both variables. This suggests that processes more correlated with elevation (e.g., drainage and redox potential) vary at a smaller scale than ones controlled by distance from the creek bank, and may be more important in controlling the microbial ecology of this habitat.

Two types of anisotropy are commonly described: geometric and zonal. In the first case, the directional semi-variograms have the same shape and sill but different range values. In its simplest form, *geometric anisotropy* is akin to elliptically shaped zones wherein the data values are correlated, i.e., the zones are "stretched" in the directions of the maximum range (Rossi et al., 1992). For example, in a river, the kind of variation expected in phytoplankton concentration between two sites 5 m apart across the current may be the same as the variation expected between two sites 50 m apart along the current – even though the variation can be modeled by spherical variograms with the same sill in two directions (Legendre and Legendre, 1998). Geometric anisotropy is measured by calculating an anisotropy ratio, which is equal to the longest range divided by the shortest range. In the example above, the anisotropy ratio would be 50/5 = 10 in the direction along the current; this means that, on average, whatever amount of variability occurs in 1 distance unit across the river, will occur over 10 distance units along the current.

Zonal anisotropy reveals directions along which an additional structure of variability is present; in this case, the sill values for the directional variograms are not the same, though the range may remain constant (Isaaks and Srivastava, 1989). An example would be a strip of land where the long axis is oriented in the direction of major environmental gradient. In this case, the total variability (the sill) encountered along the major axis maybe much greater than the variation displaced at any transect in the perpendicular direction.

5.1.5. Kriging

Besides of their ability to quantitatively describe spatial structures, variograms provide the basis for interpolation by *kriging*. Kriging is a process that uses the theoretical variogram to interpolate values for points not measured in the original experimental sampling. The results of kriging are an interpolated surface map of variable values, and kriging refers to a family of generalized least-squares regression algorithms used in the estimation process (Goovaerts, 1999). There are several different forms of kriging; the simplest are punctual and block kriging (Robertson, 1987). *Punctual kriging* is used to estimate values for exact points within the sampling unit, while *block kriging* involves estimating values for areas within the unit. (For a review of these and other kriging techniques; see Robertson, 1997; Goovaerts, 1998, 1999; Issaks and Srivastava, 1989). In addition to providing a method to estimate values at unsampled locations, geostatistics can provide a means of quantifying the uncertainty associated with each estimate (Murray, 2001), and some researchers have used this to determine confidence intervals for mapped parameters. Kriging can also be used to decompose original observations into specific spatial components, so that the spatial fluctuations can be studied at different scales (Goovaerts, 1997).

In microbial ecology, kriging has many applications, and has been used to map the distribution of microbiological and supporting environmental properties (Castrignanò et al., 2000; Goovaerts, 1998; Halvorson et al., 1994, 1995; Morris, 1999; Pennanen et al., 1999; Robertson et al., 1997). In addition, the method has proven useful for studying indicator organisms associated with fecal pollution (Beliaeff and Cochard, 1995) and for the looking at the distribution of pollutants and toxic metals in soils (Becker et al., 2006; Stein et al., 2001).

5.1.6. Additional techniques

Section 5 was intended to give a brief overview of geostatistical procedures that may be applied to microbial ecology. Readers are advised to consult reviews mentioned there, as well as textbooks by Cressie (1993), Haining (1993), and Journel and Huijbregts (1978), for more detailed discussion of the mathematical and theoretical constraints associated with the use of each technique. Before conducting any geostatistical analysis, an exhaustive exploratory data analysis should be performed (including calculating traditional univariate statistics, historgrams, etc.). Outliers can strongly influence models of spatial variation, and so a great deal of care must be taken to detect unusually large or small sample values and to modify analyses as appropriate (Rossi et al., 1992).

In addition to the methods described above, there are several advanced and nontraditional geostatistical approaches that may also be applied to the analysis of ecological data. For example, when ecological processes involved in the spatial structure of communities are interest, they may be best studied through examining the *pattern of multivariate data*. Spatial autocorrelation of such data can be described either as a complete set of individual variograms (e.g., one for each taxonomic category (Sokal and Oden, 1978)), as a correlogram/variogram of a more "synthetic variable" that compresses the full data set (e.g., principal component or factor loading scores (Cavigelli et al., 1995; Saetre and Baath, 2000; Sokal et al., 1980)), or as correlogram/ variogram of a "resemblance coefficient" (e.g., relative (dis)similarity) obtained by summation over all elements of the data. This idea has been used by several researchers (Franklin et al., 2002; Franklin and Mills, 2003; Mackas, 1984; Star and Mullin, 1981), and is especially interesting in its application to the analysis of "community fingerprinting data" such as may be obtained using fatty acid profiling or whole-community DNA analysis.

6. OTHER STATISTICAL APPROACHES

In addition to the geostatistical approaches described above, there are numerous additional methods that have been developed for the analysis of spatial patterns and autocorrelation structures. Many of these are amenable the analysis of microbial ecological data, and have been applied to various degrees to study the distribution and ecology of both bacteria and fungi. Three of these methods: *correlograms*, *time-series analysis*, and *Mantel tests* are summarized below.

6.1. Correlograms

Correlograms are, most simply, graphs that plot the autocorrelation of a variable against the spatial separation defined in *distance classes*. Two of the most common measures used are Moran's I (a product moment coefficient, like a covariance or a Pearson's correlation coefficient (Moran, 1950)) or Geary's c (a squared difference coefficient (Geary, 1954)). Moran's I is usually evaluated under the assumption that the observations were random, independent samples of a population with an unknown distribution function. I values that are significantly greater than expected occur when pairs within a distance class have scores more similar than would be expected if the variable were randomly distributed. Conversely, values significantly lower than expected indicate that scores of the variable are more dissimilar than expected by chance. Geary's c coefficient is a distance-type function; its numerator sums the squared differences between values found at the various pairs of sites being considered (Legendre and Legendre, 1998). A Geary's c correlogram varies as the reverse of Moran's I in that strong autocorrelation produces high values of I and low values of c. Results using either coefficient are presented as *correlograms*, graphic displays in which the values of the autocorrelation coefficients are plotted against distance classes. Tests of significance are obtained for each distance class by randomization processes, whereas overall significance of correlogram is modified by correction methods, such as the Bonferroni method.

After determining the significance of each correlation coefficient for each distance class, correlograms are interpreted by looking at their shape, since characteristic shapes are associated with specific types of spatial structures (Legendre and Legendre, 1998; Sokal, 1979). For example, when the spatial structure being studied is a linear gradient, the correlogram will monotonically decrease. If the spatial structure is more represented as a bull's-eye pattern (e.g., a single center bump), then the correlogram will drop off suddenly upon leaving the center of the study area, and a second change will be observed when the distance class considered is larger than the "range of influence" of the single bump. Cliff and Ord (1981) provide a good description of how to interpret correlograms and their significance. In additional, Legendre and Legendre (1998) present useful maps of simulated spatial structures and both the Moran's and Geary's correlograms that derive from them.

Moran's I and Geary's c are used to examine the distribution of a single variable and interpret its spatial structure. Correlograms can also be calculated for multivariate data using a normalized Mantel statistic r_M (Oden and Sokal, 1986; Sokal, 1986). This method is especially useful for describing the spatial structure of community assemblages by means of a multivariate similarity or dissimilarity coefficient. Thus far, the method has been used primarily by scientists analyzing vegetation data, with limited application to microbial ecology. However, Lilleskov et al. (2004) did use Mantel correlograms as part of an extensive study of the spatial distribution of ectomycorrhizal fungal (EMF) communities. The results compared favorably to those obtained via traditional Mantel tests and variogram analysis. The Mantel correlograms detected significant spatial structure that was not observed using the traditional Mantel tests, and provided results that complimented the variogram analysis of patch size.

6.2. Time-series analysis

Another approach to analyze spatial structure, which is related to geostatistics, is that of *time-series analysis* as applied to spatial ecological patterns (Platt and Denman, 1975; Rossi et al., 1992). Both geostatistics and time-series analyses utilize the covariance function, but time-series analysis focuses on transforms of the data (Rossi et al., 1992); these transformations may place practical restrictions on the input data and on subsequent interpretations. Time-series methods often require data that are continuous and evenly spaced; geostatistics can be used on irregularly spaced data and may detect the periodic as well as the nonperiodic nature of ecological parameters.

Two other related approaches, spectral analysis and wavelet analysis, have potential applications to microbial ecology. Thus far, they have been primarily used to study spatial patterns of planktonic organisms in aquatic samples (Franks, 2005; Piontkovski et al., 1997; Seuront et al., 1999; Steele and Henderson, 1992). Spectral analysis, which attempts to explain regular patterns as functions of sine waves of various periods, acts as a structure function with the distances expressed in terms of wavelengths – i.e., the frequency domain (Ford and Renshaw, 1984; Franks, 2005; Renshaw and Ford, 1984). Wavelet analysis, which offers a means of identifying and displaying spatial structure in a hierarchical manner (Bradshaw and Spies, 1992), is a similar, fairly well-developed statistical approach, but has only recently come to be used in ecology (Dale and Mah, 1998; Keitt and Urban, 2005). A main advantage of these types of techniques is that they allow for the analysis of anisotropic data, which are common in ecology. Disadvantages include the fact that they require a large data set, and that they assume the spatial pattern results from a combination of repeatable patterns (Legendre and Fortin, 1989). They do not detect other types of spatial patterns that do not involve repeatabilities. For additional discussion of these approaches, the reader is directed to Chapter 8.

6.3. Mantel and partial Mantel tests

A Mantel test (1967) is a linear regression technique in which the variables are themselves distance or dissimilarity matrices summarizing pairwise comparisons among sample locations. For example, instead of using "soil type" as a variable, the predictor might be "similarity in soil type." With a Mantel test, the operative question is, "Do samples that are similar in terms of the predictor variable also tend to be similar in terms of the dependent variable?" The Mantel test is quite flexible, and by using slight variations on the basic technique, a range of questions can be addressed (for example, see Table 2-4). For spatial analyses, the Mantel's test can be used to consider a case where the predictor variable is space itself, measured as geographic separation distance. When applied this way, the Mantel's test is one of the few spatial techniques that is explicitly geared toward hypothesis testing.

Mathematically, a Mantel test is a regression in which the variables are actually similarity or distance matrices; the Mantel statistic (r_M) is computed by determining the sum of the cross-products of the corresponding values in each of these matrices (Rossi, 1996). The partial Mantel test, as developed by Smouse et al. (1986), allows testing for the correlation between two matrices while controlling for the effect of a third matrix, and is analogous to a partial correlation. By comparing matrices in this way, it is possible to address questions such as: "do samples that are close together have similar environmental properties?" (e.g., by performing a Mantel test with a matrix of geographic distances and a matrix of environmental similarity), or "Is there a relationship between community similarity and environmental similarity, after removing the shared correlation of these variables with spatial separation?" (e.g., by performing a partial Mantel test using a matrix of community similarity and a matrix of environmental similarity, controlling for the effect of a third matrix of geographic spatial separation distances). Because Mantel's test is merely a correlation between distance matrices, and the distance matrices can be variously defined, the test can assume a variety of forms and special cases. Several example cases where a Mantel or partial Mantel's test may be useful for the analysis of spatial patterns are detailed in Table 2-4.

One advantage of Mantel's test is that, because it proceeds from a distance (dissimilarity) matrix, it can be applied to variables of different data types (e.g., categorical, rank, or interval scale). This is especially important for ecologists, who often work with categorical variables or need to relate variables on different scales. This issue is addressed by converting all the data to distance (dissimilarity) metrics, matrices of which are then used as the variables in the regression. The metrics can be univariate (e.g., similarity in soil N) or multivariate (e.g., using an index of overall soil chemistry). Because these matrices provide the foundation for the analysis (and determine the input variable in the regression), it is important that appropriate distance/dissimilarity metrics be employed. For a thorough review of distance metrics and the selection of appropriate methods for different data types (see Legendre and Legendre, 1998).

7. CLOSING REMARKS

Though spatial variation was originally considered a statistical nuisance, it is now recognized as an ecologically important feature of ecosystems. The importance of spatial heterogeneity comes from its central role in ecological theories and its practical significance in sampling theory. Even within homogeneous zones/environments, biotic processes (e.g., growth and reproduction) can produce aggregations of organisms, at various spatial and temporal scales. In microbial systems, spatial structure may arise due to biological processes and as a response to environmental gradients. Several physical, chemical, and biological variables may combine to influence the distribution of microorganisms, and the importance of each of these factors may change depending upon the system of interest and the spatial scale considered. Much can be learned about the function and stability of microbial communities by considering their spatial structure; moreover, by examining the scale at which important environmental factors operate, it may be possible to determine how these variables influence the distribution and development of stable assemblages. In order to address these types of questions, future research must include a consideration for the spatial component of the microbial ecology. This includes research specifically directed toward the analysis of spatial patterns and processes in bacterial and fungal systems, as well as a consideration of spatial structure during experimental design and statistical analysis when

Statistical Analysis of Spatial Structure

Table 2-4. Cases for the use of Mantel's tests and partial Mantel's tests in analysis of spatial structure. Includes example matrices and research questions for each matrix combination. Examples matrices represent a biological array (similarity based on overall microbial community structure), an environment array (e.g., overall similarity in soil properties), and a matrix of spatial separation distances between sampling points.

Case	Matrix	Research question
Simple Mantel's test	Matrix 1: similarity based on overall microbial community Matrix 2: similarity based on a set of environmental variables	Do sites that are environmental similar also have similar community composition?
Simple Mantel's test on geographic distance	Matrix 1: similarity based on overall microbial community Matrix 2: geographic distance (spatial dissimilarity)	Are samples that are close together otherwise similar? *Equivalent to testing for overall autocorrelation, averaged over all distances*
Mantel correlogram	Compute as above, but partition analyses into a series of discrete distance classes. Special case of test on geographical distance	Evaluates autocorrelation for distance intervals (as with tradition correlograms), but can be performed on multivariate data sets
Partial Mantel test on three matrices	Matrix 1: similarity based on overall microbial community Matrix 2: similarity based on a set of environmental variables Matrix 3: geographic distance (spatial dissimilarity)	How much of the variability in community composition can be explained by the environmental variables? Is there residual variability in community composition that is spatially structured, after removing the effects of the environmental variables? *Accounts for shared spatial structure in environmental and microbial data sets*
Partial Mantel test, repeated with multiple predictor variables	Matrix 1: similarity based on overall microbial community Matrix 2: similarity based on *individual* environmental variable Matrix 3: geographic distance (spatial dissimilarity)	Allows one to determine *which* variables are actually related to overall community structure Separately determine the contribution of each predictor variable for its pure partial effect on the dependent variable

other objectives are being addressed. In this chapter, an overview of many of the techniques needed to conduct these types of analyses has been presented.

The ability to quantitatively compare microbial communities is an important step toward more sensitive monitoring, and eventually, being able to integrate data gathered by different researchers so that overall patterns may be identified. It is important that microbial ecologists become more aware of the availability of these techniques, and consider their use when analyzing and interpreting microbial community data.

REFERENCES

Becker, J. M., T. Parkin, C. H. Nakatsu, J. D. Wilbur, and A. Konopka, 2006, Bacterial activity, community structure, and centimeter-scale spatial heterogeneity in contaminated soil, *Microb. Ecol.* **51:**220–231.

Beliaeff, B., and M. L. Cochard, 1995, Applying geostatistics to identification of spatial patterns of fecal contamination in a mussel farming area (Havre De La Vanlee, France), *Water Res.* **29:**1541–1548.

Bonham, C. D., and R. M. Reich, 1999, Influence of spatial autocorrelation on a fixed-effect model used to evaluate treatment of oil spills, *Appl. Math. Comput.* **106:**149–162.

Bradshaw, G. A., and T. A. Spies, 1992, Characterizing canopy gap structure in forests using wavelet analysis, *J. Ecol.* **80:**205–215.

Brockman, F. J., and C. J. Murray, 1997, Microbiological heterogeneity in the terrestrial subsurface and approaches for its description, in: *Microbiology of the Terrestrial Deep Subsurface*, P. S. Amy and D. H. Haldeman, eds., CRC Press, Boca Raton, FL, pp. 72–102.

Castrignanò, A., L. Giugliarini, R. Risaliti, and N. Martinelli, 2000, Study of spatial relationships among some soil physico-chemical properties of a field in central Italy using multivariate geostatistics, *Geoderma* **97:**39–60.

Cavigelli, M. A., G. P. Robertson, and M. J. Klug, 1995, Fatty-acid methyl-ester (FAME) profiles as measures of soil microbial community structure, *Plant Soil.* **170:**99–113.

Clark, P. J., and F. C. Evans, 1954, Distance to nearest neighbor as a measurement of spatial relationships in populations, *Ecology* **35:**445–453.

Cliff, A. D., and J. K. Ord, 1981, *Spatial Processes: Models and Applications*. Pion, London.

Cressie, N. A. C., 1993, *Statistics for Spatial Data*. Wiley, New York.

Dale, M. R. T., and M. J. Fortin, 2002, Spatial autocorrelation and statistical tests in ecology, *Ecoscience* **9:**162–167.

Dale, M. R. T., and M. Mah, 1998, The use of wavelets for spatial pattern analysis in ecology, *J. Veg. Sci.* **9:**805–814.

Dandurand, L. M., G. R. Knudsen, and D. J. Schotzko, 1995, Quantification of *Pythium-ultimum Var Sporangiiferum* zoospore encystment patterns using geostatistics, *Phytopathology* **85:**186–190.

Dandurand, L. M., D. J. Schotzko, and G. R. Knudsen, 1997, Spatial patterns of rhizoplane populations of *Pseudomonas fluorescens*, *Appl. Environ. Microbiol.* **63:**3211–3217.

Diggle, P. J., 1983, *Statistical Analysis of Spatial Point Patterns*. Academic Press, New York.

Dobermann, A., P. Goovaerts, and T. George, 1995, Sources of soil variation in an acid Ultisol of the Philippines, *Geoderma* **68:**173–191.

Dungan, J. L., J. N. Perry, M. R. T. Dale, P. Legendre, S. Citron-Pousty, M. J. Fortin, A. Jakomulska, M. Miriti, and M. S. Rosenberg, 2002, A balanced view of scale in spatial statistical analysis, *Ecography* **25**:626–640.

Dutilleul, P., 1993, Spatial heterogeneity and the design of ecological field experiments, *Ecology* **74**:1646–1658.

Dutilleul, P., 1998, Incorporating scale in ecological experiments: study design, in: *Ecological Scale: Theory and Applications*, D. L. Peterson and V. T. Parker, eds., Columbia University Press, New York, pp. 369–386.

Englund, E., and A. Sparks, 1991, *GEOEAS 1.2.1.: Geostatistical Environmental Assessment Software User's Guide* EPA 600/8-91/008, Environmental Monitoring Systems Laboratory, United States Environmental Protection Agency.

Escudero, A., J. M. Iriondo, and M. E. Torres, 2003, Spatial analysis of genetic diversity as a tool for plant conservation, *Biol. Conserv.* **113**:351–365.

Ettema, C. H., and D. A. Wardle, 2002, Spatial soil ecology, *Trends Ecol. Evol.* **17**:177–183.

Ford, E. D., and E. Renshaw, 1984, The interpretation of process from pattern using two-dimensional spectral-analysis – Modeling single species patterns in vegetation, *Vegetatio* **56**:113–123.

Fortin, M. J., P. Drapeau, and P. Legendre, 1989, Spatial auto-correlation and sampling design in plant ecology, *Vegetatio* **83**:209–222.

Franklin, R. B., and A. L. Mills, 2003, Multi-scale variation in spatial heterogeneity for microbial community structure in an eastern Virginia agricultural field, *FEMS Microbiol. Ecol.* **44**:335–346.

Franklin, R. B., L. K. Blum, A. C. McComb, and A. L. Mills, 2002, A geostatistical analysis of small-scale spatial variability in bacterial abundance and community structure in salt marsh creek bank sediments, *FEMS Microbiol. Ecol.* **42**:71–80.

Franks, P. J. S., 2005, Plankton patchiness, turbulent transport and spatial spectra, *Mar. Ecol. Prog. Ser.* **294**:295–309.

Geary, R. C., 1954, The contiguity ratio and statistical mapping, *Incorp. Statist.* **5**:115–145.

Gilbert, R. O., 1987, *Statistical Methods for Environmental Pollution Monitoring.* Van Nostrand Reinhold, New York.

Goovaerts, P., 1997, *Geostatistics for Natural Resources Evaluation.* Oxford University Press, New York.

Goovaerts, P., 1998, Geostatistical tools for characterizing the spatial variability of micro-biological and physico-chemical soil properties, *Biol. Fertil. Soils* **27**:315–334.

Goovaerts, P., 1999, Geostatistics in soil science: state-of-the-art and perspectives, *Geoderma* **89**:1–45.

Griffith, D. A., 1978, A spatially adjusted ANOVA model, *Geogr. Anal.* **10**:296–301.

Grundmann, G. L., and D. Debouzie, 2000, Geostatistical analysis of the distribution of NH_4^+ and NO_2^--oxidizing bacteria and serotypes at the millimeter scale along a soil transect, *FEMS Microbiol. Ecol.* **34**:57–62.

Haining, R., 1993, *Spatial Data Analysis in the Social and Environmental Sciences.* Cambridge University Press, Cambridge.

Halvorson, J. J., H. Bolton, J. L. Smith, and R. E. Rossi, 1994, Geostatistical analysis of resource islands under *Artemisia Tridentata* in the shrub-steppe, *Gt. Basin Nat.* **54**:313–328.

Halvorson, J. J., J. L. Smith, H. Bolton, and R. E. Rossi, 1995, Evaluating shrub-associated spatial patterns of soil properites in a shrub-steppe ecosystem using multiple-variable geostatistics, *Soil Sci. Soc. Am. J.* **59**:1476–1487.

Hoosbeek, M. R., A. Stein, H. van Reuler, and B. H. Janssen, 1998, Interpolation of agronomic data from plot to field scale: using a clustered versus a spatially randomized block design, *Geoderma* **81**:265–280.

Isaaks, E. H., and R. M. Srivastava, 1989, *An Introduction to Applied Geostatistics*. Oxford University Press, New York.

Journel, A. G., and C. J. Huijbregts, 1978, *Mining Geostatistics*. Academic Press, London.

Keitt, T. H., and D. L. Urban, 2005, Scale-specific inference using wavelets, *Ecology* **86**:2497–2504.

Legendre, P., 1993, Spatial autocorrelation – trouble or new paradigm, *Ecology* **74**:1659–1673.

Legendre, P., and M. -J. Fortin, 1989, Spatial pattern and ecological analysis, *Vegetatio* **80**:107–138.

Legendre, P., and L. Legendre, 1998, *Numerical Ecology*, 2nd Edition. Elsevier Scientific, Amsterdam.

Legendre, P., N. L. Oden, R. R. Sokal, A. Vaudor, and J. Kim, 1990, Approximate analysis of variance of spatially autocorrelated regional data, *J. Class.* **7**:53–75.

Legendre, P., M. R. T. Dale, M. J. Fortin, J. Gurevitch, M. Hohn, and D. Myers, 2002, The consequences of spatial structure for the design and analysis of ecological field surveys, *Ecography* **25**:601–615.

Legendre, P., M. R. T. Dale, M. J. Fortin, P. Casgrain, and J. Gurevitch, 2004, Effects of spatial structures on the results of field experiments, *Ecology* **85**:3202–3214.

Levin, S. A., 1992, The problem of pattern and scale in ecology, *Ecology* **73**:1943–1967.

Lilleskov, E. A., T. D. Bruns, T. R. Horton, D. L. Taylor, and P. Grogan, 2004, Detection of forest stand-level spatial structure in ectomycorrhizal fungal communities, *FEMS Microbiol. Ecol.* **49**:319–332.

Ludwig, J. A., and J. F. Reynolds, 1988, *Statistical ecology: a primer on methods and computing*. Wiley, New York.

Mackas, D. L., 1984, Spatial autocorrelation of plankton community composition in a continental shelf ecosystem, *Limnol. Oceanogr.* **29**:451–471.

Mantel, N., 1967, The detection of disease clustering and a generalized regression approach, *Cancer Res.* **27**:209–220.

Manz, W., G. Arp, G. Schumann-Kindel, U. Szewzyk, and J. Reitner, 2000, Widefield deconvolution epifluorescence microscopy combined with fluorescence in situ hybridization reveals the spatial arrangement of bacteria in sponge tissue, *J. Microbiol. Methods* **40**:125–134.

Miller, J. R., M. G. Turner, E. A. H. Smithwick, C. L. Dent, and E. H. Stanley, 2004, Spatial extrapolation: The science of predicting ecological patterns and processes, *Bioscience* **54**:310–320.

Moran, P. A., 1950, Notes on continuous stochastic phenomena, *Biometrika*. **37**:17–23.

Morris, S. J., 1999, Spatial distribution of fungal and bacterial biomass in southern Ohio hardwood forest soils: fine scale variability and microscale patterns, *Soil Biol. Biochem.* **31**:1375–1386.

Mottonen, M., E. Jarvinen, T. J. Hokkanen, T. Kuuluvainen, and R. Ohtonen, 1999, Spatial distribution of soil ergosterol in the organic layer of a mature Scots pine (*Pinus sylvestris* L.) forest, *Soil Biol. Biochem.* **31**:503–516.

Murray, C. J., 2001, Sampling and data analysis for environmental microbiology, in: *Manual of Environmental Microbiology*, C. J. Hurst, R. L. Crawford, G. R. Knudsen, M. J. McInerney, and L. D. Stetzenbach, eds., American Society for Microbiology Press, Washington, DC, pp. 166–177.

Nunan, N., K. Ritz, D. Crabb, K. Harris, K. J. Wu, J. W. Crawford, and I. M. Young, 2001, Quantification of the in situ distribution of soil bacteria by large-scale imaging of thin sections of undisturbed soil, *FEMS Microbiol. Ecol.* **37**:67–77.

Nunan, N., K. J. Wu, I. M. Young, J. W. Crawford, and K. Ritz, 2003, Spatial distribution of bacterial communities and their relationships with the micro-architecture of soil, *FEMS Microbiol. Ecol.* **44**:203–215.

Oberrath, R., and K. Bohning-Gaese, 2001, The Signed Mantel test to cope with autocorrelation in comparative analyses, *J. Appl. Stat.* **28**:725–736.

Oden, N. L., and R. R. Sokal, 1986, Directional autocorrelation: an extension of spatial correlograms in two directions, *Syst. Zool.* **35**:608–617.

O'Neill, R. V., and A. W. King, 1998, Homage to St. Michael; or, why are there so many books on scale?, in: *Ecological Scale: Theory and Applications*, D. L. Peterson and V. T. Parker, eds., Columbia University Press, New York, pp. 3–15.

Parkin, T. B., 1987, Soil Microsites as a source of denitrification variability, *Soil Sci. Soc. Am. J.* **51**:1194–1199.

Pennanen, T., E. Liski, E. Baath, V. Kitunen, J. Uotila, C. J. Westman, and H. Fritze, 1999, Structure of the microbial communities in coniferous forest soils in relation to site fertility and stand development stage, *Microb. Ecol.* **38**:168–179.

Pielou, E. C., 1977, *An Introduction to Mathematical Ecology*, 2nd Edition. Wiley-Interscience, New York.

Piontkovski, S. A., R. Williams, W. T. Peterson, O. A. Yunev, N. I. Minkina, V. L. Vladimirov, and A. Blinkov, 1997, Spatial heterogeneity of the planktonic fields in the upper mixed layer of the open ocean, *Mar. Ecol. Prog. Ser.* **148**:145–154.

Platt, T., and K. L. Denman, 1975, Spectral analysis in ecology, *Annu. Rev. Ecol. Syst.* **6**:189–210.

Renshaw, E., and E. D. Ford, 1984, The description of spatial pattern using two-dimensional spectral-analysis, *Vegetatio* **56**:75–85.

Ripley, B. D., 1976, The second order analysis of stationary point processes, *J. Appl. Probab.* **13**:255–266.

Ripley, B. D., 1981, *Spatial Statistics*. Wiley, New York.

Robertson, G. P., 1987, Geostatistics in ecology: interpolating with known variance, *Ecology* **68**:744–748.

Robertson, G. P., and D. W. Freckman, 1995, The spatial-distribution of nematode trophic groups across a cultivated ecosystem, *Ecology* **76**:1425–1432.

Robertson, G. P., K. M. Klingensmith, M. J. Klug, E. A. Paul, J. R. Crum, and B. G. Ellis, 1997, Soil resources, microbial activity, and primary production across an agricultural ecosystem, *Ecol. Appl.* **7**:158–170.

Rossi, J. -P., 1996, Statistical tool for soil biology. XI. Autocorrelogram and Mantel test, *Eur. J. Soil Biol.* **32**:195–203.

Rossi, R. E., D. J. Mulla, A. G. Journel, and E. H. Franz, 1992, Geostatistical tools for modeling and interpreting ecological spatial dependence, *Ecol. Monogr.* **62**:277–314.

Saetre, P., 1999, Spatial patterns of ground vegetation, soil microbial biomass and activity in a mixed spruce-birch stand, *Ecography* **22**:183–192.

Saetre, P., and E. Baath, 2000, Spatial variation and patterns of soil microbial community structure in a mixed spruce-birch stand, *Soil Biol. Biochem.* **32**:909–917.

Schlesinger, W. H., J. A. Raikes, A. E. Hartley, and A. E. Cross, 1996, On the spatial pattern of soil nutrients in desert ecosystems, *Ecology* **77**:364–374.

Schneider, D. C., 1994, *Quantitative Ecology – Spatial and Temporal Scaling.* Academic Press, San Diego.

Seuront, L., F. Schmitt, Y. Lagadeuc, D. Schertzer, and S. Lovejoy, 1999, Universal multifractal analysis as a tool to characterize multiscale intermittent patterns: example of phytoplankton distribution in turbulent coastal waters, *J. Plankton Res.* **21**:877–922.

Smith, J. L., J. J. Halvorson, and H. J. Bolton, 1994, Spatial relationships of soil microbial biomass and C and N mineralization in a semi-arid shrub-steppe ecosystem, *Soil Biol. Biochem.* **26**:1151–1159.

Smouse, P. E., J. C. Long, and R. R. Sokal, 1986, Multiple regression and correlation extensions of the Mantel test of matrix correspondence, *Syst. Zool.* **35**:627–632.

Sokal, R., 1979, Testing statistical significance of geographic variation patterns, *Syst. Zool.* **28**:227–232.

Sokal, R. R., 1986, Spatial data analysis and historical processes, in: *Data Analysis and Informatics, IV*, E. Diday, Y. Escoufier, L. Lebart, J. Pages, Y. Schertman, and R. Tomassone, eds., North Holland, Amsterdam, pp. 29–43.

Sokal, R. R., and N. L. Oden, 1978, Spatial autocorrelation in biology. 2. Some biological implications and four applications of evolutionary and ecological interest, *Biol. J. Linn. Soc.* **10**:229–249.

Sokal, R. R., J. Bird, and B. Riska, 1980, Geographic variation in *Pemphigus populicaulis* (Insecta: Aphididae) in eastern North America, *Biol. J. Linn. Soc.* **14**:163–200.

Sokal, R. R., N. L. Oden, B. A. Thomson, and J. H. Kim, 1993, Testing for regional differences in means – distinguishing inherent from spurious spatial autocorrelation by restricted randomization, *Geogr. Anal.* **25**:199–210.

Star, J. L., and M. M. Mullin, 1981, Zooplankton assemblages in three areas of the North Pacific as revealed by continuous horizontal transects, *Deep Sea Res.* **28**:1303–1322.

Steele, J. H., and E. W. Henderson, 1992, A simple-model for plankton patchiness, *J. Plankton Res.* **14**:1397–1403.

Stein, A., J. Riley, and N. Halberg, 2001, Issues of scale for environmental indicators, *Agric. Ecosyst. Environ.* **87**:215–232.

Thompson, S. K., 2002, *Sampling.* Wiley-Interscience, New York.

Troussellier, M., P. Lebaron, B. Baleux, and P. Got, 1993, Spatial-distribution patterns of heterotrophic bacterial populations in a coastal ecosystem (Thau Basin, France), *Estuar. Coast. Shelf Sci.* **36**:281–293.

Turner, M. G., 1989, Landscape ecology – the effect of pattern on process, *Ann. Rev. Ecol. Syst.* **20**:171–197.

Turner, M. G., and S. R. Carpenter, 1999, Spatial variability in ecosystem function – Introduction, *Ecosystems* **2**:383–383.

Upton, G., and B. Fingleton, 1985, *Spatial Data Analysis by Example.* Volume 1: *Point Pattern and Quantitative Data.* Wiley, Chichester, England.

van Es, H. M., and C. L. van Es, 1993, Spatial nature of randomization and its effect on the outcome of field experiments, *Agron. J.* **85**:420–428.

Walter, C., A. B. McBratney, R. A. V. Rossel, and J. A. Markus, 2005, Spatial point-process statistics: concepts and application to the analysis of lead contamination in urban soil, *Environmetrics* **16**:339–355.

Wiegand, T., and K. A. Moloney, 2004, Rings, circles, and null-models for point pattern analysis in ecology, *Oikos* **104**:209–229.

Wu, J. G., and S. A. Levin, 1994, A spatial patch dynamic modeling approach to pattern and process in an annual grassland, *Ecol. Monogr.* **64**:447–464.

Chapter 3

BACTERIAL INTERACTIONS AT THE MICROSCALE – LINKING HABITAT TO FUNCTION IN SOIL

Naoise Nunan[1], Iain M. Young[2], John W. Crawford[2], and Karl Ritz[3]
[1] *CNRS, BioEMCo, Bât. EGER, Centre INRA-INAPG, 78850 Thiverval-Grignon, France;*
[2] *SIMBIOS Center, University of Abertay Dundee, Bell Street, Dundee, DD1 1HG, UK;*
[3] *National Soil Resources Institute, Cranfield University, Silsoe, MK45 4DT, UK*

Abstract: There is a growing body of evidence that the spatial distribution of bacteria and their relationships with other soil features play a significant role in the macroscopic function of soil. In the past this has not been widely appreciated, possibly due to the difficulty of studying soils at scales that are relevant to bacterial communities. This paper reviews the evidence for the influence of microscale interactions on function at larger scales and describes recent methodological advances that allow the microscale spatial distribution of bacterial cells and bacterial activities to be quantified. Approaches for integrating the microscale into models of soil function are briefly discussed as are new techniques that have the potential to improve our understanding of microbial – habitat interactions and of how these are linked to soil function.

Keywords: bacterial spatial distribution, microscale, microhabitat, scale, biological thin sections, microsampling

1. INTRODUCTION

Patterns in the spatial distribution of organisms and the scale at which such patterns are apparent can provide information on the factors that drive individual and community development and function. Although the importance of spatial interactions in shaping ecological interactions is recognised in studies of aboveground biota, spatial aspects of soil microbial ecology have not received a great deal of attention. Traditionally the spatial heterogeneity of microbial function in soil has been viewed as a problem rather than as an intrinsic property, and there has been a tendency to homogenise samples prior to analysis. There is a *de facto* assumption associated with homogenisation that the distribution of micro-organisms within the sample (or sampled area if several samples are combined prior to analysis) is random or at least of no significance to the process of interest. Given that soils function predominantly by virtue of their spatial organisation (Harris, 1994; Young and Ritz, 2005), this is scientifically naive. However, the importance of spatial structure in soils is being increasingly recognised and patterns in the spatial distribution of soil micro-organisms and of associated microbial activity have attracted the interest of the soil scientists, with patterns at the centimetre to landscape scales being the primary focus of attention (e.g., Parkin and Shelton, 1992; Goovaerts and Chiang, 1993; Gonzalez and Zak, 1994; Velthof et al., 1996; Bergstrom et al., 1998; Bruckner et al., 1999; Stoyan et al., 2000; Ettema and Wardle, 2002; Vieublé-Gonod et al., 2005). The study of spatial patterns of soil microbial communities and activity is driven by the desire to better quantify microbial function but also to identify the factors underlying community function and dynamics (Parkin, 1993). With regard to the latter, spatial patterns have been related to landscape scale ecological gradients (Fromm et al., 1993; Robertson et al., 1997), to the patchy distribution of nutrient supply at the plot scale (Ritz et al., 2004), to decomposing plant material (Ronn et al., 1996) and, in ecosystems where plants tend to be sparse, to patterns of plant distribution (Jackson and Caldwell, 1993; Klironomos et al., 1999; Bruckner et al., 1999).

The scale at which spatial patterns occur is related to the scale at which environmental or intrinsic factors have an impact on community development. A number of studies have shown that microbial communities exhibit spatial patterns at different scales. Franklin and Mills (2003) revealed nested scales of variability in the similarity of amplified fragment length polymorphism (AFLP) profiles in samples from an agricultural soil and Nunan et al. (2002) identified spatial structure in the distribution of bacterial abundance at two scales in subsoil. With respect to the latter, the scale of spatial structure suggested that the structuring agent for the large scale pattern might be preferential flow paths while the microscale structure was thought to be related to microhabitat properties. Nested scales of variation indicate that soil micro-organisms are subjected to ecological interactions at a range of scales (Robertson and Gross, 1994).

Saetre and Bååth (2000) detected two scales of spatial structure in microbial communities measured using phospholipids fatty acid profiles in a Norway spruce-birch stand. The more important component of the community was spatially structured in large patches influenced by spruce trees whilst another component was more influenced by birch and formed smaller patches. Ritz et al. (2004) identified a highly complex set of spatial patterns in the microbial community structure of an upland grassland, with subsets of the community showing spatial patterns at scales ranging from 0.7 to >6 m whilst others were not patterned at these scales at all. Combinations of patterns such as these arise because subsets of the communities respond differently to environmental gradients or habitat properties, i.e., they interact differently with their environment (Franklin and Mills, 2003; Levin, 1992).

It has been demonstrated, in both theoretical models and experimentally, that the scale at which ecological processes occur has an effect on microbial diversity (Kerr et al., 2002; Durrett and Levin, 1997). When interactions and dispersal occur at a local level in spatially structured environments competitors can coexist, but when the scale of ecological process is larger, diversity is not maintained. Furthermore, it has been suggested that soil exhibits many features of complex adaptive systems (Crawford et al., 2005). In such systems, macroscopic properties of ecosystems may emerge from interactions among units at smaller scales and the emergent properties can feedback to alter subsequent interactions (Levin, 1998).

Thus, we are faced with the challenging prospect of having to account for mechanisms occurring at different scales and for the interaction among mechanisms across scales in order to gain a full understanding of how soil microbial communities evolve, function, and respond to external stress or change. To achieve this, measurements must therefore be made at relevant scales. Although the activity of individual microbial cells, and in particular bacterial cells, is highly localised in space, fine-scale patterns have not been described in any great detail. The lack of information concerning interactions at these scales, whether intrinsic to microbial communities or with their habitat, means that such interactions cannot be explicitly accounted for in our current models of soil microbial function and development. Whether this is because of a belief that fine-scale patterns are not pertinent for understanding ecological processes at field or landscape scales or due to technological constraints is not clear. However, a number of experimental approaches are currently being employed to study microbial-habitat interactions and the initial steps are being made with many others. The remainder of the chapter will review the different methodological approaches that have been adopted for studying microbial distribution at the microscale and the importance of such patterns for ecological processes.

2. DETECTING AND QUANTIFYING SPATIAL PATTERNS

Spatially referenced sampling schemes are generally used to establish the existence of patterns of spatial variability at the plot to landscape scales. It is not proposed to discuss optimal sampling strategies for characterizing spatial patterns at these scales as this has been discussed at great length elsewhere (Bramley and White, 1991a, b; Warrick and Myers, 1987; Chapter 8, this volume).

The size of the microscale depends on the scale at which microorganisms interact with their microhabitat and the methods of measurement employed to quantify the microscale should demonstrate a cognizance of the scales of these interactions. Different classes of soil organisms interact with their microhabitats at different scales, and such interactions are not necessarily directly confined to the microbial (μm) scale. For example, filamentous fungi transcend spatial scales across many orders-of-magnitude by virtue of the mycelium, and their effects upon soil structure and processes operate across a similarly broad range (see Chapter 7). In contrast, bacterial interactions, intrinsic to communities or with their microhabitat, tend to occur at the scale of a few micrometres at most (Harris, 1994; Chapter 1, this volume). Grundmann and Normand (2000) found that the genetic distances of the genus *Nitrobacter* at a local scale (<3 cm) were as large as those among reference strains from a range of geographical areas, suggesting that the biological and physical processes regulating diversity occur at very fine scales indeed.

2.1. Microscale distribution of bacteria and bacterial processes

Quantifying the spatial distribution of bacterial cells or bacterial activity at the microhabitat scale and in relation to other microhabitat features poses special technical challenges both with regard to sampling and to the preservation of bacterial – microhabitat relations. Two separate approaches have been adopted. One uses thin sections of soil to preserve *in situ* spatial relations, spatially referenced imaging and image analysis (Nunan et al., 2001); the other employs a microsampling technique that is based on the relationship between sample size and the frequency of occurrence of a process (Dechesne et al., 2003; Grundmann et al., 2001).

2.2. Soil thin sections

Micromorphology has a long history in soil science, pioneered by Kubiena in the first half of the twentieth century (Kubiena, 1938). The techniques adopted had their roots in geology, and although there was always an appreciation of the need to preserve soil structure (largely from a mineralogical perspective), and pore networks, there was little empathy with the delicacy of living structures in the soil. Preparation techniques tended to be such that only the most robust life forms, such as thick-walled spores, melanised hyphae, arthropod chitin, and woody roots were visible, and most microbial life was not considered. The use of micromorphological techniques for the more considered study of soil microbes began in the 1950s when soil thin sections were first used to study micro-organisms in their natural environment by Alexander and Jackson (1955). The method was subsequently modified by Jones and Griffiths (1964) and used to map the spatial distribution of bacterial colonies in soil aggregates. However, technical limitations meant that they were unable to visualise individual bacterial cells by light microscopy; resins and dehydration techniques were not refined enough, and the need for explicit fixation of biological tissue *in situ* was not fully appreciated. Pioneering work by Foster in the 1970s can be regarded as the first application of soil "histology", and here the approach was to use the emerging technology of electron microscopy to study soil bacteria at the ultrastructural scale (nanometre to micrometre), and with a crucial appreciation for the need for fixation of biological tissues. Foster produced some remarkable, and in many ways still unsurpassed, images of soil bacteria and fungi, and some pioneering illustrations of the then burgeoning concept of the rhizosphere (Foster and Rovira, 1973; Foster and Martin, 1981; Foster et al., 1983). Tippkötter et al. (1986) further explored the necessity of fixation in order to visualise biological structures in soil thin sections above the ultrastructural scale. However, this early work was largely qualitative and it was not until fluorescent stains (fluorochromes) were introduced that potential for visualising bacteria at scales appropriate for quantification emerged (Tippkötter, 1990; Altemüller and Vliet-Lanoe, 1988). Bacteria are much more clearly distinguishable from other soil features when studied with fluorescent microscopy, as a comparison of the images in Jones and Griffiths (1964) and Postma and Altemüller (1990) clearly shows. This development allowed the relationship between bacteria and soil features to be characterised, at least qualitatively (White et al., 1994; Fisk et al., 1999). With increases in computer power and the use of image analysis bacterial distributions and their relationships with soil features have been characterised in a quantitative manner (Nunan et al., 2001, 2003). The great advantages of quantitative approaches over mere qualitative observations are that the significance of the results can be tested statistically and conclusions can be incorporated into mathematical models (Fig. 3-1).

2.2.1. Preparation of biological thin sections

The preparation sequence involves sampling undisturbed cores of soil and fixing the biological component with a histological fixative to ensure that the spatial integrity of the tissues and the relations between the biological and other components of the sample are preserved (Fig. 3-2a). Samples are then stained with a fluorochrome and dehydrated. A range of fluorochromes have been used to locate bacteria, including Calcofluor white (Nunan et al., 2001), Thiazine red R. (Fisk et al., 2000) and a combination of sulfofluorescein diacetate (SFDA), a metabolic type fluorochrome, and magnesium salt of 1-anilino-8-naphthalene sulfonic acid (Mg-ANS), an adsorption type fluorochrome, for detecting the proportion of live cells (Tsuji et al., 1995). Important criteria in choice of fluorochrome are that the target organisms be adequately and uniformly stained and that the distribution of stain throughout sample be even i.e., no spatial bias. Predominantly cationic stains such as acridine orange are unlikely to be suitable candidates for spatial studies of soil bacteria. Samples are dehydrated either by chemical drying using a graded series of acetone or by air-drying. Samples are then impregnated with a resin which, after polymerisation of the resin, results in a resin-embedded block of soil and from which thin section are cut to produce slices approximately 25 µm thick. Tippkotter and Ritz (1996) tested the suitability of a range of different resins for the preparation of biological thin sections on soils with widely contrasting properties and concluded that Crystic and Palatal resins were the most suitable. The high fidelity of the procedure is suggested by the presservation of fungal hyphae spanning soil pores (Fig. 7-2) and by the presence of undisturbed bacterial colonies (Figs. 3-1a and 3-2a).

Maps of bacterial cell distributions can then be produced using microscopes fitted for epifluorescence and with a computer controlled stage and image analysis procedures (Fig. 3-1b). Due to the heterogeneity of the images of soil thin sections or of other soil preparations (Fig. 3-1a and 3-2a), the image analysis procedures required to identify bacterial cells in soil preparations are generally quite complicated and involve a great many steps (Bloem et al., 1995; Nunan et al., 2001). Using other forms of illumination such as brightfield, polarised, ultraviolet and infrared light, it is possible to categorise, identify and produce maps of soil features such as voids or organic material (Protz et al., 1992). The computer controlled stage enables one to acquire different images from the same location on the thin sections. Bacterial cell distributions in relation to other features can be quantified by superimposing the maps obtained with different illumination techniques and calculating distances between bacterial cells and features of interest (Nunan et al., 2003; Bruneau et al., 2005).

2.2.2. Conclusions from biological thin sections

Using this method the spatial distribution of individual bacterial cells has been measured in arable topsoil and subsoil (Nunan et al., 2002, 2003), and in different horizons of upland grasslands (Bruneau et al., 2005). Bacteria were found to have a patchy distribution in both topsoil and subsoil: patch sizes being greater in the topsoil (Nunan et al., 2002, 2003). The internal structure of the patches was also different: gradients of bacterial density were found in topsoil samples, whilst a mosaic of high and low values was found in subsoil samples (Nunan et al., 2003). Thus, in the topsoil samples bacterial communities appeared to develop around defined loci, possibly sources of nutrients, whilst in the subsoil this was not the case. Bacterial "hot spots" were also related to the location of pores in the subsoil (Nunan et al., 2003) and to fresh macro- and mesofaunal excremental features in upland and grassland (Bruneau et al., 2005). The spatial distribution of bacteria was more variable in samples with lower mean bacterial densities (Nunan et al., 2001, 2003; Bruneau et al., 2005). This may have been due to a patchier distribution of substrate when availability was low restricting the number of sites amenable to bacterial growth. This was particularly true in subsoil samples where colonised areas containing small colonies tended to be surrounded by vast (relative to bacterial cells or colonies) swathes of emptiness (Nunan et al., 2003). This was reflected in the very high Ripley's K function values (which give a measure of the degree of clustering as a function of distance) at small scales (Nunan et al., 2002). The different spatial distributions observed in topsoil and subsoil, allied to the fact that the microbial community structure also varies with depth (Fierer et al., 2003), suggests that mathematical models that assume that the subsoil is merely a less active version of the topsoil are unlikely to produce sensible results.

There was little evidence of biofilms (multilayered consortia of bacteria) in soil in either the arable or grassland soils (Nunan et al., 2003; Bruneau et al., 2005). It is possible that interspecies competition, nutrient limitation or both hinder the development of large colonies of morphologically similar cells (Franklin et al., 2001; Grundmann and Gourbiere, 1999). Figure 3-2a shows that several different morphotypes coexist at very fine scales.

The disadvantages of the approach are twofold: (a) no distinction is made between active and non-active cells, meaning that the functional significance of a given distribution is difficult to ascertain and specific functions cannot be attributed to bacteria, and (b) measurements are made in two dimensions whereas soil is a three-dimensional medium.

2.3. Microsampling procedure

2.3.1. Methodology

The three-dimensional spatial distribution of bacterial activity at the microscale has been measured using a microsampling procedure developed by Grundmann et al. (2001). The procedure is based on the idea that the probability of a process being detected at several different scales is linked to the spatial distribution of the process (Madden and Hughes, 1999). Microsamples of different volumes (fitting into squares of a calibrated grid with side lengths ranging from 50 to 500 µm) were dissected from an aggregate by means of a sterile scalpel. The microsamples were then transferred into wells of microculture plates containing a defined culture medium (depending on the process of interest) and incubated in the dark (Grundmann et al., 2001). Microsamples were then scored as positive or negative and the proportion of positive scores determined. Various three-dimensional theoretical spatial distributions (clustered or random) were sampled in the same way and the experimental data compared with the data from the theoretical distributions (Grundmann et al., 2001). This allows possible spatial distributions to be identified, though not the exact distributions, because microsample coordinates were not determined. The procedure was improved by replacing the computationally intensive simulation step with analytical solutions (Dechesne et al., 2003). The proportion of soil occupied by volumetric units of 50 µm side length can also be determined.

2.3.2. Conclusions from microsampling procedure

The microscale spatial distributions of NH_4^+ oxidisers and 2,4-D degraders were found to differ significantly (Dechesne et al., 2003); NH_4^+ oxidisers were found to exist in more numerous but smaller preferentially colonised than 2,4-D degraders. The different spatial distributions may have arisen because of the different natures of the substrates and how they were distributed. On the other hand, the distributions of NH_4^+ oxidisers and NO_2^- oxidisers exhibited a degree of spatial association in microsamples of 100 µm side length (Grundmann et al., 2001). The association at a very fine scale is consistent with the fact that one produces substrate for the other.

Pallud et al. (2004) found that the distribution of 2,4-D degraders was altered by percolating 2,4-D but not by water alone. Colonised patches were

Linking Habitat to Function at the Microscale 69

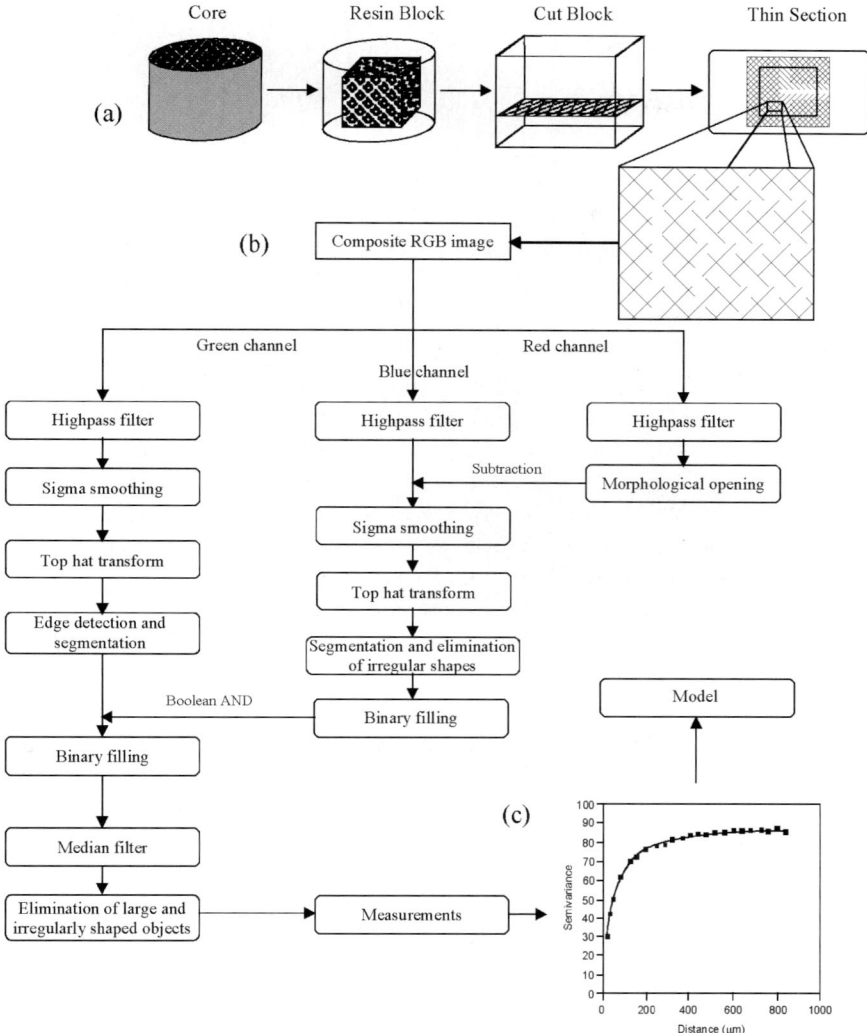

Figure 3-1. Thin section approach for study of bacterial spatial distribution at microscale. The procedure can be separated into three sections: (a) production of thin sections and visualization, (b) image analysis (of color image) for identifying bacterial cells in images and (c) quantitative analysis and informing of mathematical models. The image analysis section indicates the number of steps necessary for the recognition of cells but may be adapted depending on the image. In section (c) a semi-variogram is presented but other forms of quantitative analysis are also possible such as Ripley's K function or the distance among bacteria and other features of interest. (Adapted from Nunan et al., 2001; with permission.)

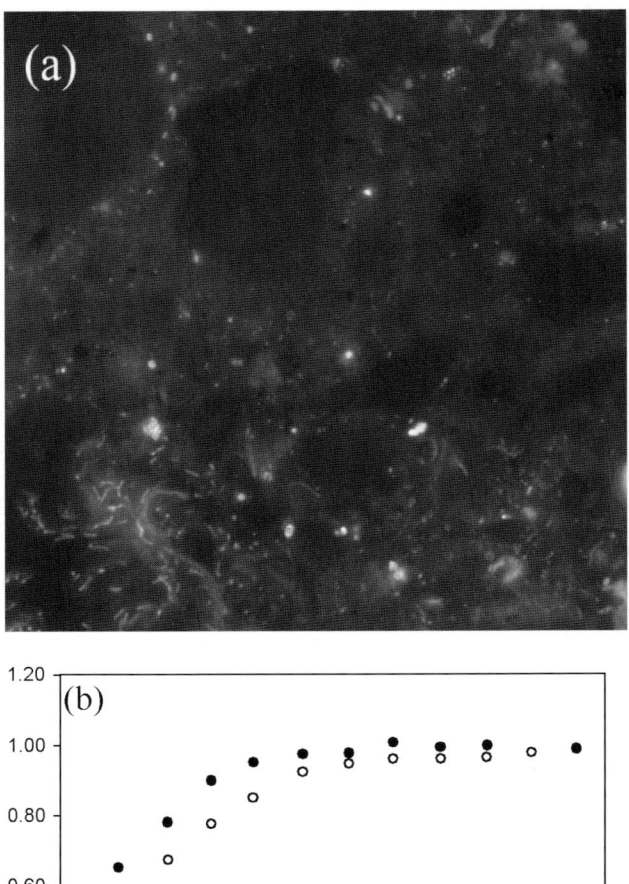

Figure 3-2. Image showing example of distribution of bacterial cells in control sample (a) and (b) semi-variograms for bacterial density in control samples (o) and after 2,4-D percolation (●). The ranges of the semi-variograms are similar but there was an increase in nugget variance (variance at scales below the scale of measurement, 50 μm). Bacteria are distributed as individuals, as pairs or in colonies. A fungal hyphum spanning a pore is also visible. The side length of the image is 100 μm. (Biological thin sections were prepared with samples obtained from Grundmann and Deschesne.)

larger after 2,4-D addition, suggesting that 2,4-D degraders spread during growth, increasing the probability of encounters with the substrate (Pallud et al., 2004). Higher 2,4-D degradation rates among samples with similar 2,4-D degrader densities were associated with those which displayed a greater dispersal of degraders within the soil volume. An analysis of the distribution of total bacterial density of two of these samples (water and 2,4-D percolation) using biological thin sections (Fig. 3-2a) indicated that, contrary to 2,4-D degraders, the scale of bacterial patterns was unaffected by 2,4-D percolation (Fig. 3-2b). Although the ranges of the semi-variograms were similar, there was an increase in nugget variance (variance at scales below the scale of measurement, 50 µm), suggesting very fine-scale changes in bacterial distribution. This corroborates the suggestion that different subcomponents of the soil bacterial populations respond differently to local environmental conditions (Franklin and Mills, 2003).

The main drawback of this method is that the relationship between microbial spatial patterns and the soil structure is not quantifiable, particularly larger scale structures such as macropores which are lost. Therefore the effect such structures might have on a given process cannot be quantified.

Thus, both approaches have shown independently that processes and interactions occurring at the bacterial scale such as interactions with local environmental gradients do not result in a random distribution of cells and of loci of activity but rather they constrain bacterial community development in such a way that quantifiable spatial patterns emerge.

3. CONSEQUENCES OF SPATIAL DISTRIBUTION

3.1. Intrinsic interactions

It is now well established that processes such as horizontal gene transfer and trait dispersal occur in soils (Top et al., 1998; Dröge et al., 1999) and there is indirect evidence that quorum sensing does so also (Chernin et al., 1998; Burmølle et al., 2003). Horizontal gene transfer among bacteria is mediated by one of three mechanisms: conjugation, transformation, or transduction (Dröge et al., 1999). Conjugation requires contact between cells for specific plasmids to be transferred from donor to recipient. Transformation is the uptake of cell-free DNA by cells and transduction is bacteriophage-mediated transfer of genetic information between cells. In soils, the conjugative transfer of plasmid-borne genes encoding 2,4-D catabolism from introduced donor strains to the indigenous bacteria has been identified as a mechanism by which the resident communities adapt to 2,4-D degradation (Top and

Springael, 2003). The effect was particularly evident in a B horizon that had no apparent ability to degrade 2,4-D without the donor strains but degraded all the 2,4-D in under 3 weeks after introduction of the donor strains (Dejonghe et al., 2000). The introduced ability to degrade was probably due to transconjugants as the donor strains were undetectable when degradation started. Soil particles are known to have the capacity to protect DNA from hydrolysis by DNases and recently it has been shown that DNA adsorption onto Montmorillonite–Humic Acids–Aluminum or Iron Hydroxypolymers does not prevent transformation of competent cells (Crecchio et al., 2005), suggesting that transformation in soil may be possible should susceptible hosts come into contact with DNA-soil particle complexes. In the rhizosphere there is potential for transduction due to the high densities of bacteria and bacteriophages. Bacteriophages were found to dominate viral communities in 6 Delaware soils, the abundance of which was between ~9×10^8 and 3×10^9 g^{-1} soil (Williamson et al., 2005).

Quorum sensing is a regulatory mechanism in which bacteria respond in a coordinated, population-density manner through the secretion and sensing of specific signalling molecules thereby accomplishing tasks which would be difficult, if not impossible, to achieve for a single bacterial cell. A concerted response is induced in the population once a threshold concentration of the signalling molecule is reached. The threshold concentration corresponds to a certain population density as the extracellular concentration of the signalling molecule is related to the population density of the secreting. This type of cell-to-cell communication was termed "quorum sensing" because a sufficient number of bacteria, the bacterial "quorum", is needed to induce or repress expression of target genes (Whitehead et al., 2001). Chernin et al. (1998) showed that the production of chitinolytic enzymes in *Chromobacterium violaceum*, a soil-borne bacterium, is regulated by the quorum sensing signaling molecule N-hexanoyl-L-homoserine lactone. The signaling molecules of non-isogenic populations have also been found to modulate the activity of other populations in wheat rhizosphere (Pierson et al., 1998).

The frequency of occurrence of these phenomena and thus their relative importance in ecosystem function is highly dependent on the spatial distribution of bacterial cells. The spatial distribution of bacterial cells will affect the likelihood of encounters among bacteria or of a bacterial quorum being reached, for example. It has been suggested that interactions via volatile organic compounds of microbial origin maybe widespread in soil (Wheatley, 2002). If this were the case, then interactions would be modulated by moisture content and one would expect a strong pore size × moisture content interaction.

Competition among micro-organisms is likely to be affected by their spatial distribution in relation to one another. The micro-structure of soils affects the spatial location of the soil biota (Nunan et al., 2003), and in so doing controls the degree to which organisms encounter each other and interact within the soil volume. Spatial isolation is believed to be one of the factors contributing to non-competitive diversity patterns and high levels of microbial diversity at small scales in soils (Zhou et al., 2002). In a theoretical model investigating the production of extracellular enzymes as a foraging strategy, Allison (2005) concluded that spatially structured environments permitted the coexistence of extracellular enzyme producing micro-organisms and micro-organisms that use the product of the enzymatic activity but without producing the enzymes themselves, "cheating" micro-organisms, despite the extra cost, and competitive disadvantage, to the enzyme producer in producing the enzymes. In homogenous environments however, interactions between the two types were altered and the cheating micro-organisms dominated. Experimental work by Treves et al. (2003) showing that less competitive micro-organisms can survive in spatially structured environments when isolated from more efficient competitors and so do not compete directly for resources, but become extinct in mixed environments, support the hypothesis that spatial isolation is an important factor in the maintenance of microbial diversity in soils. Diffusion limitations of substrate in poorly connected environments such as unsaturated soils that act to maintain a range of nutrient gradients resulting in a variety of micro-niches may be a mechanism by which high diversity is maintained (Long and Or, 2005). Furthermore, when diffusion coefficients are low the dissipation gradient length and therefore the "reach" of more competitive organisms are reduced. The diffusion pathway, which results from the interaction between moisture content and pore geometry, was also found to have a bearing on the survival of weaker species. When diffusion rates are low the diffusion pathway may be such that a weaker species intercepts nutrients that results in a nutrient depleted zone between species that inhibits invasion by the stronger species (Long and Or, 2005). The importance of diffusion pathways has been observed by electron microscopy (Chenu et al., 2001): micro-organisms were predominantly on the surface of clayey aggregates (with many pores <0.2 µm) but distributed throughout aggregates from a sandy soil (with pores ranging mostly 6–30 µm) and that glucose addition increased microbial growth throughout the sandy soil aggregates but mainly on the surface of clayey aggregates. Under conditions where diffusion rates are high these constraints to competition disappear or are reduced and this can result in a loss of diversity (Long and Or, 2005).

3.2. Bacteria – microhabitat interactions

Soil is a complex 3-D physical framework, which has variable geometry, composition and stability across scales spanning several orders of magnitude. The physical structure regulates the flow of water and nutrients in unsaturated conditions and of gases into and out of the soil matrix (Young and Ritz, 1998). The rates of these processes are heavily dependent on moisture content and fine-scale structure interactions (Young and Crawford, 2004). Small changes in moisture content can alter the process rate coefficients by orders of magnitude.

There is much evidence to suggest that this heterogeneity is not a purely random phenomenon but rather that it is ordered in an aggregated architecture (Young and Ritz, 2000). The aggregated nature of soils results in a broad range of pore sizes and a highly diverse microscale physical structure (Nunan et al., 2005). Consequently, external environmental conditions are not replicated uniformly throughout the soil and a large diversity of microhabitats can develop that are more or less suitable for bacterial growth, survival and activity. Sexstone et al. (1985) found that the O_2 distribution in water-saturated aggregates was irregular and that O_2 gradients differed among aggregates incubated under the same conditions. Their results suggest that the diffusion and consumption of O_2 is highly uneven in soil aggregates. Thus, bacteria inside an aggregate can experience radically different conditions from those situated at the periphery and from those in other aggregates of similar size.

Bacterial community structure within micro-aggregates was found to differ from that of whole micro-aggregates (Mummey and Stahl, 2004) and differences in bacterial community structure and abundance associated with different aggregate size fractions have also been reported (Ranjard et al., 2000b). Interactions between bacterial communities and soil micro-structure have been shown to affect the response of the communities to heavy metal stress (Ranjard et al., 2000a; Almås et al., 2005). Communities that were loosely bound to the exterior of aggregates were more resistant to heavy metal stress than communities adhering strongly to the interior of aggregates. The physical protection from previous heavy metal stress provided by the soil aggregates to the communities located on their interior meant that these communities were less well adapted to the presence of heavy metals.

Extracellular enzyme activity profiles associated with aggregate size fractions have also been found to differ (Marx et al., 2005). Carbohydrates showed a bimodal distribution, with activity maxima in sand and clay fractions, while acid phosphatase and leucine-aminopeptidase showed maximal activity in the clay fraction only. Kinetic studies revealed the presence of two isoenzymes in the clay fraction for all enzymes assayed as opposed to a single isoenzyme in the other aggregate size fractions for half the enzymes assayed. Although these results do not necessarily point to differences in the biological origin of the enzymes associated with the different aggregate size fractions as for example abiotic factors such as differential binding to the solid matrix may cause kinetic properties to diverge, they do suggest that the catalytic potential of soils at the microscale is highly heterogeneous. It is certainly appealing to hypothesise that the variable catalytic potential is of microbiological origin, particularly as bacterial community structure has been found to differ across aggregate size classes (Ranjard et al., 2000b). If this were the case, then a basic tenet of microbial ecology, Beijerink's principle that "everything is everywhere, the environment selects" (Beijerink, 1913) may not hold at the microscale – this in a scale-dependent concept. A variable catalytic potential modulated by the environment sensed by microbial communities may be closer to reality.

The diffusion of substrate to or metabolites from bacterial cells will greatly affect rates of activity. These processes are regulated by the soil architecture (Young and Ritz, 1998), but their importance for microbial function will also depend on how the micro-organisms are distributed at the microscale within the soil structure. Darrah et al. (1987) found that a nitrification model in which nitrifier aggregation was accounted for was better able to reproduce the different time courses of ammonium oxidation reported in the literature than a model based on a uniform distribution of cells. This difference was hypothesised to be due to a lowering of the pH in the microenvironment of bacterial clusters as a result of the nitrification process. The hypothesis was corroborated experimentally by Strong et al. (1997) who reported that nitrification became pH-limited at a higher pH in samples treated with lime (to increase the initial pH of the bulk soil) than in control samples. They concluded that acid accumulation in active microsites, because of acid diffusion into the bulk soil was slow, meant that pH measurements on bulk soil were not representative of the micro-environment pH experienced by the nitrifying bacteria and did not give a good indication of when the critical pH for nitrification limitation was reached. The location of bacterial inocula introduced into small pores (neck diameter <6 µm) was found to offer significant protection from predation by protozoa compared with inocula introduced into pores with a neck diameter of 30–60 µm (Wright et al., 1995). Recent work has shown that soil micro-organisms can act, in the presence of substrate, to alter their microhabitat towards a more porous, ordered and aggregated structure (Feeney et al., 2006). These findings have important consequences for our understanding of how physical structure regulates functional properties of soil such as microtransport processes of molecules to and from microbial communities and for the exposure of communities to environmental stresses.

There is therefore, a growing body of evidence to suggest that the location of individual bacterial cells, both in relation to other cells but also in relation to the range of micro-environments that can exist in soil, can have a significant impact on community survival, development and activity. Therefore, accounting for microscale interactions should become a primary focus of current modelling efforts.

4. MODELLING

As already stated, an advantage of being able to quantify microbial spatial distributions is that the distributions can be incorporated into mathematical models and their importance for ecological function assessed. This is by no means a trivial task. The complex structure of soil has to be adequately represented and microbial distributions incorporated into the structure. This means that the spatial relationships between microbial communities and soil structure have to be described.

Over the last decade significant advances have been made in simulating the complex spatial structure of soil (Vogel and Roth, 1998; Peat et al., 2000; Bird et al., 2000; Young et al., 2001). Some, mainly limited to fractal-based models, have attempted to link the geometry of soil structure to soil function such as hydraulic conductivity or the moisture characteristic (Young et al., 2001; Bird et al., 2000). Others have used network models in which the pore space geometry of soils is described in an idealised manner (Vogel and Roth, 2001). This is done by deriving model parameters from experimental data (Peat et al., 2000; Johnson et al., 2003) or from morphological investigations (Vogel and Roth, 2001). The impact of physical structure on the behaviour of soil biota has also been examined (Kampichler and Hauser, 1993; Young and Crawford, 2001). Many of these models suffer from the fact that they tend to use abstract, or greatly simplified, representations of structure and so cannot adequately represent fine-scale variation of soil structure or microhabitat heterogeneity.

In order to enable an explicit account microscale interactions a Markov chain approach was adopted for modelling soil structure in two dimensions (Wu et al., 2004). The model was successful in simulating structures at the microhabitat scale. The simulated structures are statistically similar ($P < 0.05$) to real structures observed in soil thin sections. This approach has already been extended to model three-dimensional structures in sandstone (Okabe and Blunt, 2004) and in soil. So far these models have been used to simulate soil structure and therefore pixels or voxels in the model can take one of two states – pore or solid. By increasing the number of states that can be assigned to a pixel or voxel, it is possible to extend the model further to include other features of soil such as the presence of microbial cells.

Because the specific location of bacterial populations in relation to other soil features is poorly understood, random or uniform distributions of populations are usually assumed when developing model of bacterial function in soil. Rappoldt and Crawford (1999) assumed a uniform distribution of potential respiration in a fractal model of soil when studying the distribution of anoxic volumes. When modelling the effects of cell clustering on ammonium oxidation, Darrah et al. (1987) assumed that clusters of nitrifying cells were uniformly distributed throughout the soil volume. In a simulation examining how the use of extracellular enzymes as a foraging strategy by soil micro-organisms is affected by competition, nutrient availability and spatial structure, Allison (2005) assumed a random distribution of micro-organisms and an even distribution of substrate inputs. These models provide important insights into soil function and demonstrate that spatial organization is an important factor influencing soil function. However, they do not explicitly account for the distribution of soil micro-organisms in relation to other soil features and therefore cannot be used for a full assessment of the importance of microscale interactions for ecological function.

If interactions among micro-organisms are important factors in regulating soil microbial function, then describing microbial activity in terms of parameters that are spatially invariant may not be adequate as this assumes that microbial communities respond to micro-environmental conditions in an identical manner throughout the soil volume. Individual-based modelling is increasingly used in population ecology in an effort to account for the impact of intra- as well as inter-specific diversity on the community structure and dynamics (Crawford et al., 2005). With individual-based modelling, individuals are assigned properties which can evolve, and interactions among individuals are accounted for. The emergence of system-level properties from the interaction among the constituent parts may be understood. Recently, individual-based modelling has been used to describe the mineralization of C and N and nitrification in laboratory incubations of soil (Ginovart et al., 2005). The model was a simplified representation of soil microbial communities in that only two types of microbial cell were considered, ammonifiers and nitrifiers, and structure is represented in a greatly simplified manner. However, the model results are promising and offer hope for future progress in this area.

5.   FUTURE DIRECTION

As the two methods presented have different, but complementary, strengths, an experimental approach combining the two may be useful and provide an immediate way forward. The microsampling technique can provide information on the spatial distribution of an activity and on how the distribution evolves with time whilst with biological thin section it is possible to measure spatial distributions in relation to other soil features.

Many techniques are now available for studying soil microbial microhabitats and the spatial arrangement of microhabitat components. A number are listed below, though the list is not exhaustive.

Secondary Ion Mass Spectrometry (SIMS) is a technique that links high resolution microscopy with isotopic analysis. SIMS involves the bombardment of a sample surface with a primary high energy ion beam. This results in secondary ions being liberated from the sample into the gas phase and dispersed in a mass spectrometer according to their energies and their mass to charge ratios. An image ("map") can be formed for any selected mass and a number of ionic species can be recorded simultaneously. The power of SIMS lies in the ability of the instrument to distinguish stable isotopes of elements with a high sensitivity, i.e., concentrations in parts per million can be detected. Furthermore, analyses can be carried out at spatial resolutions of <1 µm. SIMS has been applied to the study of biological materials e.g., in plant physiology (Lazof et al., 1992) and where ^{15}N was applied as a tracer in yeasts and plant samples (Gojon et al., 1996). SIMS has also been used to study soil minerals in thin sections (Bertrand et al., 2001). Recently, Cliff et al. (2002) were able to locate the assimilation of ^{15}N and ^{13}C to individual microbial cells in a model soil system, suggesting that SIMS shows promise as a tool for studying microhabitat heterogeneity and microbial metabolism in combination. Such an approach could be complemented by the use of fluorescent *in situ* hybridization (FISH) with specific nucleic probes (Macnaughton et al., 1996), thus enabling the distribution of genetic information and associated activity to be compared. However, non-specific staining in the case of FISH and the frequency of occurrence might make all but the commonest of functions difficult to locate, bearing in mind the large amount of unoccupied space in soil.

The use of x-ray tomography is now widely used in studies of soil heterogeneity and has been reviewed (Young et al., 2001). Recent developments in x-ray microtomography and nano-tomography have enabled the description and quantification of soil structure at the microhabitat scale (Nunan et al., 2005; Feeney et al., 2006) and the spatial arrangement of bacteria and soil particles in colloidal flocculates has also been described (Thieme et al., 2003).

Environmental scanning electron microscope (ESEM) represents several important advances in scanning electron microscopy. Whereas conventional scanning electron microscopy requires a relatively high vacuum in the specimen chamber to prevent atmospheric interference with primary or secondary electrons, an ESEM may be operated with a poor vacuum (up to 10 Torr of vapour pressure, or 176 of an atmosphere) in the specimen chamber, so called "wet mode" imaging where water is the imaging gas (Callow et al., 2003). This means that pre-processing of samples such as dehydration and fixation that can distort sample is not necessary. Furthermore, it is not necessary to make non-conductive samples conductive by coating with a conductant (Callow et al., 2003). Samples can therefore be imaged before and after modification as damage to the sample is limited. Environmental scanning electron microscopy was used to show that the colonization of sand grain surfaces by *Pseudomonas aeruginosa* PG201 was relatively sparse and well distributed rather than in the form of multilayered biofilms (Holden et al., 2002).

A more complete understanding of the impact of microscale properties on macroscopic behaviour can only be attained if these techniques and others not listed here are fully exploited. This means that findings should be incorporated mathematical models that are linked to suitable models of physical processes. In this way, it should be possible to search for scaling laws from first principals that relate the microscopic to function at larger scales and therefore to link habitat to function.

REFERENCES

Alexander, F. E. S., and R. M. Jackson, 1955, Preparation of sections for study of soil microorganisms, in: *Soil Zoology*, D. K. M. Kevan, ed., Butterworth. London, pp. 433–440.

Allison, S. D., 2005, Cheaters, diffusion and nutrients constrain decomposition by microbial enzymes in spatially structured environments, *Ecol. Lett.* **8**:626–635.

Almås, A. R., J. Mulder, and L. R. Bakken, 2005, Trace metal exposure of soil bacteria depends on their position in the soil matrix, *Environ. Sci. Technol.* **39**:5927–5932.

Altemüller, H., and B. Vliet-Lanoe, 1988, Soil thin section fluorescence microscopy, *Devel. Soil Sci.* **19**:565–579.

Beijerink, M. W., 1913, De infusies en de ontdekking der backteriën. Jaarboek van de Koninklijke Akademie v. Wetenschappen, Müller, Berlin.

Bergstrom, D. W., C. M. Monreal, J. A. Millette, and D. J. King, 1998, Spatial dependence of soil enzyme activities along a slope, *Soil Sci. Soc. Am. J.* **62**:1302–1308.

Bertrand, I., N. Grignon, P. Hinsinger, G. Souche, and B. Jaillard, 2001, The use of secondary ion mass spectrometry coupled with image analysis to identify and locate chemical elements in soil minerals: the example of phosphorus, *Scanning* **23**:279–291.

Bird, N. R. A., E. Perrier, and M. Rieu, 2000, The water retention function for a model of soil structure with pore and solid fractal distributions, *Eur. J. Soil Sci.* **51**:55–63.

Bloem, J., M. Veninga, and J. Shepherd, 1995, Fully-automatic determination of soil bacterium numbers, cell volumes, and frequencies of dividing cells by confocal laser-scanning microscopy and image-analysis, *Appl. Environ. Microbiol.* **61**:926–936.

Bramley, R. G. V., and R. E. White, 1991a, An analysis of variability in the activity of nitrifiers in a soil under pasture. 1. Spatially dependent variability and optimum sampling strategy, *Aust. J. Soil Res.* **29**:95–108.

Bramley, R. G. V., and R. E. White, 1991b, An analysis of variability in the activity of nitrifiers in a soil under pasture. 2. Some problems in the geostatistical analysis of biological soil properties, *Aust. J. Soil Res.* **29**:109–122.

Bruckner, A., E. Kandeler, and C. Kampichler, 1999, Plot-scale spatial patterns of soil water content, pH, substrate-induced respiration and N mineralization in a temperate coniferous forest, *Geoderma* **93**:207–223.

Bruneau, P. M. C., D. A. Davidson, I. C. Grieve, I. M. Young, and N. Nunan, 2005, The effects of soil horizons and faunal excrement on bacterial distribution in an upland grassland soil, *FEMS Microbiol. Ecol.* **52**:139–144.

Burmølle, M., L. H. Hansen, G. Oregaard, and S. J. Sørensen, 2003, Presence of N-acyl homoserine lactones in soil detected by a whole-cell biosensor and flow cytometry, *Microb. Ecol.* **45**:226–236.

Callow, J. A., M. P. Osborne, M. E. Callow, F. Baker, and A. M. Donald, 2003, Use of environmental scanning electron microscopy to image the spore adhesive of the marine alga *Enteromorpha* in its natural hydrated state, *Coll. Surf. B.* **27**:315–321.

Chenu, C., J. Hassink, and J. Bloem, 2001, Short-term changes in the spatial distribution of microorganisms in soil aggregates as affected by glucose addition, *Biol. Fertil. Soils* **34**:349–356.

Chernin, L. S., M. K. Winson, J. M. Thompson, S. Haran, B. W. Bycroft, I. Chet, P. Williams, and G. S. A. B. Stewart, 1998, Chitinolytic activity in *Chromobacterium violaceum*: substrate analysis and regulation by quorum sensing, *J. Bacteriol.* **180**:4435–4441.

Cliff, J. B., D. J. Gaspar, P. J. Bottomley, and D. D. Myrold, 2002, Exploration of inorganic C and N assimilation by soil microbes with time-of-flight secondary ion mass spectrometry, *Appl. Environ. Microbiol.* **68**:4067–4073.

Crawford, J. W., J. A. Harris, K. Ritz, and I. M. Young, 2005, Towards an evolutionary ecology of life in soil, *Trends Ecol. Evol.* **20**:81–87.

Crecchio, C., P. Ruggiero, M. Curci, C. Colombo, G. Palumbo, and G. Stotzky, 2005, Binding of DNA from *Bacillus subtilis* on montmorillonite-humic acids-aluminum or iron hydroxylpolymers: Effects on transformation and protection against Dnase, *Soil Sci. Soc. Am. J.* **69**:834–841.

Darrah, P. R., R. E. White, and P. H. Nye, 1987, A theoretical consideration of the implications of cell clustering for the prediction of nitrification in soil, *Plant Soil.* **99**:387–400.

Dechesne, A., C. Pallud, D. Debouzie, J. P. Flandrois, T. M. Vogel, J. P. Gaudet, and G. L. Grundmann, 2003, A novel method for characterizing the microscale 3D spatial distribution of bacteria in soil, *Soil Biol. Biochem.* **35**:1537–1546.

Dejonghe, W., J. Goris, S. El Fantroussi, M. Hofte, P. De Vos, W. Verstraete, and E. M. Top, 2000, Effect of dissemination of 2,4-dichlorophen-oxyacetic acid (2,4-D) degradation plasmids on 2,4-D degradation and on bacterial community structure in two different soil horizons, *Appl. Environ. Microbiol.* **66**:3297–3304.

Dröge, M., A. Puhler, and W. Selbitschka, 1999, Horizontal gene transfer among bacteria in terrestrial and aquatic habitats as assessed by microcosm and field studies, *Biol. Fertil. Soils* **29**:221–245.

Durrett, R., and S. Levin, 1997, Allelopathy in spatially distributed populations, *J. Theoret. Biol.* **185**:165–171.

Ettema, C. H., and D. A. Wardle, 2002, Spatial soil ecology, *Trends Ecol. Evol.* **17**:177–183.

Feeney, D. S., J. C. Crawford, T. J. Daniell, P. D. Hallett, N. Nunan, K. Ritz, M. Rivers, and I. M. Young, 2006, 3D micro-organisation of the soil-root-microbe system. *Microb. Ecol.* **52**:151–158.

Fierer, N., J. P. Schimel, and P. A. Holden, 2003, Variations in microbial community composition through two soil depth profiles, *Soil Biol. Biochem.* **35**:167–176.

Fisk, A. C., S. L. Murphy, and R. L. Tate, 1999, Microscopic observations of bacterial sorption in soil cores, *Biol. Fertil. Soils* **28**:111–116.

Foster, R., and J. Martin, 1981, In situ analysis of soil components of biological origin, *Soil Biochem.* **5**:75–111.

Foster, R., and A. Rovira, 1973, The rhizosphere of wheat roots studied by electron microscopy of ultra-thin sections. "Modern methods in the study of microbial ecology", *Bull. Ecol. Res. Comm. Sweden.* **17**:93–95.

Foster, R., A. Rovira, and T. Cock, 1983, *Ultrastructure of the Root–Soil Interface.* American Phytopathological Society, St. Paul, Minnesota.

Franklin, R. B., and A. L. Mills, 2003, Multi-scale variation in spatial heterogeneity for microbial community structure in an eastern Virginia agricultural field, *FEMS Microbiol. Ecol.* **44**:335–346.

Franklin, R. B., J. L. Garland, C. H. Bolster, and A. L. Mills, 2001, Impact of dilution on microbial community structure and functional potential: comparison of numerical simulations and batch culture experiments, *Appl. Environ. Microbiol.* **67**:702–712.

Fromm, H., K. Winter, J. Filser, R. Hantschel, and F. Beese, 1993, The influence of soil type and cultivation system on the spatial distributions of the soil fauna and microorganisms and their interactions, *Geoderma* **60**:109–118.

Ginovart, M., D. Lopez, and A. Gras, 2005, Individual-based modelling of microbial activity to study mineralization of C and N and nitrification process in soil, *Nonlinear Anal. Real World Appl.* **6**:773–795.

Gojon, A., N. Grignon, P. Tillard, P. Massiot, F. Lefebvre, M. Thellier, and C. Ripoll, 1996, Imaging and microanalysis of N-14 and N-15 by SIMS microscopy in yeast and plant samples, *Cell. Mol. Biol.* **42**:351–360.

Gonzalez, O. J., and D. R. Zak, 1994, A geostatistical analysis of soil properties in a secondary tropical dry forest, St-Lucia, West-Indies, *Plant Soil* **163**:45–54.

Goovaerts, P., and C. N. Chiang, 1993, Temporal persistence of spatial patterns for mineralizable nitrogen and selected soil properties, *Soil Sci. Soc. Am. J.* **57**:372–381.

Grundmann, G. L., and F. Gourbiere, 1999, A micro-sampling approach to improve the inventory of bacterial diversity in soil, *Appl. Soil Ecol.* **13**:123–126.

Grundmann, G. L., and P. Normand, 2000, Microscale diversity of the genus *Nitrobacter* in soil on the basis of analysis of genes encoding rRNA, *Appl. Environ. Microbiol.* **66**:4543–4546.

Grundmann, G. L., A. Dechesne, F. Bartoli, J. P. Flandrois, J. L. Chasse, and R. Kizungu, 2001, Spatial modeling of nitrifier microhabitats in soil, *Soil Sci. Soc. Am. J.* **65**:1709–1716.

Harris, P. J., 1994, Consequences of the spatial distribution of microbial communities in soil, in: *Beyond the Biomass*, K. Ritz, J. Dighton, and K. E. Giller, eds., British Society of Soil Science, Wiles-Sayce, London, pp. 239–246.

Holden, P. A., M. G. LaMontagne, A. K. Bruce, W. G. Miller, and S. E. Lindow, 2002, Assessing the role of *Pseudomonas aeruginosa* surface-active gene expression in hexadecane biodegradation in sand, *Appl. Environ. Microbiol.* **68:**2509–2518.

Jackson, R. B., and M. M. Caldwell, 1993, The scale of nutrient heterogeneity around individual plants and its quantification with geostatistics, *Ecology* **74:**612–614.

Johnson, A., I. M. Roy, G. P. Matthews, and D. Patel, 2003, An improved simulation of void structure, water retention and hydraulic conductivity in soil with the Pore-Cor three-dimensional network, *Eur. J. Soil Sci.* **54:**477–489.

Jones, D., and Griffiths, E., 1964, The use of thin sections for the study of soil microorganisms, *Plant Soil* **20:**232–240.

Kampichler, C., and M. Hauser, 1993, Roughness of soil pore surface and its effect on available habitat space of microarthropods, *Geoderma* **56:**223–232.

Kerr, B., M. A. Riley, M. W. Feldman, and B. J. M. Bohannan, 2002, Local dispersal promotes biodiversity in a real-life game of rock-paper-scissors, *Nature* **418:**171–174.

Klironomos, J. N., M. C. Rillig, and M. F., Allen, 1999, Designing below-ground field experiments with the help of semi-variance and power analyses, *Appl. Soil Ecol.* **12:**227–238.

Knorr, W., I. C. Prentice, J. I. House, and E. A. Holland, 2005, Long-term sensitivity of soil carbon turnover to warming, *Nature* **433:**298–301.

Lazof, D., R. W. Linton, R. J. Volk, and T. W. Rufty, 1992, The application of SIMS to nutrient tracer studies in plant physiology, *Biol. Cell.* **74:**127–134.

Levin, S. A., 1992, The problem of pattern and scale in ecology, *Ecology* **73:**1943–1967.

Levin, S. A., 1998, Ecosystems and the biosphere as complex adaptive systems, *Ecosystems* **1:**431–436.

Long, T., and D. Or, 2005, Aquatic habitats and diffusion constraints affecting microbial coexistence in unsaturated porous media, *Water Resour. Res.* **41:**W08408.

Macnaughton, S. J., T. Booth, T. M. Embley, and A. G. O'Donnell, 1996, Physical stabilization and confocal microscopy of bacteria on roots using 16S rRNA targeted, fluorescent-labeled oligonucleotide probes, *J. Microbiol. Methods* **26:**279–285.

Madden, L. V., and G. Hughes, 1999, An effective sample size for predicting plant disease incidence in a spatial hierarchy, *Phytopathology* **89:**770–781.

Marx, M. C., E. Kandeler, M. Wood, N. Wermbter, and S. C. Jarvis, 2005, Exploring the enzymatic landscape: distribution and kinetics of hydrolytic enzymes in soil particle-size fractions, *Soil Biol. Biochem.* **37:**35–48.

Mummey, D. L., and P. D. Stahl, 2004, Analysis of soil whole- and inner-microaggregate bacterial communities, *Microb. Ecol.* **48:**41–50.

Nunan, N., K. Ritz, D. Crabb, K. Harris, K. J. Wu, J. W. Crawford, and I. M. Young, 2001, Quantification of the in situ distribution of soil bacteria by large-scale imaging of thin sections of undisturbed soil, *FEMS Microbiol. Ecol.* **37:**67–77.

Nunan, N., K. Wu, I. M. Young, J. W. Crawford, and K. Ritz, 2002, In situ spatial patterns of soil bacterial populations, mapped at multiple scales, in an arable soil, *Microb. Ecol.* **44:**296–305.

Nunan, N., K. J. Wu, I. M. Young, J. W. Crawford, and K. Ritz, 2003, Spatial distribution of bacterial communities and their relationships with the micro-architecture of soil, *FEMS Microbiol. Ecol.* **44:**203–215.

Nunan, N., K. Ritz, M. Rivers, D. S. Feeney, and I. M. Young, 2006, Investigating microbial micro-habitat structure using x-ray computed tomography, *Geoderma* **133**:398–407.
Okabe, H., and M. J. Blunt, 2004, Prediction of permeability for porous media reconstructed using multiple-point statistics, *Phys. Rev. E* **70**:066135.
Pallud, C., A. Dechesne, J. P. Gaudet, D. Debouzie, and G. L. Grundmann, 2004, Modification of spatial distribution of 2,4-dichloro-phenoxyacetic acid degrader microhabitats during growth in soil columns, *Appl. Environ. Microbiol.* **70**:2709–2716.
Parkin, T. B., 1993, Spatial variability of microbial processes in soil – a review, *J. Environ. Qual.* **22**:409–417.
Parkin, T. B., and D. R. Shelton, 1992, Spatial and temporal variability of carbofuran degradation in soil, *J. Environ. Qual.* **21**:672–678.
Peat, D. M. W., G. P. Matthews, P. J. Worsfold, and S. C. Jarvis, 2000, Simulation of water retention and hydraulic conductivity in soil using a three-dimensional network, *Eur. J. Soil Sci.* **51**:65–79.
Pierson, E. A., D. W. Wood, J. A. Cannon, F. M. Blachere, and L. S. Pierson, 1998, Inter-population signaling via N-acyl-homoserine lactones among bacteria in the wheat rhizosphere, *Mol. Plant Microbe Interact.* **11**:1078–1084.
Postma, J., and H. J. Altemüller, 1990, Bacteria in thin soil sections stained with the fluorescent brightener Calcofluor White M2R, *Soil Biol. Biochem.* **22**:89–96.
Protz, R., S. J. Sweeney, and C. A. Fox, 1992, An application of spectral image-analysis to soil micromorphology. 1. Methods of analysis, *Geoderma* **53**:275–287.
Ranjard, L., S. Nazaret, F. Gourbiere, J. Thioulouse, P. Linet, and A. Richaume, 2000a, A soil microscale study to reveal the heterogeneity of Hg(II) impact on indigenous bacteria by quantification of adapted phenotypes and analysis of community DNA fingerprints, *FEMS Microbiol. Ecol.* **31**:107–115.
Ranjard, L., F. Poly, J. Combrisson, A. Richaume, F. Gourbiere, J. Thioulouse, and S. Nazaret, 2000b, Heterogeneous cell density and genetic structure of bacterial pools associated with various soil microenvironments as determined by enumeration and DNA fingerprinting approach (RISA), *Microb. Ecol.* **39**:263–272.
Rappoldt, C., and J. W. Crawford, 1999, The distribution of anoxic volume in a fractal model of soil, *Geoderma* **88**:329–347.
Ritz, K., W. McNicol, N. Nunan, S. Grayston, P. Millard, D. Atkinson, A. Gollotte, D. Habeshaw, B. Boag, C. D. Clegg, B. S. Griffiths, R. E. Wheatley, L. A. Glover, A. E. McCaig, and J. I. Prosser, 2004, Spatial structure in soil chemical and microbiological properties in an upland grassland, *FEMS Microbiol. Ecol.* **49**:191–205.
Robertson, G. P., and K. L. Gross, 1994, Assessing the heterogeneity of belowground resources: quantifying pattern and scale, in: *Exploitation of Environmental Heterogeneity by Plants: Ecophysiological Processes Above- and Belowground*, M. M. Caldwell, and R. W. Pearcy (eds.), Academic Press, New York, pp. 237–253.
Robertson, G. P., K. Klingensmith, M. Klug, E. Paul, J. Crum, and B. Ellis, 1997, Soil resources, microbial activity, and primary production across an agricultural ecosystem, *Ecol. Appl.* **7**:158–170.
Ronn, R., B. S. Griffiths, F. Ekelund, and S. Christensen, 1996, Spatial distribution and successional pattern of microbial activity and micro-faunal populations on decomposing barley roots, *J. Appl. Ecol.* **33**:662–672.
Saetre, P., and E. Bååth, 2000, Spatial variation and patterns of soil microbial community structure in a mixed spruce-birch stand, *Soil Biol. Biochem.* **32**:909–917.

Sexstone, A., N. Revsbech, T. Parkin, and J. Tiedje, 1985, Direct measurement of oxygen profiles and denitrification rates in soil aggregates, *Soil Sci. Soc. Am. J.* **49**:645–651.

Stoyan, H., H. De Polli, S. Bohm, G. P. Robertson, and E. A. Paul, 2000, Spatial heterogeneity of soil respiration and related properties at the plant scale, *Plant Soil* **222**:203–214.

Strong, D. T., P. W. G. Sale, and K. R. Helyar, 1997, Initial soil pH affects the pH at which nitrification ceases due to self-induced acidification of microbial microsites, *Aus. J. Soil Res.* **35**:565–570.

Thieme, J., G. Schneider, and C. Knochel, 2003, X-ray tomography of a microhabitat of bacteria and other soil colloids with sub-100 nm resolution, *Micron.* **34**:339–344.

Tippkötter, R., 1990, Staining of soil microorganisms and related materials with fluorochromes, in: *Soil Micromorphology: A Basic and Applied Science*, L. A. Douglas (ed.), Elsevier, Amsterdam, pp. 605–611.

Tippkötter, R., and K. Ritz, 1996, Evaluation of polyester, epoxy and acrylic resins for suitability in preparation of soil thin sections for in situ biological studies, *Geoderma* **69**:31–57.

Tippkötter, R., K. Ritz, and J. F. Darbyshire, 1986, The preparation of soil thin sections for biological studies, *J. Soil Sci.* **77**:681–690.

Top, E. M., and D. Springael, 2003, The role of mobile genetic elements in bacterial adaptation to xenobiotic organic compounds, *Curr. Opin. Biotechnol.* **14**:262–269.

Top, E. M., P. Van Daele, N. De Saeyer, and L. J. Forney, 1998, Enhancement of 2,4-dichlorophenoxyacetic acid (2,4-D) degradation in soil by dissemination of catabolic plasmids, *Antonie Van Leeuwenhoek Int. J. Gen. Mol. Microbiol.* **73**:87–94.

Treves, D. S., B. Xia, J. Zhou, and J. M. Tiedje, 2003, A two-species test of the hypothesis that spatial isolation influences microbial diversity in soil, *Microb. Ecol.* **45**:20–28.

Tsuji, T., Y. Kawasaki, S. Takeshima, T. Sekiya, and S. Tanaka, 1995, A new fluorescence staining assay for visualizing living microorganisms in soil, *Appl. Environ. Microbiol.* **61**:3415–3421.

Velthof, G. L., S. C. Jarvis, A. Stein, A. G. Allen, and O. Oenema, 1996, Spatial variability of nitrous oxide fluxes in mown and grazed grasslands on a poorly drained clay soil, *Soil Biol. Biochem.* **28**:1215–1225.

Vieublé-Gonod, L., J. Chadoeuf, and C. Chenu, 2006, Spatial distribution of microbial 2,4-Dichlorophenoxy acetic acid mineralization from field to microhabitat scales, *Soil Sci. Soc. Am. J.* **70**:64–71.

Vogel, H. J., and K. Roth, 1998, A new approach for determining effective soil hydraulic functions, *Eur. J. Soil Sci.* **49**:547–556.

Vogel, H. J., and K. Roth, 2001, Quantitative morphology and network representation of soil pore structure, *Adv. Water Resour.* **24**:233–242.

Warrick, A. W., and D. E. Myers, 1987, Optimization of sampling locations for variogram calculations, *Water Resour. Res.* **23**:496–500.

Wheatley, R. E., 2002, The consequences of volatile organic compound mediated bacterial and fungal interactions, *Antonie van Leeuwenhoek Int. J. Gen. Mol. Microbiol.* **81**:357–364.

White, D., E. A. Fitzpatrick, and K. Killham, 1994, Use of stained bacterial inocula to assess spatial-distribution after introduction into soil, *Geoderma* **63**:245–254.

Whitehead, N. A., A. M. L. Barnard, H. Slater, N. J. L. Simpson, and G. P. C. Salmond, 2001, Quorum-sensing in gram-negative bacteria, *FEMS Microbiol. Rev.* **25**:365–404.

Williamson, K. E., M. Radosevich, and K. E. Wommack, 2005, Abundance and diversity of viruses in six Delaware soils, *Appl. Environ. Microbiol.* **71**:3119–3125.

Wright, D. A., K. Killham, L. A. Glover, and J. I. Prosser, 1995, Role of pore-size location in determining bacterial-activity during predation by protozoa in soil, *Appl. Environ. Microbiol.* **61**:3537–3543.

Wu, K. J., N. Nunan, J. W. Crawford, I. M. Young, and K. Ritz, 2004, An efficient Markov chain model for the simulation of heterogeneous soil structure, *Soil Sci. Soc. Am. J.* **68**:346–351.

Young, I. M., and J. W. Crawford, 2001, Protozoan life in a fractal world, *Protist* **152**:123–126.

Young, I. M., and J. W.Crawford, 2004, Interactions and self-organization in the soil-microbe complex, *Science* **304**:1634–1637.

Young, I. M., and K. Ritz, 1998, Can there be a contemporary ecological dimension to soil biology without a habitat? *Soil Biol. Biochem.* **30**:1229–1232.

Young, I. M., and K. Ritz, 2000, Tillage, habitat space and function of soil microbes, *Soil Till. Res.* **53**:201–213.

Young, I. M., and K. Ritz, 2005, The habitat of soil microbes, in: *Biological Diversity and Function in Soils*, R. D. Bardgett, M. B. Usher, and D. W. Hopkins eds., Cambridge University Press, Cambridge, pp. 31–43.

Young, I. M., J. W. Crawford, and C. Rappoldt, 2001, New methods and models for characterising structural heterogeneity of soil, *Soil Till. Res.* **61**:33–45.

Zhou, J. Z., B. C. Xia, D. S. Treves, L. Y. Wu, T. L. Marsh, R. V. O'Neill, A. V. Palumbo, and J. M. Tiedje, 2002, Spatial and resource factors influencing high microbial diversity in soil, *Appl. Environ. Microbiol.* **68**:326–334.

Chapter 4

SPATIAL DISTRIBUTION OF BACTERIA AT THE MICROSCALE IN SOIL

Arnaud Dechesne[1], Céline Pallud[2], and Geneviève L. Grundmann[3]
[1]*Institute of Environment and Resources, Technical University of Denmark, Kgs. Lyngby, Denmark;* [2]*Soil and Environmental Biogeochemistry, Department of Geological and Environmental Sciences, Stanford University, Stanford, CA 94305, USA;* [3]*Laboratoire d'Ecologie Microbienne, Université de Lyon1, Villeurbanne cedex, France*

Abstract: After the discovery of the tremendous bacterial diversity in soil at all spatial scales, numerous studies have been motivated by the fact that soil represents a very large reservoir of various genes. Nevertheless, the organization of bacterial cells at the microscale in the soil fabric has been overlooked, although all functional interactions appearing at the ecosystem level initially intervene at the scale of the bacterial cells. Many microbiological processes are based on encounters between cells, and between cells and substrates, between cells and surfaces. This chapter provides insight into the microscale spatial distribution of bacteria in soil, with a special emphasis on the concepts of microcolonies and microhabitats as structuring elements for these patterns.

Keywords: bacterial diversity, soil, spatial organization, microscale, microhabitat

1. INTRODUCTION

There is great interest by microbial ecologists in trying to understand and quantify microbially driven processes at scales of natural environments (Parkin, 1993). Soils are complex multiphase environments, with structural and geochemical heterogeneities taking place over spatial scales ranging from nanometres to kilometres. Individual bacterial cells function at a scale relevant to their size, the microscale, and their combined localized activities affect the environment up to the global scale. However, microbial activity measurements are usually carried out at field or ecosystem scales, and the small spatial scale at which individuals and populations are actually living and interacting is rarely considered (Franklin et al., 2002). Studies at the microscale could provide us with a description of factors controlling microbial activities, and eventually result in improved predictive models for microbial soil processes (Wachinger et al., 2000). Similarly, a better knowledge of the processes occurring at the microscale is a prerequisite to understanding bacterial community structures; the bacterial community at a given site, at a given time, is the integrated result of migration, clonal reproduction, and lateral gene transfer. Bacterial survival and activities are dependent on contact between the bacterial cells and one or more components of the soil ecosystem (e.g., carbon source, electron acceptor, other bacteria). Due to soil chemical and structural heterogeneities, contact between bacterial cells and any components of the soil ecosystem is under the dependence of the respective distributions of two, or more, partners (Daane et al., 1996, 1997). The abundance of each constituent of the pairing, however, is not indicative of the encounter probability (Holden and Firestone, 1997). Therefore, microscale spatial distribution of bacteria in soil plays a functional role in soil microbial ecology, and the complex relationships between bacterial localization, bacterial activity, and the physico-chemical environment in soil is in need of further investigation.

A key factor in soil function is the probability for bacteria to encounter other bacteria or other components of the soil ecosystem. Bacteria located close to each other will be more likely to interact than bacteria separated by several millimetres. Therefore, the distance separating bacterial cells, or bacterial microhabitats, has to be explored. Furthermore, bacteria, electron donors, and electron acceptors can move or be transported through soil, provided that a water film is present (Gammack et al., 1992). The spatial and temporal flow paths for solutes and bacteria are mainly determined by the connectivity and composition of the porous network, as well as by water content. As a consequence, bacteria situated several millimetres away from each other would be able to interact if they are located in connected pores and if conditions are favourable for bacterial movement. Conversely, bacteria entrapped in unconnected micropores will not be able to interact unless the soil structure is modified (e.g., by the action of the soil microfauna). The microscale soil structure as well as the possibilities of bacterial movements thus have to be considered.

This chapter provides an insight into the microscale spatial distribution of bacteria in soil, with a special emphasis on the concepts of microcolonies and microhabitats as structuring elements of bacterial spatial distribution. Advantages of different methodological approaches to study the spatial distribution of bacteria are presented. The stability of the microhabitats in space and time, and their bacterial diversity, are discussed from an ecological point of view. This text will mainly deal with mineral soils and will not detail the influence of plant roots on the spatial distribution of bacteria. That topic is discussed in by Knudsen and Dandurand in Chapter 5.

2. METHODOLOGICAL APPROACHES

Study of the microscale spatial distribution of bacteria in soils is made especially challenging by specific sampling and visualization difficulties caused by the structural and geochemical complexity of the soil.

2.1. Microscopy

The detection of bacteria in complex samples such as soil thin sections using optical microscopy typically requires that bacteria are stained. Several dyes have been proposed to visualize bacterial cells in soil using epifluorescence microscopy (Li et al., 2003). The first descriptions of the bacterial microlandscape in soil include those by Jones and Griffiths (1964), Gray et al. (1968), and Hattori (1973). Their early results indicated that bacterial cells were not evenly distributed in the soil porosity and that colonies were a characteristic feature of bacterial communities in soils.

Refinements of the staining step have been proposed that allow specific subgroups of microbial communities to be identified. Some dyes, such as the sulfofluoroscein diacetate (SFDA) for example (Tsuji et al., 1995), only stain living microorganisms. Fluorescent *in situ* hybridization (FISH) labelling, based on the detection of a fluorescently labelled DNA or RNA probe hybridized to bacterial nucleic acids at the cell scale, is also a promising technique. With FISH, a given functional gene or bacteria belonging to a given phylogenetic group can be imaged. Whereas FISH has been used for studying bacteria in water and activated sludge (e.g., Lee et al., 1999), its application in soil is still limited due to technical difficulties such as the autofluorescence of soil minerals and organic materials (Li et al., 2003). In soil, these techniques were used for enumerating specific bacterial groups, rather than for studying their spatial patterns (Wachinger et al., 2000; Richardson et al., 2002). Future efforts need to be carried out in these directions since both FISH and living cell staining could provide a functional viewpoint on the spatial distribution of bacteria, potentially uncovering the links between spatial pattern and bacterial diversity and activity.

In 1994, a novel molecular marker, the green fluorescent protein (GFP) became available (Chalfie et al., 1994). It allows the monitoring of bacterial cells introduced into complex environments, against background of large numbers of microbial populations, using epiflorescence microscopy (Errampalli et al., 1999). The use of GFP-labelled bacteria has greatly improved our description of the rhizoplane colonization (Tombolini et al., 1999) as well as their community structure (Bloemberg et al., 2000), but it has rarely been used to specifically study the spatial pattern of bacteria in bulk soil (Dechesne et al., 2005). Moreover, GFP-labelled bacteria have to be introduced into soils, and it is supposed that introduced and indigenous bacteria have different spatial distributions (Recorbet et al., 1995; Li et al., 2003; Dechesne et al., 2005). This clearly allows tracing bacteria in soil but cannot be directly representative of the spatial behaviour of autochtonous bacteria.

One of the main interests of microscopy is that the relationship between bacteria and their geochemical and physical environment can be uncovered. Electron microscopy, in that respect, can provide information on the materials surrounding bacterial cells (Kilbertus, 1980; Lünsdorf et al., 2000). X-ray tomography has further been used to uncover the three-dimensional structure of bacterial microhabitats (Thieme et al., 2003). Recently, Nunan et al. (2001) developed a technique for determining the number and *in situ* spatial distribution of bacterial cells over scales ranging from micrometres to centimetres in mineral soils. This technique is based on the acquisition and computer analysis of composite microscopic images of soil thin sections. Quantified results are obtained, which are suitable for a range of statistical analyses. The soil structure may also be taken into account to study potential relationships between bacterial spatial distribution and the presence of large pores (Nunan et al., 2003).

2.2. Approaches based on soil microsampling

Microscopic techniques for studying microscale bacterial distributions suffer some limitations. First, microscopy usually implies sample preparation, which may affect the soil structure and modify the location of bacterial cells. Furthermore, with the exception of the confocal laser scanning microscopy, none of the microscopic techniques permit a three-dimensional description of bacterial spatial distribution (Li et al., 2003). Moreover, indigenous bacteria may be difficult to observe because they are sometimes entrapped into soil structures and thus difficult to stain (Li et al., 2003). To overcome some of the limitations associated with microscopic approaches, a specific method to quantitatively characterize the three-dimensional spatial pattern of the zones in soil colonized by bacteria has been proposed (Grundmann et al., 2001;

Dechesne et al., 2003), combining a specific microsampling strategy with a data analysis method. Any given bacterial type may be targeted, provided that a presence test is available and applicable to soil microsamples. So far, only viable bacteria have been studied, because the presence tests were performed by cultivating bacteria in soil microsamples. This approach allows investigating bacterial distributions at the microhabitat scale, with a spatial resolution corresponding to the minimum soil sample size, which is typically larger than 50 µm in diameter. It provides the microbial ecologist with spatial descriptors of bacterial distribution, such as the average size of colonized patches and their average number per gram of soil, and enables him/her to compare different situations and check for statistical differences between distributions. Statistical methods developed in other disciplines were also applied to microsampling data (e.g., geostatistics in Grundmann and Debouzie (2000)) to describe the spatial distributions of nitrifiers along a soil transect at the millimetre scale.

2.3. Approaches based on soil fractionation

Soils are heterogeneous porous media, composed of individual aggregates. Two types of habitats for the microorganisms have been described in soil aggregates, based on aggregate structure and on the water retention in each site of the aggregate (Hattori, 1967). The inner compartment corresponds to micropores having a diameter of 2–6 µm, where water is retained by capillarity and where substrate access relies on diffusion. The outer compartment corresponds to the inter-aggregate porosity and to the particle surfaces in contact with solutes during convective flow. In the case of unsaturated soil, convective–dispersive transport is limited to only a fraction of the liquid-filled pores (mobile water), while immobile water participates through diffusion. As a consequence, outer surfaces of soil aggregates tend to sustain oxidizing conditions, while aggregate interiors are more reducing. These two habitats are characterized by different water retention, substrate availability, reducing/oxidizing conditions, or susceptibility to predation, factors known to directly influence the dynamics and survival of bacterial populations (Ranjard and Richaume, 2001). Based on the fact that the microorganisms located in the inner compartment of stable aggregates are retained by capillarity, and thus are more difficult to extract, Hattori (1967) developed a method to separate the microorganisms located in the two soil compartments. This method consists of successive washings of the soil and has been improved by Nishio et al. (1968), Nishio and Furusaka (1970), and more recently by Ranjard et al. (1997), to obtain a more representative and reproducible extraction of the outer compartment bacteria. Recently, a UV irradiation-based procedure was developed to specifically isolate inner-microaggregate bacterial communities (Mummey and Stahl, 2004).

Another approach is based on the physical fractionation of the soil in order to separate aggregates of various stability and size (<0.1 μm to several millimetres), which are considered as different bacterial habitats (Jocteur-Monrozier et al., 1991; Mendes and Bottomley, 1998; Poly et al., 2001; Sessitsch et al., 2001). The main drawback of these methods is that the overall context of each aggregate to its neighbours is lost.

3. CELLS, MICROCOLONIES AND MICROHABITATS AS STRUCTURING ELEMENTS

3.1. Microcolonies versus isolated cells

Early microscopic studies on soil samples revealed that bacteria are unevenly distributed in the soil porosity, with large portions of soil being devoid of bacteria (Jones and Griffiths, 1964; Hattori, 1973). The fraction of available surface effectively colonized by bacteria is estimated to be approximately 0.17% and 0.02%, for organic and sand grain surfaces, respectively (Hissett and Gray, 1976). Bacteria in soil can either be found as free living or as attached cells, and a bacterium is likely to experience the two lifestyles during its existence (Nikin and Kunc, 1988). The ratio of free to settled cells depends on the ecological situation and especially on the water regimen of the soil (Nikin and Kunc, 1988). Nevertheless, differences in activity between attached bacteria and their suspended counterparts have been observed in many studies, possibly because cell attachment modifies the substrate supply to the bacterial cells (e.g., Harms and Zehnder, 1994).

Attached bacteria may either be isolated or form groups of contiguous cells, which are referred to as "microcolonies" due to their small size compared to colonies typically observed on solid culture media. An early study (Hissett and Gray, 1976) using fluorescent sera, showed that *Bacillus subtilis* cells in soil generally occurred as very small colonies, averaging 4–5 cells on organic particles and only 2 on mineral particles. Accordingly, most of the microscopic studies identified a low number of cells *per* microcolony (Jones and Griffiths, 1964; Kilbertus, 1980; Foster, 1988; Nunan et al., 2001). This observation may be linked to the oligotrophic status of soil habitats, which may not support the growth of large numbers of bacteria. The limited size of hydrated microsites in unsaturated soils also potentially restricts the development of microcolonies (Or et al., 2007).

A colony is typically formed after division of one or several parent cells when the local conditions are favourable. Nevertheless, microcolonies were observed to be either constituted by cells of identical or of different morphologies (Nikin and Kunc, 1988; Harris, 1994; Nunan et al., 2003). It is thus probable, although conclusive evidence is missing, that microcolonies in soil do not systematically consist of one clone but might rather represent a microcommunity. Such "diverse" microcolonies would then have originated from the movement of various bacterial strains to the same site, followed by their growth.

In specific conditions, such as following an experimental glucose amendment to soil samples, bacterial structures larger than microcolonies have been observed at the surface of soil aggregates (Chenu et al., 2001; Nunan et al., 2003). In a clayey soil, Chenu et al. (2001) described monolayered biofilms covering areas of 20–300 µm in diameter on the external surfaces of soil aggregates. The origin and significance of isolated cells are unclear. An isolated cell could result from active or passive cell movements (Gammack et al., 1992) or from previous microcolonies that have been subjected to predation or mortality. As emphasized by Harris (1994), the important question is the extent to which an isolated cell can develop further. If conditions are suitable, a single cell could develop into a microcolony. However, isolated bacteria may already be in a non-viable state and indicate the failure of new habitat colonization. Skinner (1976) concluded that the growth of a single cell into a colony is subject to severe limitation. On plant leaves, isolated and aggregated cells have different ecological traits, for example, aggregated cells resist desiccation stresses better (Monier and Lindow, 2003).

3.2. Concept of bacterial microhabitat

Bacterial microhabitat is proposed as one of the major structuring elements of their spatial distribution in soil. Metting (1992) defined the microhabitat as the volume of soil whose physico-chemical status influences the behaviour of bacterial cells that, in turn, alter or control the physical and chemical characteristics of the environment within that space. This functional definition is purposely not too rigid because the microhabitat may fluctuate in time and space. Depending on the type of microorganisms, as well as on the electron donors and acceptors used by the resident bacteria, the microhabitat size may vary (Metting, 1992). Harris (1994) argued that, the term "community" only makes sense at the microhabitat scale because this is the scale at which microorganisms are truly interacting and benefiting from each other.

Size and spatial limits of some microbial microhabitats have been assessed by microscopic observations. Lünsdorf et al. (2000), using electron microscopy, observed that bacteria were able to interact with clay minerals to form "clay hutch"-like structures enclosing one or few bacterial cells. These hutches were in average 3.7 µm in width at their base and 2.9 µm in height. The authors speculated that the clay hutches may represent a "minimal nutritional sphere" and consequently a minimal soil microhabitat. Organic and inorganic nutrients recruited during the development of the hutches could subsequently be accessed by bacteria, while the entire matrix may function as a diffusion barrier, hindering the loss of nutrients. X-ray tomography, with 45 nm resolution, has been used to obtain information about the three-dimensional spatial arrangement of associations of bacteria and colloids (Thieme et al., 2003). The microhabitat described was approximately 6 µm in diameter and consisted of the association between colloidal soil particles and a dozen rod-shaped bacteria. It is probable that many other types of bacterial microhabitats exist in soil, which may or may not result in the formation of specific soil microstructures. The absence of any detectable feature makes the determination of microhabitat sizes conspicuously difficult. Nevertheless microhabitat extension can be related to the soil matric potential as water will influence connectivity between microhabitats (Focht, 1992).

The temporal limits of the microhabitats may be even more difficult to evaluate than the spatial limits. It is probable that if the majority of the bacteria located in a microhabitat are heterotrophs, its time limit would be associated with the exhaustion of the energy sources (Nikin and Kunc, 1988). Nevertheless, this time span may be extended if some bacteria are capable of surviving in a dormant state to face adverse conditions.

4. CELL AND MICROHABITAT MICROSCALE DISTRIBUTIONS AND THEIR DETERMINANTS

Having recognized the cell, microcolony and microhabitats as structuring elements in soil bacterial ecology, the question of their spatial distribution remains to be addressed. Is this distribution random, or are, for example, bacterial microhabitats organized in hot spots? In the case of non-randomness, is it possible to identify factors contributing to the shaping of these spatial distributions?

4.1. Spatial distribution

The two-dimensional method based on microscopy and image analysis developed by Nunan and collaborators (2001) provides us with quantified descriptions of bacterial distributions in soil at the cell scale. The bacteria appear to be distributed in an aggregated way, with some cells forming microcolonies and other being more loosely clustered (Nunan et al., 2001). The authors attributed this aggregation to bacterial growth, with more aggregation in the topsoil, where growth is considered more intense, than in the subsoil. This aggregated pattern was quantified using geostatistics, leading to the identification of lengths of spatial autocorrelation varying between 240 and 1,560 µm in the topsoil and 0–990 µm in the subsoil (Nunan et al., 2002). This suggests that there are relatively large patches of soil that harbour more cells than their surroundings. The existence of an aggregated distribution of bacteria was confirmed for specific bacteria communities studied at the millimetre scale. Grundmann and Debouzie (2000) used geostatistics to study the one-dimensional distribution of NH_4^+- and NO_2^--oxidizers and of various serotypes along a 10 cm transect sampled every millimetre: the presences of NH_4^+- and NO_2^--oxidizers were autocorrelated with ranges of 4 and 2 mm, respectively.

At a similar spatial resolution, three-dimensional spatial descriptors have been estimated for NH_4^+-oxidizers, NO_2^--oxidizers, and 2,4-D degraders microhabitat distributions using microsampling and a dedicated data analysis method (Grundmann et al., 2001; Dechesne et al., 2003; Pallud et al., 2004). The spatial pattern of NO_2^--oxidizer and of NH_4^+-oxidizer distribution at the abundance of 10^6 cells g^{-1} suggested that their microhabitats occurred as randomly distributed patches of, respectively, 250 and <180 µm diameter, whose centers were on average separated by 375 µm and 55–95 µm for the two groups of nitrifiers (Grundmann et al., 2001; Dechesne et al., 2003). At an abundance of 10^2 cells g^{-1}, 2,4-D degrader microhabitats occurred as 100–300 µm diameter patches that were dispersed in the soil, with an average distance between patch centers estimated to be 900 µm. When bacterial abundance of 2,4-D degraders increased up to 10^6 cells g^{-1}, their microhabitats formed a very dense network of coalescing patches of 500–2,250 µm in diameter (Pallud et al., 2004). As a consequence of these observations, the sampling size of a few 100 µm^3, which is far below the soil sample size commonly used, seems to be consistent with the spatial distribution of nitrifiers and 2,4-D degraders and is probably relevant to functional spatial units.

4.2. Physico-chemical controls on bacterial distributions

Studies focusing on the spatial pattern of bacteria at subcentimetre scales indicate non-random and, most often, aggregated distributions. Therefore, some factors must shape the spatial pattern of bacteria. The mode of bacterial growth by cell division contributes to the formation of microcolonies. Nevertheless, all the microhabitats are not equally favourable to bacterial growth or survival, and bacterial movements can result in the colonization of new microhabitats or in cell movements between microhabitats. Bacteria interactions with soil surfaces have been overlooked and, apart from bacteria abundance on particulate organic carbon (Parkin, 1987; Wachinger et al., 2000), it has not been demonstrated yet whether bacteria are located at random or in a deterministic manner on specific surfaces. Two important features of soil microhabitats are well documented: their protective role and their nutritional status. In soil, bacteria may be subjected to predation by microfauna, which mainly includes protozoa. In the microhabitats described by Lündsdorf et al. (2000) and Thieme et al. (2003), mineral particles play a role in protection against predation. Pores with a diameter <3 μm are suggested to act as a soil protective space (Postma and Van Veen, 1990) because they are too narrow to be accessed by the predators of bacteria. The relevance of such a protective space is confirmed by the observation that the predation of protozoa on introduced bacteria is marked in the outer fraction of soil but insignificant in the inner fraction (Vargas and Hattori, 1990). The concept of protective pores would explain why, according to observations of soil thin sections, most of the indigenous bacteria are located in relatively small pores (1–3 μm in diameter) (Kilbertus, 1980). Additionally, the protective microhabitats provided by clays would account for the increased bacterial survival in clay-amended soils (England et al., 1993).

The availability of nutrients is one of the major structuring factors of the spatial pattern of bacteria. In the case of heterotrophic bacteria, one of the main carbon sources is the particulate organic matter (Foster, 1988). Hissett and Gray (1976) showed that 64% of the bacteria were associated with organic particles, even though these represent only 15% of the soil volume. Around large organic particles, such as decomposing wheat straw, strong gradients of bacterial abundance ranging over several millimetres have been observed (Gaillard et al., 1999), suggesting that bacterial distribution is affected by soluble substrates that have diffused from the wheat straw. The experimental addition of soluble substrate was also shown to affect the distribution of bacteria at the microscale. Whereas indigenous bacteria in soil are usually organized in small microcolonies, the addition of highly labile organic carbon induced apparition of bacterial biofilms (Chenu et al.,

2001; Nunan et al., 2003) on the outer part of soil aggregates, while no marked bacterial growth was observed inside aggregates (Chenu et al., 2001; Nunan et al., 2003).

Soils are nutrient-poor media, in which the inter-aggregate porosity constitutes the major channel by which substrate may become available to bacteria (Bundt et al., 2001). Accordingly, preferential flow paths may be considered as biological hot spots in soil, characterized by high bacterial abundances and activities (Pivetz and Steenhuis, 1995; Bundt et al., 2001). The accessibility of substrate in preferential flow paths probably explains why Nunan et al. (2003) observed that the density of bacteria in subsoil samples was greatest close to large pores. Conversely, such an association with pores was not significant in topsoil samples, probably because nutrient supply is more spatially homogenous.

NH_4^+-oxidizers, NO_2^--oxidizers, and 2,4-D degraders distribution in the inner and outer compartments has been evaluated using soil columns before and after ammonium or 2,4-D supply. Interestingly, the two groups of nitrifiers that act successively in the nitrification process were preferentially located in different compartments. The NO_2^--oxidizers were found primarily in the inner compartment while the NH_4^+-oxidizers and the 2,4-D degraders were preferentially located in the outer compartment (Fig. 4-1). The relative distribution of bacteria between inner and outer compartments was the same

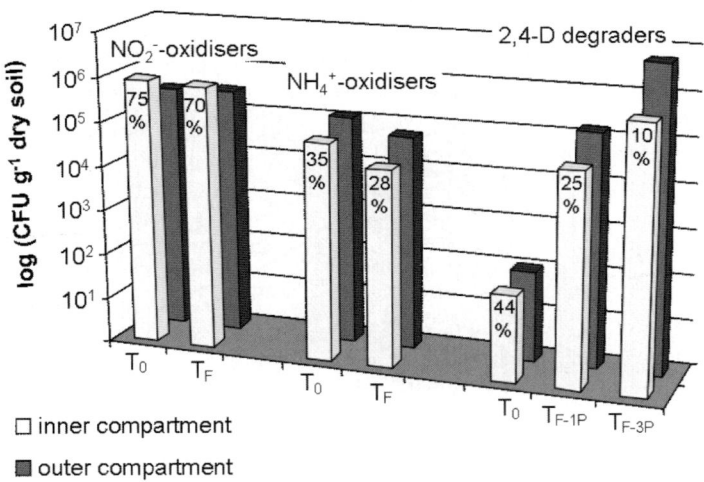

Figure 4-1. Densities of NO_2^--oxidizers, NH_4^+-oxidizers and 2,4-D degraders in the inner and outer compartments of a soil before any treatment (T_0), and after NH_4^+ supply (T_F) or after one (T_F-1P) or three (T_F-3P) 2,4-D applications in flow through soil columns. The percentages indicate the proportion of bacteria located in the inner compartment.

in the original soil and in the soil after substrate addition. The differential location of the two groups of nitrifiers (Fig. 4-1) could be a result of the different locations of their respective substrates, and could be looked upon as a way of optimizing substrate interception. During convective transport, NH_4^+ adsorption takes place primarily in the outer compartment. It is then an advantage for the NH_4^+-oxidizers to preferentially occupy the outer compartment. Conversely, NO_2^--oxidizers were preferentially situated in the inner compartment. They were spatially associated with NH_4^+ oxidizers, as shown by their frequent simultaneous presence in microsamples, and more spatially spread (Grundmann and Debouzie, 2000; Grundmann et al., 2001). Their relative distribution in the soil compartments should allow efficient NO_2^- uptake by creating a NO_2^- concentration gradient from the outer compartment toward the inner compartment, the latter acting as a NO_2^- sink and thus diverting NO_2^- from convective transport in the water flow. A functional spatial scheme of nitrifiers distribution is proposed (Fig. 4-2). Results obtained on nitrite oxidizers, ammonium oxidizers and 2,4-D degraders (Fig. 4-1) indicated that partitioning of the communities between outer and inner compartments, is not random, but probably determined by substrate location, and could eventually be related to functional characteristics.

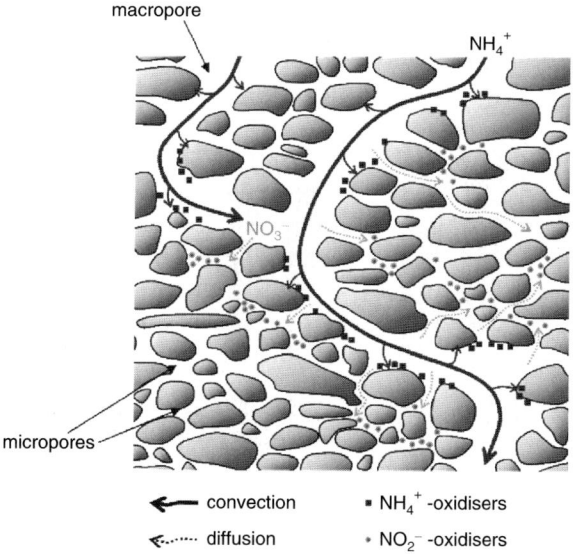

Figure 4-2. Scheme of the relative spatial distribution of NH_4^+-oxidizers and NO_2^--oxidizers in the inner and outer compartments of soil.

4.3. Microscale bacterial movements in soil

Bacterial spatial isolation controls several microbial interactions, such as competition (Treves et al., 2003), and has an influence on the population genetics of bacteria (Cho and Tiedje, 2000; Papke and Ward, 2004). In soil, isolation depends on local water content and bacterial movements.

Following movements of indigenous bacteria is quite challenging because, unlike introduced bacteria, they cannot be tracked by tagging. Therefore, in this matter, the microbial ecologist has to rely mostly on indirect evidence. Zvyagintsev (1962) observed that relatively few soil bacteria (<1%) are motile, suggesting that bacterial movements in soil are limited or rely on passive mechanisms, such as diffusion or transport with water flow (Gammack et al., 1992). Lünsdorf et al. (2000) suggested that bacteria may migrate within their microhabitats but rarely observed bacteria outside the clay hutches. Indeed, observing the migration of bacteria from one microhabitat to another may be unlikely since migration times may be short compared to the period of occupation of the microhabitats. Nevertheless, it is possible to indirectly detect movements of indigenous bacteria; two clone mates at different locations in soil, for example, is an indication that at least one of them has moved from the original location of the parent cell. It is not exceptional to detect two clone mates separated by several millimetres (Vogel et al., 2003) or even several kilometres (Fulthorpe et al., 1998; Vilas Boas et al., 2002). These observations, made on a variety of bacterial taxa, would indicate that bacterial movements in soil are not rare and that large distances may be travelled.

Experiments on soil columns showed that substrate addition triggers significant spatial modification in the microhabitat spatial structure of indigenous and introduced bacteria within a few days. For example, upon 2,4-D percolation, the estimated average diameter of the patches of soil colonized by 2,4-D degrading bacteria increased from 0.1 to >0.5 mm (Pallud et al., 2004). Such a spatial spreading cannot solely be due to the growth of 2,4-D degraders in stationary microcolonies; bacterial movements, probably active since the dominant 2,4-D degraders were motile, contributed to the colonization of new microhabitats. Similar spatial spreading following substrate amendment was also demonstrated for a GFP-tagged *Pseudomonas* introduced in soil columns (Dechesne et al., 2005). A dispersal of cells was also mentioned by El Balkhi et al. (1978), who showed that after percolation with a carbon solution, individual cells rather than microcolonies were found in the soil fabric.

Overall, experimental results indicate that microhabitats are not stable in time. For a better understanding of potential encounters between bacteria and between substrates and bacteria, the spatial isolation of bacteria and the extent of movement from one microhabitat to another must be addressed. They could differ as a function of bacterial type, bacterial activity, and/or soil type, and could influence degradation efficiency and gene transfer. For example, 97% of the 2,4-D applied to a flow-through soil column was degraded in the case of dispersed degrader populations, whereas for the same abundance, only 74% of the 2,4-D was degraded in the case of a less dispersed population (Pallud et al., 2004).

5. BACTERIAL DIVERSITY AT THE MICROSCALE

5.1. Microscale distribution of bacterial diversity

On different occasions, it has been shown that bacterial diversity in small soil samples (1 g, one to few aggregates or micro-pieces of soil) was very large, covering complex communities of pesticide degraders (Dunbar et al., 1997; Vallaeys et al., 1997), or different genotypes of the same species or genus: *Rhodopseudomonas* (Oda et al., 2003), *Nitrobacter* (Grundmann and Normand, 2000), *Agrobacterium* (Vogel et al., 2003), and *Myxococcus xanthus* (Vos and Velicer, 2006). The diversity observed in small samples could represent the same diversity as the one of larger sample surface or volume (Felske and Akkermans, 1998), or could cover the same genetic distances as in larger samples (Grundmann and Normand, 2000; Vogel et al., 2003). At small scales, it does not seem that genetic distances are related to spatial distances, as shown in a study on *Agrobacterium* spp. carried out on 1 cm^3 of undisturbed soil, where the spatial coordinates of microsamples (500 µm diameter) from which *Agrobacteria* (55 isolates) were exhaustively isolated, were recorded. Identical ARDRA patterns were found close-by as well as 10 mm apart (Vogel et al., 2003). At this small scale, clones were spatially overlapping, probably due to cell movements. Similarly, *M. xanthus* genotypes are not clustered at the centimetre scale (Vos and Velicer, 2006). In a sediment studied with a lower spatial resolution, the genetic similarity of the *Rhodopseudomas* population decreased with distance, although identical *Rhodopseudomonas* genotypes were found in the most distant samples (1 m) (Oda et al., 2003). Interestingly, the different genotypes exhibited distinct ecological traits, suggesting that the pattern of selective pressure imposed in the microhabitats was increasingly different with the distance (Oda et al., 2003). The existence of a spatial structure in soil bacterial communities at scales larger than the centimetre is supported by experiments on the dominant bacterial populations, studied by molecular fingerprinting (Noguez et al., 2005).

Rius et al. (2001), studying *Pseudomonas stutzeri*, concluded that the exceptionally high genetic diversity in a strongly clonal population structure could be the result of niche-specific selection that occurs during colonization and adaptation to a wide range of microenvironments. Studies of the genetic structure of populations of free-living bacteria combined with spatial structure are scarce in soil. McArthur et al. (1988) found that habitat variability was correlated with genetic variability in soil-borne *B. cepacia*. The results obtained on 1.6 km^2, suggested a pattern of micro-geographical adaptation of clones due to selection.

A taxa–area relationship, linking the number of species in an area and the size of that area was repeatedly observed in plant and animal communities over large scales. Such a law was described at a scale of centimetres to hundreds of metres for bacteria in salt marsh, based on the decay of community similarity with distance (Horner-Devine et al., 2004). Following results by Vogel et al. (2003), the power–law relationship does not seem to apply at the small scale for bacterial communities.

5.2. Microscale dynamics of bacterial diversity

The processes of emergence and maintenance of bacterial diversity have been explored in laboratory experiments (Arber, 1995; Rainey et al., 2000). The influence of spatial structures on adaptive radiation in experimental systems has been demonstrated, giving insight into potential mechanisms for establishment of micro-diversity, but a gap between results of laboratory experiments and diversity data in the field must be filled.

As detailed in the previous sections, both laboratory experiments and isolate collection indicate that genotypes have the potential for physical spreading within the soil matrix. A phenomenon of "spatial spreading", consisting of a rapid colonization of new microhabitats by a bacterial population or community, upon the onset of favourable conditions (high soil water content and availability of soluble substrate) has been described (Pallud et al., 2004; Dechesne et al., 2005). This suggests that during changes in bacterial density due to variations in nutrient availability, a spread of cells will occur, followed by retreat due to cell death, leaving some cells alive in some locations (Fig. 4-3). Thus, following a new growth event, there would be new cell settlements originating from the cells that did not die following the previous decline period. This cyclic invading process would give a bacterial genotype the opportunity to invade the soil fabric a little further at each cycle. It would thus explain the shuffling of clones observed at the microscale and the strong micro-diversity observed at this same small scale (Vogel et al., 2003). The degree of spatial isolation within microbial communities, which is known to influence several microbial interactions, such

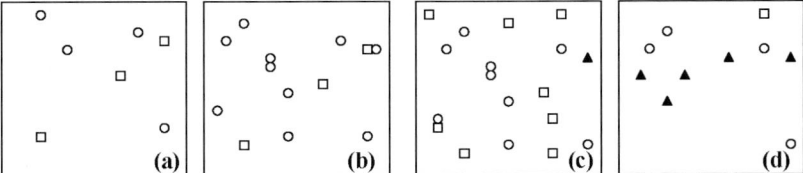

Figure 4-3. Two-dimensional scheme of invasion/decline cycles of bacteria inducing shuffling of genotypes in the soil space. Symbols represent four genotypes in space: (**a**) initial spatial distribution; (**b**) invasion of genotype noted ○, and one encounter and gene transfer between the two genotypes, new genotype noted ▲; (**c**) invasion of genotype noted □ and encounters with other genotypes; (**d**) recess of two genotypes and invasion by the new genotype.

as competition (Treves et al., 2003), and thus the population genetics of bacteria (Cho and Tiedje, 2000; Papke and Ward, 2004), would then fluctuate following the cycles of invasion and recess. For example, the period of invasions has been estimated to increase the chances of cell-to-cell contact, which is a prerequisite to conjugal gene transfers (Dechesne et al., 2005).

6. CONCLUSION

Describing and understanding the spatial distribution of bacterial cells and bacterial diversity, particularly in soil, remains a constant challenge in soil microbial ecology. Despite the difficulty of the task, new results are regularly obtained, which contribute to a better description and understanding of microbial spatial patterns and their ecological relevance. Nevertheless, most of the data available so far are limited to qualitative descriptions of bacterial distributions. There is a need of more quantitative surveys, which would enable ecological hypotheses to be tested such as encounter probability, species–area relationships, etc.

Future research should focus spatial studies on specific groups of bacteria, eventually in parallel, in order to decrease the complexity of the system under investigation and thus to uncover more easily the causes and consequences of bacterial structures in soil. These groups may either be defined phylogenetically or functionally. The first attempts in this direction are promising (Grundmann and Debouzie, 2000; Bent et al., 2003; Oda et al., 2003; Vogel et al., 2003).

Studying the spatial distribution of bacteria and its dynamics in simplified environments may also help in testing hypotheses or focusing on specific processes. For example, a sterile sand matrix has been used to examine the role of spatial isolation on the dynamics of two competing strains (Treves et al., 2003), and to observe the colonization dynamics of a *lux*-tagged strain, in the presence of a percolating substrate (Yarwood et al., 2002). The understanding of spatial mechanisms at the small scale in soil would probably benefit from laboratory experiments.

The knowledge of spatial distribution characteristics may have wide applications. One interest has been demonstrated through the calculation of encounter probabilities in soil between an introduced strain and a sub-community (Dechesne et al., 2005). It was shown that stimulating spatial spread could increase encounters by a factor of one hundred. Large applications could be found in the frame of bioremediation, risk assessment, and predictive modelling.

REFERENCES

Arber, W., 1995, The generation of variation in bacterial genomes, *J. Mol. Evol.* **40**:7–12.

Bent, S. J., C. L. Gucker, Y. Oda, and L. J. Forney, 2003, Spatial distribution of *Rhodopseudomonas palustris* ecotypes on a local scale, *Appl. Environ. Microb.* **69**:5192–5197.

Bloemberg, G. V., A. H. Wijfjes, G. E. Lamers, N. Stuurman, and B. J. Lugtenberg, 2000, Simultaneous imaging of *Pseudomonas fluorescens* WCS365 populations expressing three different autofluorescent proteins in the rhizosphere: new perspectives for studying microbial communities, *Mol. Plant Microbe Interact.* **13**:1170–1176.

Bundt, M., F. Widmer, M. Pesaro, J. Zeyer, and P. Blaser, 2001, Preferential flow paths: biological 'hot spots' in soils, *Soil Biol. Biochem.* **33**:729–738.

Chalfie, M., Y. Tu, G. Euskirchen, W. W. Ward, and D. C. Prasher, 1994, Green fluorescent protein as a marker for gene expression, *Science* **263**:802–805.

Chenu, C., J. Hassink, and J. Bloem, 2001, Short-term changes in the spatial distribution of microorganisms in soil aggregates as affected by glucose addition, *Biol. Fert. Soils* **34**:349–356.

Cho, J. C., and J. M. Tiedje, 2000, Biogeography and degree of endemicity of fluorescent *Pseudomonas* strains in soil, *Appl. Environ. Microb.* **66**:5448–5456.

Daane, L. L., J. A. Molina, E. C. Berry, and M. J. Sadowsky, 1996, Influence of earthworm activity on gene transfer from *Pseudomonas fluorescens* to indigenous soil bacteria, *Appl. Environ. Microb.* **62**:515–521.

Daane, L. L., J. A. E. Molina, and M. J. Sadowsky, 1997, Plasmid transfer between spatially separated donor and recipient bacteria in earthworm-containing soil microcosms, *Appl. Environ. Microb.* **63**:679–686.

Dechesne, A., C. Pallud, D. Debouzie, J. P. Flandrois, T. M. Vogel, J. P. Gaudet, and G. L. Grundmann, 2003, A novel method for characterizing the microscale 3D spatial distribution of bacteria in soil, *Soil Biol. Biochem.* **35**:1537–1546.

Dechesne, A., C. Pallud, F. Bertolla, and G. L. Grundmann, 2005, Impact of the microscale distribution of a *Pseudomonas* strain introduced into soil on potential contacts with indigenous bacteria, *Appl. Environ. Microb.* **71**:8123–8131.

Dunbar, J., S. White, and L. Forney, 1997, Genetic diversity through the looking glass: effect of enrichment bias, *Appl. Environ. Microb.* **63**:1326–1331.

El Balkhi, M., F. Mangenot, J. Proth, and G. Kilbertus, 1978, Influence de la percolation d'une solution de saccharose sur la composition qualitative et quantitative de la microflore bactérienne d'un sol, *Soil Sci. Plant Nutr.* **24**:15–25.

England, L. S., H. Lee, and J. T. Trevors, 1993, Bacterial survival in soil: effect of clays and protozoa, *Soil Biol. Biochem.* **25**:525–531.

Errampalli, D., K. Leung, M. B. Cassidy, M. Kostrzynska, M. Blears, H. Lee, and J. T. Trevors, 1999, Applications of the green fluorescent protein as a molecular marker in environmental microorganisms, *J. Microbiol. Methods* **35**:187–199.

Felske, A., and A. D. L. Akkermans, 1998, Spatial homogeneity of abundant bacterial 16S rRNA molecules in grassland soils, *Microbial Ecol.* **36**:31–36.

Focht, D. D., 1992, Diffusional constraints on microbial processes in soil, *Soil Sci.* **154**:300–307.

Foster, R. C., 1988, Microenvironments of soil microorganisms, *Biol. Fert. Soils* **6**:189–203.

Franklin, R. B., L. K. Blum, A. C. McComb, and A. L. Mills, 2002, A geostatistical analysis of small-scale spatial variability in bacterial abundance and community structure in salt marsh creek bank sediments, *FEMS Microbiol. Ecol.* **42**:71–80.

Fulthorpe, R. R., A. N. Rhodes, and J. M. Tiedje, 1998, High levels of endemicity of 3-chlorobenzoate-degrading soil bacteria, *Appl. Environ. Microb.* **64**:1620–1627.

Gaillard, V., C. Chenu, S. Recous, and G. Richard, 1999, Carbon, nitrogen and microbial gradients induced by plant residues decomposing in soil, *Eur. J. Soil Sci.* **50**:567–578.

Gammack, S. M., E. Paterson, J. S. Kemp, M. S. Cresser, and K. Killham, 1992, Factors affecting the movement of microorganisms in soils, in: *Soil Biochemistry*, Vol. 7, G. Stotzky and J. Bollag, eds., Marcel Dekker, New York, pp. 263–305.

Gray, T. R. G., P. Baxby, I. R. Hill, and M. Goodfellow, 1968, Direct observation of bacteria in soil, in: *The Ecology of Soil Bacteria*, T. Gray and D. Parkinson, eds., Liverpool University Press, Liverpool, UK, pp. 171–192.

Grundmann, G. L., and D. Debouzie, 2000, Geostatistical analysis of the distribution of NH_4^+ and NO_2^--oxidizing bacteria and serotypes at the millimeter scale along a soil transect, *FEMS Microbiol. Ecol.* **34**:57–62.

Grundmann, G. L., and P. Normand, 2000, Microscale diversity of the genus *Nitrobacter* in soil on the basis of analysis of genes encoding rRNA, *Appl. Environ. Microb.* **66**:4543–4546.

Grundmann, G. L., A. Dechesne, F. Bartoli, J. P. Flandrois, J. L. Chasse, and R. Kizungu, 2001, Spatial modeling of nitrifier microhabitats in soil, *Soil Sci. Soc. Am. J.* **65**:1709–1716.

Harms, H., and A. J. Zehnder, 1994, Influence of substrate diffusion on degradation of dibenzofuran and 3-chlorodibenzofuran by attached and suspended bacteria, *Appl. Environ. Microb.* **60**:2736–2745.

Harris, P. J., 1994, Consequences of the spatial distribution of microbial communities in soil, in: *Beyond the Biomass*, K. Ritz, et al., eds., Wiley, Chichester, UK, pp. 239–246.

Hattori, T., 1967, Microorganisms and soil aggregates as their microhabitat, *Bull. Inst. Agr. Res. Tohoku Univ.* **18**:159–193.

Hattori, T., 1973, *Microbial Life in the Soil*, Marcel Dekker, New York.

Hissett, R., and T. R. G. Gray, 1976, Microsites and time changes in soil microbe ecology, in: *The Role of Terrestrial and Aquatic Organisms in Decomposition Process*, J. Anderson, and A. MacFadyen, eds., Blackwell, Oxford, pp. 23–39.

Holden, P. A., and M. K. Firestone, 1997, Soil microorganisms in soil cleanup: how can we improve our understanding? *J. Environ. Qual.* **26**:32–40.

Horner-Devine, M. C., K. M. Carney, and B. J. M. Bohannan, 2004, An ecological perspective on bacterial biodiversity, *Proc. Roy. Soc. Lond. B Bio.* **271**:113–122.

Jocteur-Monrozier, L., J. N. Ladd, R. W. Fitzpatrick, R. C. Foster, and M. Rapauch, 1991, Physical properties, mineral and organic components and microbial biomass content of size fractions in soils of contrasting aggregation, *Geoderma* **50**:37–62.

Jones, D., and E. Griffiths, 1964, The use of thin soil sections for the study of soil microorganisms, *Plant Soil* **20**:232–240.

Kilbertus, G., 1980, Etude des microhabitats contenus dans les agrégats du sol. Leur relation avec la biomasse bactérienne et la taille des procaryotes présents, *Rev. Ecol. Biol. Sol* **17**:543–557.

Lee, N., P. H. Nielsen, K. H. Andreasen, S. Juretschko, J. L. Nielsen, K. H. Schleifer, and M. Wagner, 1999, Combination of fluorescent *in situ* hybridization and microautoradiography – a new tool for structure-function analyses in microbial ecology, *Appl. Environ. Microb.* **65**:1289–1297.

Li, Y., W. A. Dick, and O. H. Tuovinen, 2003, Evaluation of fluorochromes for imaging bacteria in soil, *Soil Biol. Biochem.* **35**:737–744.

Lünsdorf, H., R. W. Erb, W. R. Abraham, and K. N. Timmis, 2000, 'Clay hutches': a novel interaction between bacteria and clay minerals, *Environ. Microbiol.* **2**:161–168.

McArthur, J. V., D. A. Kovacic, and M. H. Smith, 1988, Genetic diversity in natural populations of a soil bacterium across a landscape gradient, *Proc. Natl. Acad. Sci. USA* **85**:9621–9624.

Mendes, I. C., and P. J. Bottomley, 1998, Distribution of a population of *Rhizobium leguminosarum* bv. trifolii among different size classes of soil aggregates, *Appl. Environ. Microb.* **64**:970–975.

Metting, F. B., 1992, Structure and physiological ecology of soil microbial communities, in: *Soil Microbial Ecology: Application in Agricultural and Environmental Management*, F. Metting, ed., Marcel Dekker, New York, pp. 3–25.

Monier, J. M., and S. E. Lindow, 2003, Differential survival of solitary and aggregated bacterial cells promotes aggregate formation on leaf surfaces, *Proc. Natl. Acad. Sci. USA* **100**:15977–15982.

Mummey, D. L., and P. D. Stahl, 2004, Analysis of soil whole- and inner-microaggregate bacterial communities, *Microbial Ecol.* **48**:41–50.

Nikin, D. I., and F. Kunc, 1988, Structure of microbial soil associations and some mechanisms of their autoregulation, in: *Soil Microbial Associations*, V. Vancura and F. Kunc, eds., Elsevier, Amsterdam, pp. 157–190.

Nishio, M., and C. Furusaka, 1970, The distribution of nitrifying bacteria in soil aggregates, *Soil Sci. Plant Nutr (Tokyo)* **16**:24–29.

Nishio, M., T. Hattori, and C. Furusaka, 1968, The growth of bacteria in sterilized soil aggregates, *Rep. Inst. Agr. Res. Tohoku Univ.* **19**:37–43.

Noguez, A. M., H. T. Arita, A. E. Escalante, L. J. Forney, F. Garcia-Oliva, and V. Souza, 2005, Microbial macroecology: highly structured prokaryotic soil assemblages in a tropical deciduous forest, *Global Ecol. Biogeogr.* **14**:241–248.

Nunan, N., K. Ritz, D. Crabb, K. Harris, K. J. Wu, J. W. Crawford, and I. M. Young, 2001, Quantification of the *in situ* distribution of soil bacteria by large-scale imaging of thin sections of undisturbed soil, *FEMS Microbiol. Ecol.* **37**:67–77.

Nunan, N., K. Wu, I. M. Young, J. W. Crawford, and K. Ritz, 2002, *In situ* spatial patterns of soil bacterial populations, mapped at multiple scales, in an arable soil, *Microbial Ecol.* **44**:296–305.

Nunan, N., K. J. Wu, I. M. Young, J. W. Crawford, and K. Ritz, 2003, Spatial distribution of bacterial communities and their relationships with the micro-architecture of soil, *FEMS Microbiol. Ecol.* **44**:203–215.

Oda, Y., B. Star, L. A. Huisman, J. C. Gottschal, and L. J. Forney, 2003, Biogeography of the purple nonsulfur bacterium *Rhodopseudomonas palustris*, *Appl. Environ. Microb.* **69**:5186–5191.

Or, D., B. F. Smets, J. M. Wraith, A. Dechesne, and S. P. Friedman, 2007, Physical constraints affecting bacterial habitats and activity in unsaturated porous media – A review, *Adv. Water. Res.* **30**:1505–1527.

Pallud, C., A. Dechesne, J. P. Gaudet, D. Debouzie, and G. L. Grundmann, 2004, Modification of spatial distribution of 2,4-dichloro-phenoxyacetic acid degrader microhabitats during growth in soil columns, *Appl. Environ. Microb.* **70**:2709–2716.

Papke, R. T., and D. M. Ward, 2004, The importance of physical isolation to microbial diversification, *FEMS Microbiol. Ecol.* **48**:293–303.

Parkin, T. B., 1987, Soil microsites as a source of denitrification variability, *Soil Sci. Soc. Am. J.* **51**:1194–1199.

Parkin, T. B., 1993, Spatial variability of microbial processes in soil – a review, *J. Environ. Qual.* **22**:409–417.

Pivetz, B. E., and T. S. Steenhuis, 1995, Soil matrix and macropore biodegradation of 2,4-D, *J. Environ. Qual.* **24**:564–570.

Poly, F., L. Ranjard, S. Nazaret, F. Gourbiere, and L. J. Monrozier, 2001, Comparison of *nifH* gene pools in soils and soil microenvironments with contrasting properties, *Appl. Environ. Microb.* **67**:2255–2262.

Postma, J., and J. A. Van Veen, 1990, Habitable pore space and survival of *Rhizobium leguminosarum* biovar *trifolii* introduced into soil, *Microbial Ecol.* **19**:149–161.

Rainey, P. B., A. Buckling, R. Kassen, and M. Travisano, 2000, The emergence and maintenance of diversity: insights from experimental bacterial populations, *Trends Ecol. Evol.* **15**:243–247.

Ranjard, L., and A. S. Richaume, 2001, Quantitative and qualitative microscale distribution of bacteria in soil, *Res. Microbiol.* **152**:707–716.

Ranjard, L., A. Richaume, L. Jocteur-monrozier, and S. Nazaret, 1997, Response of soil bacteria to Hg(II) in relation to soil characteristics and cell location, *FEMS Microbiol. Ecol.* **24**:321–331.

Recorbet, G., A. Richaume, and L. Jocteur Monrozier, 1995, Distribution of a genetically-engineered *Escherichia coli* population introduced into soil, *Lett. Appl. Microbiol.* **21**:38–40.

Richardson, R. E., C. A. James, V. K. Bhupathiraju, and L. Alvarez Cohen, 2002, Microbial activity in soils following steam treatment, *Biodegradation* **13**:285–295.

Rius, N., M. C. Fuste, C. Guasp, J. Lalucat, and J. G. Loren, 2001, Clonal population structure of *Pseudomonas stutzeri*, a species with exceptional genetic diversity, *J. Bacteriol.* **183**:736–744.

Sessitsch, A., A. Weilharter, H. Gerzabek M, H. Kirchmann, and E. Kandeler, 2001, Microbial population structures in soil particle size fractions of a long-term fertilizer field experiment, *Appl. Environ. Microb.* **67**:4215–4224.

Skinner, F. A., 1976, Methods in soil examination, in: *Microbiology in Agriculture, Fisheries and Food*, F. A. Skinner and J. G. Carr, eds., Academic Press, London, pp. 19–35.

Thieme, J., G. Schneider, and C. Knochel, 2003, X-ray tomography of a microhabitat of bacteria and other soil colloids with sub-100 nm resolution, *Micron* **34**:339–344.

Tombolini, R., D. J. Van Der Gaag, B. Gerhardson, and J. K. Jansson, 1999, Colonization pattern of the biocontrol strain *Pseudomonas chlororaphis* MA 342 on barley seeds visualized by using green fluorescent protein, *Appl. Environ. Microb.* **65**:3674–3680.

Treves, D. S., B. Xia, J. Zhou, and J. M. Tiedje, 2003, A two-species test of the hypothesis that spatial isolation influences microbial diversity in soil, *Microbial Ecol.* **45**:20–28.

Tsuji, T., Y. Kawasaki, S. Takeshima, T. Sekiya, and S. Tanaka, 1995, A new fluorescence staining assay for visualizing living microorganisms in soil, *Appl. Environ. Microb.* **61**:3415–3421.

Vallaeys, T., F. Persello-carteaux, N. Rouard, C. Lors, G. Laguerre, and G. Soulas, 1997, PCR-RFLP analysis of 16S rRNA, *tfdA* and *tfdB* genes reveals a diversity of 2,4-D degraders in soil aggregates, *FEMS Microbiol. Ecol.* **24**:269–278.

Vargas, R., and T. Hattori, 1990, The distribution of protozoa among soil aggregates, *FEMS. Microbiol. Lett.* **74**:73–78.

Vilas Boas, G., V. Sanchis, D. Lereclus, M. V. Lemos, and D. Bourguet, 2002, Genetic differentiation between sympatric populations of *Bacillus cereus* and *Bacillus thuringiensis*, *Appl. Environ. Microb.* **68**:1414–1424.

Vogel, J., P. Normand, J. Thioulouse, X. Nesme, and G. L. Grundmann, 2003, Relationship between spatial and genetic distance in *Agrobacterium* spp. in 1 cubic centimeter of soil, *Appl. Environ. Microb.* **69**:1482–1487.

Vos, M., and G. J. Velicer, 2006, Genetic population structure of the soil bacterium *Myxococcus xanthus* at the centimeter scale, *Appl. Environ. Microb.* **72**:3615–3625.

Wachinger, G., S. Fiedler, K. Zepp, A. Gattinger, M. Sommer, and K. Roth, 2000, Variability of soil methane production on the micro-scale: spatial association with hot spots of organic material and Archaeal populations, *Soil Biol. Biochem.* **32**:1121–1130.

Yarwood, R. R., M. L. Rockhold, M. R. Niemet, J. S. Selker, and P. J. Bottomley, 2002, Noninvasive quantitative measurement of bacterial growth in porous media under unsaturated-flow conditions, *Appl. Environ. Microb.* **68**:3597–3605.

Zvyagintsev, D. G., 1962, Adsorption of microorganisms to soil particles, *Soviet Soil Sci.* 140–144.

Chapter 5

ANALYSIS OF SPATIAL PATTERNS OF RHIZOPLANE COLONIZATION

Guy R. Knudsen and Louise-Marie Dandurand
Department of Plant, Soil, and Entomological Sciences, University of Idaho, Moscow, ID 83844-2339, USA

Abstract: Natural populations and habitats are spatially heterogeneous: organisms and the resources they use are not uniformly distributed over space or time, but instead are found in different degrees of aggregation. The rhizosphere, that region of soil surrounding plant roots where microbial activity is influenced by the root, is one of the most microbiologically active habitats on earth. The rhizoplane, the innermost boundary of the rhizosphere, is the surface of the plant root including root hairs, and is a hot spot of plant–microbe activity. Bacteria and fungi on the rhizoplane are favorably positioned to intercept root exudates, and rhizoplane sites are points of root interaction with plant pathoens as well as beneficial microbes. Microbial populations in the rhizosphere differ quantitatively and qualitatively from those in the bulk soil, since microbial numbers generally are higher, and different populations are represented. Our emphasis in this chapter is on descriptive and quantitative aspects of the spatial associations of plant roots and rhizoplane microbes. Physiological and spatial heterogeneity have important implications for the ecology of rhizoplane microbial populations and communities. Although the published literature contains a great deal of information on microbial colonization of roots, there is a need for an understanding of how spatial associations of rhizoplane and rhizosphere microbes evolve over time. Spatial statistical analysis provides a mechanism to explore processes that generate different patterns of organisms over time, and to determine the sensitivity of pattern to variations in these processes. Geostatistics provides a quantitative assessment of spatial distributions which maintains spatial integrity of data, and is able to analyze spatial autocorrelation based on direction and distance between samples. We describe the use of geostatistics to evaluate rhizoplane colonization and changes in spatial distribution of microbes associated with growing roots, and some implications for root infection by plant pathogens, biological control of plant diseases by beneficial microbes, and bacterial gene exchange events on the rhizoplane. We also discuss methods for observing and quantifying rhizosphere and rhizoplane habitats, including novel developments in molecular biology and microscopy which hold great promise for these types of studies.

Keywords: microbial ecology, spatial statistics, geostatistics, biological control, rhizosphere

1. INTRODUCTION

The rhizosphere is one of the most microbiologically active habitats on earth. "Rhizosphere" is most commonly defined as that region of soil surrounding plant roots, in which activity of the microbiota is affected by the presence of the root. Conversely, microbial activity in the rhizosphere also has profound effects on the root itself, although the scale of the different interactions may vary. The rhizoplane, which may be considered the innermost boundary of the rhizosphere, is the surface of the plant root (including root hairs), and is a hot spot of plant-microbe activity. Microbes associated with plant roots modify the rhizosphere environment by producing extracellular enzymes and plant growth factors, and by forming parasitic or mutualistic relationships with the plant. Populations of microbes in the rhizosphere differ quantitatively and qualitatively from those in the bulk soil: their numbers generally are higher, and different populations commonly are represented. This "rhizosphere effect" reflects the influence of plant roots on the size and composition of the surrounding soil microbial community. One numerical measure of rhizosphere effect is the rhizosphere/soil (R/S) ratio, which compares the total number of microbes in a rhizosphere habitat to the number in the root-free soil. R/S ratios range from as low as 5–100 or more, but most frequently are from 10 to 20 (Curl and Truelove, 1986; Gray and Parkinson, 1968; Katznelson, 1965). Not surprisingly, the extent of the rhizosphere effect is difficult to accurately measure, and can vary with the methodology used and the specific microbes that are looked at. Operationally, researchers have sometimes made a crude estimate of the rhizosphere, as that soil which is loosely adherent to a root which has been carefully removed from soil, i.e., a distance of up to several millimeters from the root surface. Occasionally the term "endorhizosphere" has been used to include the root interior as part of the rhizosphere, however a good case has been made for eliminating the use of that term (Kloepper and Beauchamp, 1992).

A number of biochemical gradients exist in the rhizosphere, which can significantly affect microbial activity. Because growing roots exude a number of compounds including amino acids, organic acids, hormones, vitamins, and sugars, there usually is a significant concentration of these compounds available near the rhizoplane, and this concentration decreases with increasing distance from the root. Because of the respiratory activity of both the root and rhizosphere microbes, there is typically an increasing gradient of oxygen away from the root, a increasing gradient of CO_2 towards from the root, and a decreasing pH gradient towards the root (due to carbonic acid produced by CO_2 dissolution in soil water). Plant roots remove water from soil, so that there usually is a gradient of increasing soil moisture that increases moving away from the root, except that during rain or irrigation events the channels made by roots may enhance infiltration of water into soil, thus reversing the gradient at least temporarily.

Because it is a two-dimensional entity, the rhizoplane is more easily defined and observed than is the rhizosphere. Bacteria and fungi present on the rhizoplane are the most favorably positioned to intercept root exudates; sites on the rhizoplane are also points of root infection by a variety of plant pathogens and mutualistic microbes (e.g., *Rhizobium* and its relatives, *Frankia*, mycorrhizal fungi), as well as interactions with beneficial associative microbes (e.g., *Azotobacter* and others). In this chapter, we will focus on descriptive and quantitative aspects of the spatial associations of plant roots and rhizoplane microbes. We will emphasize spatial statistical analysis of rhizoplane colonization activity, the changes in spatial distribution of microbes associated with growing roots, and some implications for root infection by plant pathogens, biological control of plant pathogens by beneficial bacteria and/or fungi, and bacterial gene exchange events which may be enhanced by rhizoplane nutritional and surface influences. We will also discuss methods for observing and quantifying the rhizosphere and rhizoplane habitat, including some relatively recent developments that hold great promise for these types of studies.

2. RHIZOPLANE SPATIAL ASSOCIATIONS

Biological organisms and their controlling variables rarely are distributed in a random or in a uniform way, since the environment is spatially structured by various energy inputs that result in patchy structures or gradients (Legendre and Fortin, 1989). The rhizoplane is a good example of this, since energy input is largely due to root exudates, and certain zones of roots produce more exudate than others (Rovira, 1956). Although only 1–2% of a root system may be colonized by microbes, space can be a limiting factor (Baker, 1981; Lockwood, 1981). It is generally believed that the zone just behind the root tip is the site of maximum root exudation. As roots elongate through soil, cells are sloughed off from the root cap, and root hairs and cortical cells are abraded from the root surface. Deposition from roots provides a readily used energy source for microbes, and spatiotemporal variation in the components of rhizodeposition over the root surface greatly influences variability in the composition of the rhizoplane community (Whipps, 1990). Sites may be preferentially colonized by some rhizoplane microbes, and thus no longer be available to others (Cook, 1993). As a result, microbial distributions typically are neither random nor uniform. The tendency for both leaf- and root-surface microbial populations to conform to lognormal or similar frequency distributions has been noted (Bahme and Schroth, 1987; Loper et al., 1984; Newman and Bowen, 1974), although there has been less attention to mechanisms of population development that might lead to such distributions. Later, we will discuss some significant limitations to the use of frequency distributional analysis in microbial ecology.

3. SPATIOTEMPORAL AVAILABILITY OF SITES FOR ROOT PATHOGENS AND BIOLOGICAL CONTROL AGENTS

Physiological and spatial heterogeneity have important implications for the ecology of rhizoplane microbial populations and communities. As noted by Martin and English (1997), the structural, physical, and biological complexity of the soil environment in which pathogens interact with plant roots constrains disease control options, including biological control. As seeds germinate and roots subsequently elongate, the spatial and temporal availability of infection courts is constantly changing. For example, Deacon and Donaldson (1993) described zoospores of phytopathogenic Oomycete fungi as "homing agents" or "site-selection agents," because their motility is linked to receptor functions for detecting environmental signals. They described the zoospore homing response as a sequence requiring two factors: a chemotactic stimulus, and a suitable surface on which zoospores can orient (Deacon and Donaldson, 1993). Further, they pointed out that zoospores can precisely locate root tips, wounds, or even individual root cells, so that understanding the homing response, and factors that may modify it, is central to understanding zoosporic fungi, and for attempts to control them.

Many or most bacteria associated with the exterior surfaces of plants are nonpathogenic. Among those are a number of organisms that have been found to be antagonistic to bacterial and fungal plant pathogens, thus potentially acting as biological control (biocontrol) agents. Similarly, colonization dynamics of biocontrol agents on seeds and roots, in the presence of an indigenous rhizosphere microbial community, will determine how well these potential infection courts are protected. Successful manipulation of rhizoplane microflora to enhance native or introduced beneficial microorganisms, depends on knowledge of their ecological associations over time and space (Schroth and Hancock, 1981; Stanghellini and Rasmussen, 1989). Although the published literature contains a great deal of information on microbial colonization of roots, there is a need for an understanding of how spatial associations of rhizoplane and rhizosphere microbes evolve over time.

4. EXPERIMENTAL METHODOLOGY FOR INVESTIGATING RHIZOSPHERE MICROBIAL DYNAMICS

A number of special considerations arise when sampling microbes from plant surfaces. Parberry et al. (1981), Hirano and Upper (1983, 1986), Kinkel et al. (1995), and Kloepper and Beauchamp (1992) have reviewed some of the difficulties inherent in sampling the phyllosphere (leaf surface) and the rhizosphere; these include the nonnormal distribution of microorganisms, selection of sampling strategy, and scale of sample and sample unit. Nonspatial traditional methods such as dilution plating are inadequate for addressing heterogeneity in rhizoplane populations, so that the ability of more recently developed molecular and microscopic methods to address heterogeneity in environmental samples is especially exciting. Several of these methods enhance our ability to observe, *in situ*, many aspects of microbial physiology that previously could only be demonstrated with pure cultures under highly controlled conditions. Various new strategies also are available for detection of microbial genes and gene products in natural habitats.

There are two general approaches to the quantitative estimation of microbial populations in the rhizosphere or phyllosphere: direct methods such as visualization techniques, and traditional indirect methods such as dilution plating of root or leaf washings, or most probable number (MPN) estimates (Cochran, 1950). Indirect methods have been reported to recover only about 0.01–10% of the numbers of bacteria enumerated by direct counts (Campbell and Porter, 1982; Colwell et al., 1985; Richaume et al., 1993). However, direct counts are difficult to obtain from an opaque medium such as soil. Also, they may overestimate the number of bacteria in soil because it is difficult to differentiate between living and dead cells unless using a viability stain. Direct counts are highly labor intensive if large numbers of samples need to be processed. Another disadvantage with direct counts is that different groups of organisms are difficult to identify unless their morphologies differ.

5. MARKER AND REPORTER GENES TO TRACK RHIZOPLANE MICROBES

Although several methods have been used to study the occurrence and distribution of fungi in natural soils (Green and Jensen, 1995), few methods have allowed quantitative evaluation of population dynamics and proliferation. For example, in previous studies we quantified influences of temperature, soil matric potential, nutrient source, and antagonistic bacteria, on the hyphal growth and biocontrol efficacy of pelletized *Trichoderma harzianum* (Knudsen et al., 1990, 1991a, b). However, in those studies we were unable to differentiate the hyphal growth of the fungal agent from that of indigenous *Trichoderma* strains in natural soils (Knudsen et al., 1990, 1991a, b). The use of dilution plating for numerical estimation of fungal populations does not differentiate among the different propagules (hyphal fragments, conidia, chlamydospores) that may generate colonies when plated on agar, and thus is an unreliable estimate of fungal biomass and active physiological status (Lumsden et al., 1990). The use of mutant strains resistant to specific fungicides may partially overcome problems related to nonspecific recovery, but this method does not allow for *in situ* monitoring of growth dynamics and survival structures of introduced *Trichoderma* strains, or differentiation of introduced *Trichoderma* strains from indigenous strains. There are similar constraints to the use of antibiotic-marked bacterial strains.

Recently, genetic engineering of bacteria and fungi with marker genes has provided useful tools for detecting and monitoring them in natural environments (Green and Jensen, 1995; Lo et al., 1998; Bae and Knudsen, 2000). For example, the selectable hygromycin B phosphotransferase (hygB) gene, coding for resistance to this antibiotic, has been used to detect fungal biocontrol agents in the rhizosphere and phyllosphere (Lo et al., 1998). The β-glucuronidase (GUS) marker gene also is a promising tool for ecological studies of soil- and plant-associated microbes, because of the low background activity of GUS in fungi and plants, the relative ease and sensitivity of detection (Roberts et al., 1989), and the apparent lack of influence of GUS expression on biocontrol efficacy or other activity (Thrane et al., 1995). Green et al. (2001) monitored activity of *T. harzianum* on seeds and in the rhizosphere of different plant species, using a GUS transformant strain. They were able to directly observe microhabitats supporting metabolic activity of

T. harzianum, and to correlate GUS activity of *T. harzianum* with biomass of the root pathogen *Pythium ultimum* in infected roots. However, some GUS activity may be present in unsterile systems or natural soils. For example, *Aspergillus niger* has some indigenous GUS activity (Thrane et al., 1995). Therefore, for study of growth patterns of an introduced fungus in natural ecosystems, this reporter gene system may be less useful.

The green fluorescent protein (GFP) of the jellyfish *Aequorea victoria* also has been developed both as a marker and a reporter for gene expression (Chalfie et al., 1994). The GFP gene has been successfully cloned and expressed in a variety of different organisms including plants, animals, fungi, and bacteria. GFP requires only UV or blue light and oxygen to induce green fluorescence. An exogenous substrate, such as GUS requires, is not needed for the detection of GFP, thus avoiding problems related to cell perrmeability and substrate uptake (Sheen et al., 1995). Expression of cloned GFP has been reported in several organisms (Cormack et al., 1997; Sheen et al., 1995). GFP was shown to be a useful tool for studying host-fungal pathogen interactions in vivo (Spelling et al., 1996) and has been used to assess colonization and dispersal of *Aureobasidium pullulans* in the phyllosphere (Vanden Wymelenberg et al., 1997). We generated a stable transformant of the fungal biocontrol agent *T. harzianum* expressing both GFP and GUS phenotypes (Bae and Knudsen, 2000). Presence of the GFP and GUS marker genes in the cotransformant strain allowed differentiation of introduced *Trichoderma* from indigenous strains. For example, Fig. 5-1 shows a sample of leaf debris colonized by the *Trichoderma* GFP transformant ThzID1-M3, along with indigenous fungi, bacteria, and protozoa. Under epifluorescence illumination, hyphae of the transformant are easily identifiable and distinguishable from the indigenous microflora. GFP activity of the transformant also was a useful tool for nondestructive monitoring of hyphal growth *in situ* (Bae and Knudsen, 2001; Orr and Knudsen, 2004). Formation of green-fluorescing conidiophores and conidia was observed within the first three days of incubation in soil, followed by formation of terminal and intercalary chlamydospores and subsequent disintegration of older hyphal segments. By capturing images viewed with epifluorescence, and subsequent computer image analysis of the captured images, we were able to quantify the hyphal growth and biomass dynamics of the introduced fungus (Orr and Knudsen, 2004). This methodology, combining GFP, epifluorescence microscopy, and digital image analysis, should be highly applicable to quantification of fungal spatial dynamics in soil and on the rhizoplane.

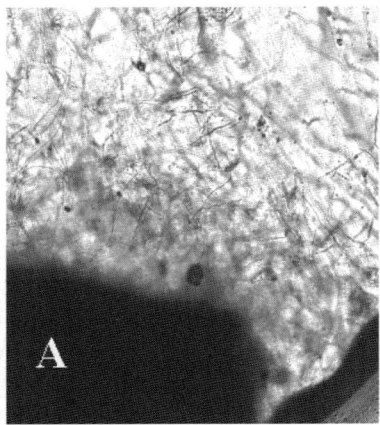
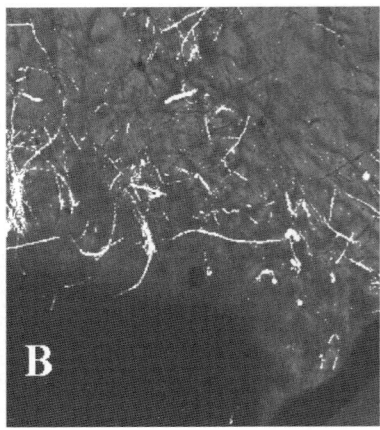

Figure 5-1. Transmitted light (**A**) and fluorescence (**B**) micrographs, identical field of view, of a sample of natural soil and plant debris, colonized by indigenous microbes and *T. harzianum* ThzID1 M3. Fluorescence of hyphae and spores of ThzID1 M3 is visible (green fluorescence was electronically filtered to white in this micrograph).

6. ADVANCES IN MICROSCOPY

A number of methods to quantify metabolic activity and/or biomass of individual microbes, populations, or microbial communities, involve direct microscopic observation of cells. These include simple measurement of cell size and size distributions, formation of microscopically observable metabolic products, fluorochromes used as "physiological stains," and genetically based marker and reporter systems. Most of the wide range of potential microscopic techniques have been attempted for visualization of microbes on plant surfaces or in soil, including brightfield, phase contrast and differential interference, epifluorescence and confocal scanning laser microscopy, scanning and transmission electron microscopy (e.g., Dandurand and Knudsen, 1994; Dandurand et al., 1995, 1997), and atomic force microscopy (Fendorf et al., 1997). In general, direct visualization for quantitative purposes can be difficult because of opacity of the substratum and the small size of microorganisms. Spatial associations can be difficult to describe because of the short working distance and the high magnification needed to see them. Phase contrast or interference microscopy of fresh material is possible, but associated soil and root material can give confusing background for observation of microorganisms. Fluorochromes and other stains are commonly used to observe microbes on plant surfaces, but autofluorescence of plant tissue can be a problem when using fluorochromes to detect microbes on plant surfaces.

6.1. Example: use of confocal microscopy in conjunction with GFP-marked *T. harzianum*

Scanning confocal laser microscopy (SCLM) potentially provides several advantages over conventional light or standard epifluorescence microscopy for visualization of microbes. A confocal microscope combines fluorescence microscopy with electronic image analysis to obtain three-dimensional images. Confocal laser microscopy has proven to be a powerful tool for examining the structure, organization, and physiology of microbial cells on surfaces, among other uses (Huang et al., 1995; McFeters et al., 1995). The shallow depth of field (as little as 0.5–1.5 µm) of confocal microscopes allows information to be collected from well-defined optical sections, rather than from most of the specimen as in conventional microscopy. Thus, out-of-focus fluorescence is eliminated, resulting in increased contrast and clarity. Effectively, the sample can be optically sectioned, and stacks of optical sections taken at successive focal planes (i.e., a "Z-series") can be reconstructed to produce a focused view of the sample. For example, a standard epifluorescence micrograph of our GFP-transformed isolate ThzID1-M3 growing in soil is shown in Fig. 5-2A. Some hyphae and conidia are visible. Next, using a Leica TCS-NT confocal scanning laser microscope, a Z-series from this same sample was captured by taking a series of vertical sections at approximately 50 µm intervals, resulting in a much more detailed view of the fungal hyphae in the sample (Fig. 5-2B). The ability to quantify microbial spatial distributions in the third ("Z") dimension suggests the possibility of three-dimensional spatial statistical analysis as well, although we are unaware of any reports of this to date.

7. STATISTICAL METHODOLOGY FOR INVESTIGATING RHIZOPLANE MICROBIAL DYNAMICS

Natural populations and habitats are spatially heterogeneous. Organisms and their resources are usually not uniformly distributed over space or time, but are found in different degrees of aggregation. Typically, natural environments are spatially structured by various energy inputs that result in patchy structures or gradients (Legendre and Fortin, 1989). The rhizoplane is a good

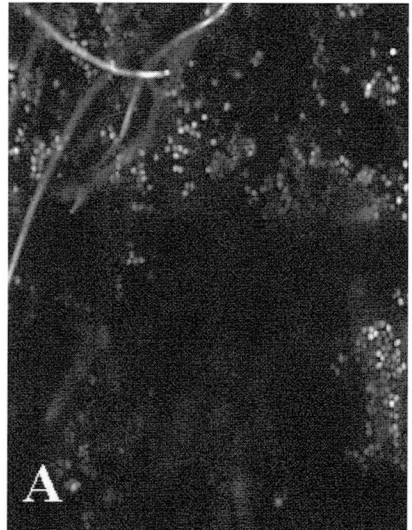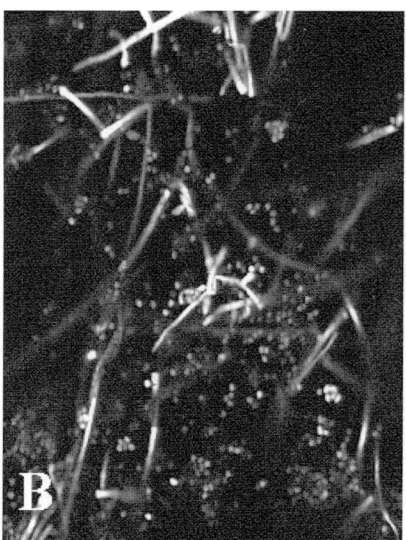

Figure 5-2. SCLM image captures of GFP fluorescence of *Trichoderma harzianum* ThzID1-M3 growing in soil. Image captured at a single focal plane (**A**), compared to a composite Z-series (10 images, taken at 50 µm intervals) (**B**). The Z-series results in a more detailed three-dimensional view of the fungal hyphae and spores in the soil sample.

example of this, since energy input is largely due to root deposition. Spatial variability of nutrient deposition from seeds as well as the spatial presence of roots may influence sites of colonization by rhizoplane microbes, and the quantity and composition of root deposition varies along the root surface. Sites may be preferentially colonized by some microbes, such as bacteria in cell junctions (Rovira, 1956), or zoospore encystment in the zone of root cell elongation (Hickman and Ho, 1966; Mitchell and Deacon, 1986). Although the tendency for both phyllosphere and rhizosphere microbial populations to conform to lognormal or similar frequency distributions has been noted (Hirano et al., 1982; Loper et al., 1984; Newman and Bowen, 1974), the actual spatial variability of these populations is less well documented. Appropriate characterization of the spatial variability inherent in soil- and plant-associated microbial communities will allow us to improve quantification of the organisms and processes under study.

Spatial statistical analysis provides a mechanism to explore processes that generate different patterns of organisms over time, and determines the sensitivity of patterns to variations in these processes. Spatial analysis is defined here as quantitative evaluation of variations or changes in spatial orientation of an entity or population within a defined area or volume. Such an analysis requires that the spatial integrity (spatial coordinate framework) of the observations be maintained.

Several theoretical studies have demonstrated the importance of spatial heterogeneity for population and metapopulation persistence and abundance (Hanski, 1991; Harrison and Quinn, 1989). Hirano and Upper (1993) demonstrated changes in frequency distributions of epiphytic bacterial populations on populations of habitats (individual leaflets) within a plant canopy. However, despite continuing demonstration of the theoretical and practical importance of spatial processes to ecology, relatively few mechanistic studies of the factors that create heterogeneous spatial distributions and the processes by which they occur in nature are available. As pointed out by Hirano and Upper (1993), an understanding of factors that regulate plant surface microbial population dynamics must address the mechanisms that underlie variability.

7.1. Sampling considerations

The choice of sampling strategy, scale of sample, and sample unit can all have a significant effect on the results of an experiment. The type of sampling strategy used in sampling the rhizosphere and the phyllosphere is often dependent on the question that is being asked. It is generally recommended that samples be taken at random so that the assumption of independence of samples for standard statistical tests (e.g., analysis of variance) is met. However, sampling of microbes from the rhizoplane, in many cases, cannot be done randomly. For example, colonization is often determined by taking root segments at a prespecified distance (e.g., 1 cm segments 2 cm below the soil line). Another example, which can result in autocorrelated data, is when roots are sequentially sampled in order to assess the root colonizing ability of a microbe. In these situations, one must verify that the data meet the assumption of the statistical analysis prior to conducting the tests. When data are autocorrelated, the knowledge of a variable's value at some location gives the observer some prior knowledge of the value the variable will take at another location. Thus, each observation does not bring one full degree of freedom. Autocorrelation sometimes causes results of standard statistical tests to be declared significant when they are not (Legendre, 1993; Robertson, 1987). When spatial, as opposed to temporal, autocorrelation is present, spatial dependency among observations can be removed by using trend surface analysis or spatial variate differencing so that the usual statistical tests can be done (Legendre, 1993). For further information on correcting standard statistical methods when autocorrelation is present the reader is referred to Legendre (1993).

Franklin and Mills (2003) discussed the need for multiscale analysis studies to more fully characterize the spatial variability of microbial systems. The selection of sample scale and sample unit is based on biological features of plants, reduction of sample variance, ease of processing and other methodological constraints. When epiphytic bacterial populations were sampled, a high level of variability was reported among sample units at every scale investigated (leaf segments, leaflets, or whole leaves) (Kinkle et al., 1995). In contrast, variability of rhizobacteria was significantly less when whole roots were sampled compared to root segments (Kloepper et al., 1991). Thus, for sampling of plant-surface microflora, Kinkel et al. (1995) suggested that selection of sample scale and unit should be based not only on the assumption that the variance is scale dependent, but may be more appropriately based on methodological and biological considerations. Although sample variance may be reduced when sampling whole root systems rather than root segments, this sample scale may not always be feasible (e.g., when sampling mature plants).

7.2. Advantages and shortcomings of frequency distribution indices

The tendency for both phyllosphere and rhizosphere microbial populations to conform to lognormal distribution (i.e., the logarithm of bacteria per leaf was normally distributed) has been noted (Bahme and Schroth, 1987; Hirano et al., 1982; Hirano and Upper, 1983, 1986; Loper et al., 1984; Newman and Brown, 1974), and appropriate transformations routinely are used. However, Kinkel et al. (1995) and Ishimaru et al. (1991) have reported that the distribution of epiphytic bacteria was not consistently described by the lognormal distribution. The appropriateness of the lognormal distribution should be verified before its application to a data set. Frequency distribution methods (e.g., indices of dispersion) are commonly based on mean/variance ratios, which do not provide reliable interpretations of spatial structure, since information on the location of each sample site is ignored. Although such indices are useful for estimation of the population mean, they do not maintain spatial integrity of samples, making spatial analysis impossible (Jumars et al., 1977; Nicot et al., 1984; Sawyer, 1989; Schotzko and Knudsen, 1992). A second drawback to the use of frequency distribution analyses is that they assume independence of observations (Legendre and Fortin, 1989; Nicot et al., 1984). However, the existence of spatial structure implies that the assumption of independence is not met, because any ecological phenomenon located at a given sampling point may have an influence on other points close by, or some distance away (Legendre, 1993; Legendre and Fortin, 1989).

7.3. Use of geostatistics for analysis of rhizoplane microbial populations

A true spatial statistical analysis, such as geostatistics, provides a mechanism to explore processes that generate different patterns of organisms over time, and to determine the sensitivity of pattern to variations in these processes. Spatial analysis is defined here as any analysis that quantitatively evaluates variations or changes based on spatial orientation within a defined area or volume. Unlike frequency analysis, spatial analysis requires that the spatial integrity of observations be maintained; i.e., spatial coordinates are recorded for each sample point. One method for spatial analysis, geostatistics, provides a quantitative assessment of spatial distributions that maintains the spatial integrity of data, and is able to analyze the degree of association (auto-correlation) based on direction and distance between samples (Isaaks and Srivastava, 1989; Trangmar et al., 1975; see Chapter 2 for additional details on geostatistical analysis).

Geostatistical analysis determines the degree of autocorrelation among samples based on the direction and distance between them (Isaaks and Srivastava, 1989; Trangmar et al., 1975) and can be used to interpolate values between measured points based on the degree of autocorrelation encountered (Robertson, 1987). Geostatistics detects spatial dependence among neighboring samples and defines the degree of dependence by giving quantifiable parameters. For readers interested in an introductory text on geostatistical theory and practice, Isaaks and Srivastava's (1989) text is highly recommended. Other recommended reading includes Rossi et al. (1992), Legendre (1993), and Robertson (1987). Geostatistics has proven highly applicable to biological systems; for example, geostatistics has been used effectively to evaluate insect spatial distributions (Kemp et al., 1989; Schotzko and Smith, 1991) and a plant/insect spatial simulation model (Knudsen and Schotzko, 1991; Schotzko and Knudsen, 1992), as well as plant disease patterns (Chellemi et al., 1988; Johnson et al., 1991; Nicot et al., 1984) and patterns of zoospore encystment on roots (Dandurand et al., 1995, 1997).

Standardized covariograms (or alternatively, variograms) allow comparison in changes in spatial structure which are independent of the population variance. Covariograms are plots of the covariance of sample pairs against the distance (lag) between sample points (Isaaks and Srivastava, 1989). Statistical models (linear, spherical, etc.) can be fitted to covariograms and allow the investigator to relate observed structures to hypothesis-generating processes (Legendre, 1993). With aggregated patterns, the covariance is smaller at shorter distances, increasing with distance between points, and leveling off at the population covariance. This typifies a spherical model. With a gradient, the covariance is smaller at shorter distances, increasing with distance between points, but does not level off. Gradients typify a linear model and could be expected if, for example, densities of a bacteria are higher at one end of a root than another. With random patterns, there are no consistent changes in covariance with distance, thus the covariogram consists of points varying about the population covariance. Three key aspects of a covariogram are: (1) the sill, (2) localized discontinuity ("nugget" or y intercept), and (3) the range. The sill is the point at which the covariance no longer increases or in other words, becomes random. Localized discontinuity is a measure of variation below the sampling scale, random variation, measurement error, or all three. The range (range of spatial dependence) is the distance to the sill. Modeling of a covariogram is concerned with the response between the localized discontinuity and the sill and reveals the spatial structure. Spatial dependence is the degree of association of any two points at a given separation distance revealed by the spatial structure. Geostatistics detects spatial dependence among neighboring samples and defines the degree of dependence by giving quantifiable parameters. Some advantages of geostatistical analysis as a methodology may be summarized as follows: Geostatistics is independent of the relationship between the mean and variance; geostatistics maintains the spatial integrity of locations of samples and uses the variation between points to evaluate spatial dependence; geostatistics assesses spatial dependence quantitatively and can be used to compare spatial dependence at different points in time, or at the same point in time under different conditions.

7.4. Geostatistical analysis of rhizoplane microbial populations: some examples

7.4.1. Encystment of zoospores of *Pythium ultimum*

Spatial variability of exudates from seeds and roots may influence sites of colonization of plant pathogens. Chemotaxis of fungal zoospores and growth of mycelia towards roots is highly regulated by root exudates (Cunningham and Hagedorn, 1962; Dandurand and Menge, 1994; Hickman and Ho, 1966; Mitchell and Deacon, 1986; Paulitz, 1991; Royle and Hickman, 1964). Zoospores are an important infectious propagule for *Pythium* spp. The greatest accumulation of zoospores has been reported to be at approximately 2.5 mm behind the root tip (zone of cell elongation), where a major portion

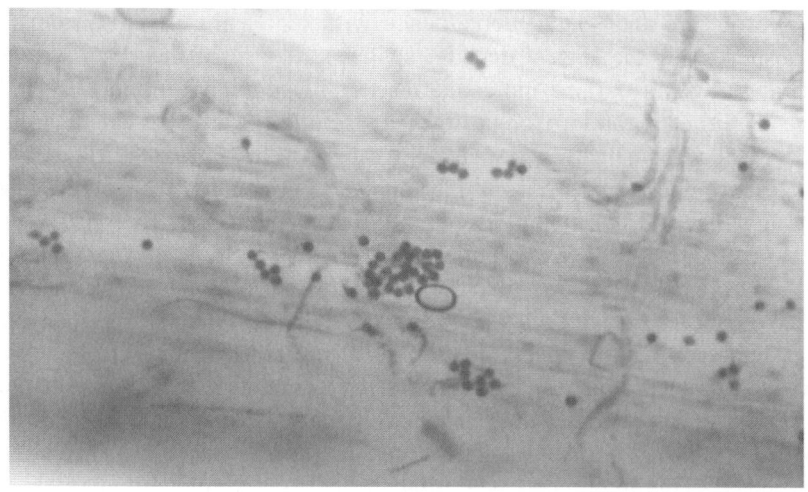

Figure 5-3. Light micrograph of *Pythium ultimum* var. sporangiiferum zoospores encysted on the pea rhizoplane (zoospores stained with Trypan blue, photographed at 250×).

of diffusable compounds are exuded (Mitchell and Deacon, 1986; Rovira, 1973; Royle and Hickman, 1964). Disease may depend, at least in part, on access to spatially important entry sites for pathogens. If these are blocked by biocontrol agents or other microbes, then infection and disease may be reduced. For example, Dandurand and Menge (1992, 1994) found that prior colonization of citrus roots by *Fusarium solani* reduced numbers of encysted zoospores of *Phytophthora parasitica* and *P. citrophthora*, and reduced *P. parasitica* populations.

Dandurand et al. (1995) evaluated effects of zoospore density and root age on zoospore encystment patterns on pea roots. Roots were exposed *in situ* to different densities of zoospore suspensions of *P. ultimum* var. sporangiiferum, then removed and gently washed. Entire roots were stained with Trypan blue and zoospore counts were made microscopically using an optical grid (unit size = 83 × 83 µm) over the entire visible surface of the root (Fig. 5-3).

Spatial coordinates and numbers of zoospores were recorded for each sample unit, and spatial statistics were calculated using a geostatistics computer program. Distinctive spatial organization of encysted zoospores developed with changes in inoculum density. At low densities, cysts were either randomly or uniformly distributed over the root surface, whereas at high inoculum densities, encysted zoospores were highly aggregated. When 2-, 3-, or 5-day-old seedlings were exposed to zoospores, spatial analysis indicated no effect of root age on spatial patterns. Results did show an increase in spatial structure and spatial dependence with increasing inoculum density of zoospores. At the lowest density, spatial structure of the cysts was not evident; flat covariograms indicated absence of aggregation and were indicative of either a random or uniform distribution. However, at the intermediate and high densities, the covariograms reveal aggregated patterns of encystment. These observations that encysted zoospores were aggregated on roots suggested two hypotheses: First, that the source of chemoattractant (e.g., root exudates) is itself spatially patterned, resulting in aggregates of encysted zoospores. Second, that root exudates act only as general signals, so that zoospores initially encyst randomly. Then, as the density of encysted zoospores reaches some threshold, signals are released from them, resulting in autoaggregation. Our results from these experiments would appear to support the latter hypothesis, since, if autoaggregation were not a factor in the encystment process, similar patterns at low and high zoospore densities would be expected. This hypothesis is similar to the dual-component chemotactic process proposed by Thomas and Peterson (1990), and may illustrate one way in which spatial statistical analysis may complement other more physiologically based studies.

7.4.2. Geostatistical analysis of spatial patterns of the biocontrol agents *Pseudomonas fluorescens* and *Trichoderma harzianum*, independently and in combination

The variable success of biocontrol agents in controlling diseases may, in part, be caused by a lack of understanding of the spatial partitioning of resources in the rhizoplane. Spatial analysis of biocontrol agents may increase our predictive ability for effective biocontrol agents. For example, colonization patterns of bacteria were reported by Fukui et al. (1994); although it was observed that two strains colonized various parts of sugar beet seeds, a quantitative analysis of the spatial patterns of the two strains was not made, and conclusions derived from spatial patterns of the two bacteria based on observation only are difficult to interpret. Nonpathogenic rhizoplane colonizers (biocontrol agents) at or near infection courts may be well positioned to modify the zoospore encystment process, and subsequent root infection. For example, the frequency distribution of cucumber root sections without encysted zoospores of *P. aphanidermatum* was higher for roots treated with biocontrol bacteria than for untreated roots (Zhou and Paulitz, 1993). A quantitative analysis of spatial patterns is particularly important to determine whether biocontrol agents change the spatial patterns of pathogens in predictable and consistent ways.

P. ultimum var. sporangiiferum is an important pea damping-off and root rot pathogen. There are numerous examples of seed and/or root protection from *Pythium* spp. using biocontrol bacteria or fungi (Hadar et al., 1984; Harman et al., 1980; Lifshitz et al., 1986; Loper, 1988; Nelson et al., 1988; Parke, 1990; Tedla and Stanghellini, 1992). Dandurand and Knudsen (1994) showed a reduction in disease caused by another zoosporic plant pathogen, *Aphanomyces* root rot, when peas were treated with *P. fluorescens* and/or *T. harzianum* isolate ThzID1. In one study (L. M. Dandurand, 1994), spatial patterns of *P. ultimum* var. sporangiiferum were compared in the presence or absence of *T. harzianum* ThzID1. Numbers of encysted zoospores per unit root area and encystment patterns were quantified as described above. Mean numbers of encysted zoospores did not differ between roots that were colonized by *T. harzianum* and untreated roots. However, the spatial pattern of zoospore encystment was significantly affected by treatment with *T. harzianum* (Fig. 5-4).

Without *T. harzianum*, a relatively lower proportion of the total variation was spatially structured (Fig. 5-4), indicating greater aggregation without *T. harzianum*. Additionally, the presence of *T. harzianum* increased the range over which there was autocorrelation. In other words, in the presence of *T. harzianum*, the zoospores were less aggregated, and the aggregates were more diffuse. These results suggest that the potential role of *T. harzianum* as a biocontrol agent may not be simply in reducing pathogen access to the root, but instead may be as a determinant of where zoospores are able to encyst. Probably *T. harzianum* did not affect the magnitude of the homing response, because average zoospore density on the root surface was approximately the same in the presence or absence of *T. harzianum*. However, *T. harzianum* had a strong effect in mediating the aggregation process, so that zoospores were less aggregated and aggregates were more diffuse. The ecological significance of aggregation is not known, e.g., whether aggregation plays a role in germination of zoospores or the infection process (perhaps through a localized inoculum density threshold effect).

Although geostatistical analysis cannot say specifically what the mechanism of biocontrol activity of *T. harzianum* is, it can give credence to hypotheses about mechanisms. As an example, Mandeel and Baker (1991) observed that potential infection courts on the rhizoplane can be protected by an agent that actively competes for these sites, and they suggested that this mechanism has more potential impact on biocontrol efficiency than does rhizosphere nutrient competition. Mandeel and Baker (1991) further observed that biocontrol efficiency values are influenced by spatial relationships, especially

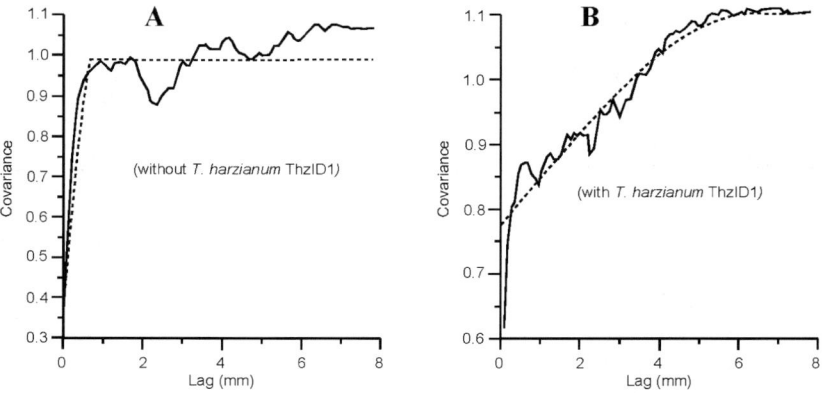

Figure 5-4. Covariograms of *Pythium ultimum* var. sporangiiferum zoospore encystment patterns in the absence (**A**) or presence (**B**) of *T. harzianum* ThzID1 (gnotobiotic system).

relative proximities of the pathogen to penetration sites compared to proximity of the biocontrol agent. If space is indeed a mechanism in biocontrol efficacy, than proliferation of the biocontrol agent would be expected to generate a change in observed spatial patterns of the pathogen population, as our preliminary results have indicated. Spatial statistical analysis can provide quantitative answers about whether spatial attributes of a biocontrol agent will change or not change in the presence of indigenous microbes. Effective development of spatially rigorous analysis techniques will provide a necessary framework for evaluating the effects of mechanisms that are studied at the genetic and biochemical level.

In another set of experiments, we tested whether the biocontrol agent, *P. fluorescens* 2-79RN10 (which produces a phenazine antibiotic), influenced spatial patterns of *T. harzianum* ThzID1 on the rhizoplane of pea. In other words, is there evidence that antibios is involved in spatial partitioning of the rhizoplane? Seeds were inoculated with *T. harzianum* either alone or in combination with *P. fluorescens* 2-79RN10, or *P. fluorescens* B46, a non-phenazine-producing strain of 2-79RN10. Roots of 3-day-old seedlings grown under gnotobiotic conditions were observed using scanning electron microscopy (Fig. 5-5). Spatial structure (aggregation) of *T. harzianum* was evident along the entire root, but was not affected by the presence of either *P. fluorescens* 2-79RN10 or *P. fluorescens* B46. However, whether this phenomenon would be the same in a natural soil could not be addressed at that time, due to the impossibility of distinguishing the added *T. harzianum* from other indigenous fungi. The availability of a GFP transformant of the fungus, as described above, may help alleviate this technical difficulty in future experiments.

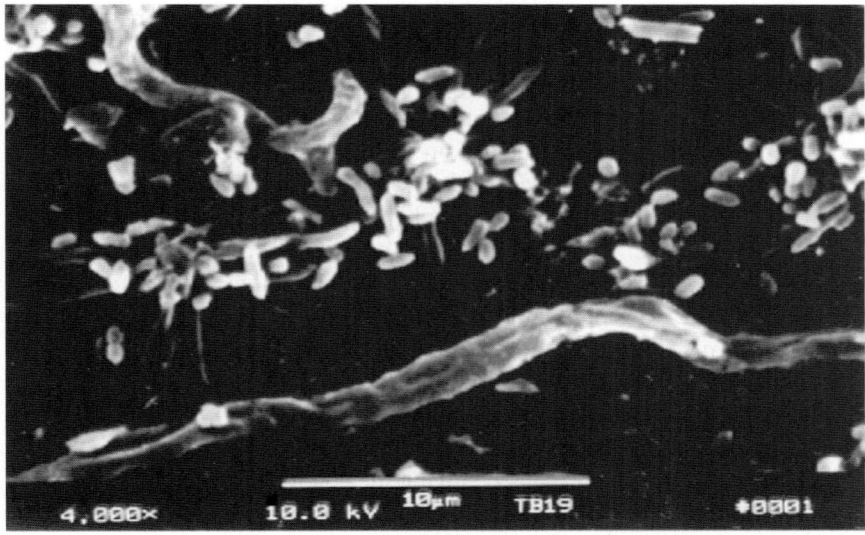

Figure 5-5. Scanning electron micrograph of *Trichoderma harzianum* ThzID1 hyphae and cells of *Pseudomonas fluorescens* 2-79RN10, on the pea rhizoplane.

8. POTENTIAL INFLUENCE OF BACTERIAL POPULATION SPATIAL STRUCTURE ON GENE TRANSFER ON THE RHIZOPLANE

The ability of bacteria to transfer plasmid-borne genes between species and even genera is well known. for example, conjugative or "self-transmissible" plasmids belonging to the IncP and IncW groups have been shown to be transferred among a wide range of bacteria including soil inhabitants. Some nonconjugative plasmids may be mobilized and transferred in the presence of conjugative plasmids. The frequency of plasmid transfer events in natural habitats remains poorly understood. Mathematical models have been used to describe and predict kinetics of conjugative plasmid transfer, and have taken the form of analytical (Levin et al., 1979) and simulation (Knudsen et al., 1988) models, however these models were developed for well-mixed (i.e., homogeneous) systems. A previous comparison of plasmid pMON5003 mobilization rates in broth culture with rates on root surfaces suggested that numbers of transconjugants formed on root surfaces were 10- to 100-fold lower than in the liquid medium (Sudarshana, 1995). Also, another previous study (Dandurand et al., 1997) showed highly aggregated patterns of rhizoplane bacterial populations, and geostatistically quantified those patterns.

We hypothesized that observed lower plasmid transfer rates on the rhizoplane are due, at least in part, to the more limited chances for cell-to-cell contact imposed by the spatial structure of rhizoplane bacterial populations. To address this hypothesis, in a preliminary study we modeled effects of bacterial population spatial structure on gene transfer frequencies. A grid of 1,617 spatial cells was simulated, with each cell 2.5 µm in size. A population of recipient cells was initialized in a highly aggregated spatial pattern within the grid. Then, equal numbers of donor cells (containing the conjugative plasmid) and cells containing the mobilizable plasmid were initialized either in randomly selected grid locations, or in highly aggregated patterns. Aggregates were formed using a cellular automaton routine. Geostatistical analysis of the simulated patterns showed similar patterns (nugget, sill, range) to those experimentally observed by Dandurand et al. (1997) for *P. fluorescens* cells on pea roots. Simulation results indicated that significantly higher numbers of transconjugants (cells receiving the conjugative and/or mobilizable plasmids) were formed when plasmid donor cells were initialized in aggregates, compared to randomly distributed cells. Thus, although it has been shown that parental growth kinetics can strongly affect plasmid transfer frequencies, our experimental and simulation results also suggest a strong influence of spatial structure of surface-colonizing populations on gene transfer. Further investigation of the quantitative relationship between pattern (e.g., spatial structure which can be quantified using spatial statistics) and process (e.g., plasmid transfer kinetics) is promising.

9. CONCLUDING REMARKS

Historically, analysis of microbial processes usually has been conducted separately from analysis of microbial population and community patterns. In many cases, the connection between process and pattern has been ignored, in part due to the unavailability of appropriate tools. Geostatistical analysis provides one such tool, with potential to serve as the critical link in evaluating the process-pattern cycle.

REFERENCES

Bae, Y. S., and G. R. Knudsen, 2000, Cotransformation of *Trichoderma harzianum* with β-glucuronidase and green fluorescent protein genes provides a useful tool for monitoring fungal growth and activity in natural soils. *Appl. Environ. Microbiol.* **66**:810–815.

Bae, Y. S., and G. R. Knudsen, 2001, Influence of a fungus-feeding nematode on growth and biocontrol efficacy of *Trichoderma harzianum*. *Phytopathology* **91**:301–306.

Bahme, J. B., and M. N. Schroth, 1987, Spatial-temporal colonization patterns of a rhizobacterium on underground organs of potato. *Phytopathology* **77**:1093–1100.

Baker, R., 1981, Ecology of the fungus *Fusarium*, competition, in: *Fusarium: Diseases, Biology, and Taxonomy*, P. E. Nelson, T. A. Nelson, T. A. Tousson, and R. J. Cook, eds., Penn State University Press, University Park, PA, pp. 245–258.

Campbell, R., and R. Porter, 1982, Low-temperature scanning electron microscopy of microorganisms in soil. *Soil Biol. Biochem.* **14**:241–245.

Chalfie, M., Tu, Y., Euskirchen, G., Ward, W. W., and Prasher, D. C., 1994, Green fluorescent protein as a marker of gene expression. *Science* **263**:802–805.

Chellemi, D. O., K. G. Rohrbach, R. S. Yost, and R. M. Sonoda, 1988, Analysis of the spatial pattern of plant pathogens and diseased plants using geostatistics. *Phytopathology* **78**:221–226.

Cochran, W. G., 1950, Estimation of bacterial densities by means of the most probable number. *Biometrics* **March**:105–116.

Colwell, R. R., P. R. Brayton, D. J. Grimes, D. B. Roszak, S. A. Huq, and L. M. Palmer, 1985, Viable but non-culturable *Vibrio cholerae* and related pathogens in the environment: implications for release of genetically engineered microorganisms. *Biol. Tech.* **3**:817–820.

Cook, R. J., 1993, Making greater use of introduced microorganisms for biological control of plant pathogens. *Ann. Rev. Phytopathol.* **31**:53–80.

Cormack, B. P., G. Bertram, M. Egerton, N. A. R. Gow, S. Falkow, and A. J. P. Brown, 1997, Yeast-enhanced green fluorescent protein (yEGFP): a reporter of gene expression in *Candida albicans*. *J. Gen. Microbiol.* **143**:303–311.

Cunningham, J. L., and Hagedorn, D. J., 1962, Attraction of *Aphanomyces euteiches* zoospores to pea and other plant roots. *Phytopathology* **52**:616–618.

Curl, E. A., and B. Truelove, 1986, *The Rhizosphere*. Springer, Berlin, p. 288.

Dandurand, L. M., and G. R. Knudsen, 1994, Spatial description of encysted zoospores on roots: Influence of sampling intensity. *Phytopathology* **84**:1154.

Dandurand, L. M., and Menge, J. A., 1992, The influence of *Fusarium solani* on citrus root rot caused by *Phytophthora parasitica* and *Phytophthora citrophthora*. *Plant Soil.* **144**:13–21.

Dandurand, L. M., and Menge, J. A., 1994, Influence of *Fusarium solani* on chemotaxis of zoospores of *Phytophthora parasitica* and *Phytophthora citrophthora* and on distribution of ^{14}C in citrus tissues and root exudate. *Soil Biol. Biochem.* **26**:75–79.

Dandurand, L. M., Knudsen, G. R., and Schotzko, D. J., 1995, Quantification of *Pythium ultimum* var. sporangiiferum zoospore encystment patterns using geostatistics. *Phytopathology* **85**:186–190.

Dandurand, L. M., D. J. Schotzko, and G. R. Knudsen, 1997, Spatial patterns of rhizoplane populations of *Pseudomonas fluorescens*. *Appl. Environ. Microbiol.* **63**:3211–3217.

Deacon, J. W., and S. P. Donaldson, 1993, Molecular recognition in the homing responses of zoosporic fungi, with special reference to *Pythium* and *Phytophthora*. *Mycol. Res.* **97**:1153–1171.

Fendorf, S. E., Li, G., Morra, M. J., and Dandurand, L. M., 1997, High-resolution non-destructive imaging of bacterial cells in pure cultures and mineral suspensions. *Soil Sci. Soc. J.* **61**:109–115.

Franklin, R. B., and A. L. Mills, 2003, Multi-scale variation in spatial heterogeneity for microbial community structure in an eastern Virginia agricultural field. *FEMS Microbiol. Ecol.* **44**:335–346.

Fukui, R., E. I. Poinar, P. H. Bauer, M. N. Schroth, M. Hendson, X. -L. Wang, and J. G. Hancock, 1994, Spatial colonization patterns and interaction of bacteria on inoculated sugar beet seed. *Phytopathology* **84**:1338–1345.

Gray, T. R. G., and D. Parkinson, 1968, *The Ecology of Soil Bacteria*. University of Toronto Press, Toronto, p. 681.

Green, H., and D. F. Jensen, 1995, A tool for monitoring *Trichoderma harzianum*: II. The use of a GUS transformant for ecological studies in the rhizosphere. *Phytopathology* **85**:1436–1440.

Green, H., N. Heiberg, K. Lejbølle, and D. Funck Jensen, 2001, The use of a GUS transformant of *Trichoderma Harzianum*, strain T3a, to study metabolic activity in the spermosphere and rhizosphere related to biocontrol of *Pythium* damping-off and root rot. *Eur. J. Plant Pathol.* **107**:349–359.

Hadar, Y., G. E. Harman, and A. G. Taylor, 1984, Evaluation of *Trichoderma koningii* and *Trichoderma harzianum* from New York soils for biological control of seed rot caused by *Pythium* spp. *Phytopathology* **74**:106–110.

Hanski, I., 1991, Single-species metapopulation dynamics: concepts, models and observations. *Biol. J. Linn. Soc.* **42**:73–88.

Harman, G. E., Chet, I., and Baker, R., 1980, *Trichoderma harzianum* effects on seed and seedling disease induced in radish and pea by *Pythium* spp. or *Rhizoctonia solani*. *Phytopathology* **70**:1167–11

Harrison, S., and J. F. Quinn, 1989, Correlated environments and the persistence of metapopulations. *Oikos* **56**:293–298.

Hickman, C. J., and H. H. Ho, 1966, Behavior of zoospores in plant-pathogenic phycomycetes. *Annu. Rev. Phytopathol.* **4**:195–220.

Hirano, S. S., and C. D. Upper, 1983, Ecology and epidemiology of foliar bacterial plant pathogens. *Annu. Rev. Phytopathol.* **21**:243–269.

Hirano, S. S., and C. D. Upper, 1986, Temporal, spatial and genetic variability of leaf-associated bacterial populations, in: *Microbiology of the Phyllosphere*, N. J. Fokkema and J. Van Den Heuvel, eds., Cambridge University Press, New York, pp. 235–241.

Hirano, S. S., and C. D. Upper, 1993, Dynamics, spread, and persistence of a single genotype of *Pseudomonas syringae* relative to those of its conspecifics on populations of snap bean leaflets. *Appl. Environ. Microbiol.* **59**:1082–1091.

Hirano, S. S., E. V. Nordheim, D. C. Arny, and C. D. Upper, 1982, Lognormal distribution of epiphytic bacterial populations on leaf surfaces. *Appl. Environ. Microbiol.* **44**:695–700.

Huang, C. -T., F. P. Yu, G. A. McFeters, and P. S. Stewart, 1995, Non-uniform spatial patterns of respiratory activity within biofilm during disinfection. *Appl. Environ. Microbiol.* **61**:2252–2256.

Isaaks, E. H., and R. H. Srivastava, 1989, *Applied Geostatistics*, Oxford University Press, Oxford.

Ishimaru, C., K. M. Eskridge, and A. K. Vidaver, 1991, Distribution analysis of naturally occurring epiphytic populations of *Xanthomonas campestris* p.v. phaseoli on dry beans. *Phytopathology* **81**:262–268.

Johnson, D. A., J. R. Alldredge, J. R. Allen, and R. Allwine, 1991, Spatial pattern of downy mildew in hop yards during severe and mild disease epidemics. *Phytopathology* **81**:1369–1374.

Jumars, P. A., D. Thistle, and M. L. Jones, 1977, Detecting two-dimensional spatial structure in biological data. *Oecologia* **8**:109–123.

Katznelson, H., 1965, Nature and importance of the rhizosphere, in: *Ecology of Soil Borne Plant Pathogens – Prelude to Biological Control*, K. J. Baker and W. C. Snyder, eds., University of California Press, Berkeley, CA, pp. 187–209.

Kemp, W. P., T. M. Kalaris, and W. F. Quimby, 1989, Rangeland grasshopper (Orthoptera: Acrididae) spatial variability: Macroscale population assessment. *J. Econ. Entomol.* **82**: 1270–1276.

Kinkle, L. L., M. Wilson, S. E. Lindow, 1995, Effect of sampling scale on the assessment of epiphytic bacterial populations. *Microb. Ecol.* **29**:283–297.

Kloepper, J. W., and C. J. Beauchamp, 1992, A review of issues related to measuring colonization of plant roots by bacteria. *Can. J. Microbiol.* **29**:1219–1232.

Knudsen, G. R., and L. Bin, 1990, Effects of temperature, soil moisture and wheat bran on growth of *Trichoderma harzianum* from alginate pellets. *Phytopathology* **80**:724–727.

Knudsen, G. R., and D. J. Schotzko, 1991, Simulation of Russian wheat aphid movement and population dynamics on preferred and non-preferred host plants. *Ecol. Model.* **57**:117–131.

Knudsen, G. R., M. V. Walter, L. A. Prorteous, V. J. Prince, J. L. Armstrong, and R. J. Seidler, 1988, Predictive model of conjugative plasmid transfer in the rhizsphere and phyllosphere. *Appl. Environ. Microbiol.* **54**:343–347.

Knudsen, G. R., D. J. Eschen, L. M. Dandurand, and L. Bin, 1991a, Potential for biocontrol of *Sclerotinia sclerotiorum* through colonization of sclerotia by *Trichoderma harzianum*. *Plant Dis.* **75**:466–469.

Knudsen, G. R., D. J. Eschen, L. M. Dandurand, and Z. G. Wang, 1991b, Method to enhance growth and sporulation of pelletized biocontrol fungi. *Appl. Environ. Microbiol.* **57**:2864–2867.

Legendre, P., 1993, Spatial autocorrelation: trouble or new paradigm? *Ecology* **74**:1659–1673.

Legendre, P., and M. -J. Fortin, 1989, Spatial pattern and ecological analysis. *Vegetatio* **80**:107–138.

Levin, B. R., F. M. Stewart, and V. A., Rice, 1979, The kinetics of conjugative plasmid transmission: fit of a simple mass action model. *Plasmid* **2**:247–260.

Lifshitz, R., M. T. Windham, and R. Baker, 1986, Mechanism of biological control of pre-emergence damping-off of pea by seed treatment with *Trichoderma* spp. *Phytopathology* **76**:720–725.

Lo, C. -T., E. B. Nelson, C. K. Hayes, and G. E. Harman, 1998, Ecological studies of transformed *Trichoderma harzianum* strain 1295-22 in the rhizosphere and on the phylloplane of creeping bentgrass. *Phytopathology* **88**:129–136.

Lockwood, L. J., 1981, Exploitation competition, in: *The Fungal Community*, D. T. Wicklow and G. C. Carroll, eds., Marcell Dekker, New York, pp. 319–349.

Loper, J. E., 1988, Role of fluorescent siderophore production in biological control of *Pythium ultimum* by a *Pseudomonas fluorescens* strain. *Phytopathology* **78**:166–172.

Loper, J. E., Suslow, T. V., and Schroth, M. N., 1984, Lognormal distribution of bacterial populations in the rhizosphere. *Phytopathology* **74**:1454–1460.

Lumsden, R. D., J. P. Carter, J. M. Whipps, and J. M. Lynch. 1990. Comparison of biomass and viable propagule measurements in the antagonism of *Trichoderma harzianum* against *Pythium ultimum*. *Soil Biol. Biochem.* **22**:187–194.

McFeters, G. A., F. P. Yu, B. H. Pyle, and P. S. Stewart. 1995. Physiological methods to study disinfection. *J. Ind. Microbiol.* **15**:333–338.

Mandeel, Q., and R. Baker. 1991. Mechanisms involved in biological control of *Fusarium* wilt on cucumber with strains of non-pathogenic *Fusarium oxysporum*. *Phytopathology* **81**:462–469.

Martin, F. N., and J. T. English. 1997. Population genetics of soilborne fungal plant pathogens. *Phytopathology* **87**:446–447.

Mitchell, R. T., and Deacon, J. W. 1986. Differential (host-specific) accumulation of zoospores of *Pythium* on roots of graminaceous and non-graminaceous plants. *New Phytol.* **102**:113–122.

Nelson, E. B., G. E. Harman, and G. T. Nash. 1988. Enhancement of *Trichoderma*-induced biological control of *Pythium* seed rot and pre-emergence damping-off of peas. *Soil Biol. Biochem.* **20**:145–150.

Newman, E. I., and H. J. Bowen. 1974. Patterns of distribution of bacterial on root surfaces. *Soil Biol. Biochem.* **6**:205–209.

Nicot, P. C., D. I. Rouse, and B. S. Yandell. 1984. Comparison of statistical methods for studying spatial patterns of soilborne plant pathogens in the field. *Phytopathology* **74**:1399–1402.

Orr, K. A., and G. R. Knudsen. 2004. Use of green fluorescent protein and image analysis to quantify proliferation of *Trichoderma harzianum* in nonsterile soil. *Phytopathology* **94**:1383–1389.

Parberry, I. H., J. F. Brown, and V. J. Bofinger, 1981, Statistical methods in the analysis of phylloplane populations, in: *Microbial Ecology of the Phylloplane*, J. P. Blakeman, ed., Academic Press, London, pp. 47–65.

Parke, J. L. 1990. Population dynamics of *Pseudomonas cepacia* in the pea spermosphere in relation to biocontrol of *Pythium*. *Phytopathology* **80**:1307–1311.

Richaume, A., D. Steinberg, L. Jocteur-Monrozier, and G. Faurie, 1993. Differences between direct and indirect enumeration of soil bacteria: the influence of soil structure and cell location. *Soil Biol. Biochem.* **25**:641–643.

Roberts, I. N., R. P. Oliver, P. J. Punt, and C. A. M. J. J. van den Hondel. 1989. Expression of *Escherichia coli*. β-glucuronidase gene in industrial and phytopathogenic filamentous fungi. *Curr. Genet.* **15**:177–180.

Robertson, G. P. 1987. Geostatistics in ecology: interpolating with known variance. *Ecology* **68**:744–748.

Rossi, R. E., D. J. Mulla, A. G. Journel, and E. H. Franz. 1992. Geostatistical tools for modeling and interpreting ecological spatial dependence. *Ecol. Monogr.* **62**:277–314.

Rovira, A. D. 1956. A study of the development of the root surface microflora during the initial stages of plant growth. *J. Appl. Bacteriol.* **19**:72–79.

Rovira, A. D. 1973. Zones of exudation along plant roots and spatial distribution of microoorganisms in the rhizosphere. *Pestic. Sci.* **4**:361–366.

Royle, D. J., and Hickman, C. J. 1964. Analysis of factors governing in vitro acccumulation of zoospores of *Pythium aphanidermatum* on roots. I. Behavior of zoospores. *Can. J. Microbiol.* **10**:151–162.

Sawyer, A. J. 1989. Inconstancy of Taylor's B: simulated sampling with different quadrant sizes and spatial distributions. *Res. Popul. Ecol.* **31**:11–24

Schroth, M. N., and J. G. Hancock, 1981. Selected topics in biological control. *Ann. Rev. Microbiol.* **35**:453–476.

Schotzko, D. J., and G. R. Knudsen. 1992. Use of geostatistics to evaluate a spatial simulation of Russian wheat aphid (Homoptera: Aphididae) movement behavior on preferred and nonpreferred hosts. *Environ. Entomol.* **21**:1271–1282.

Sheen, J., Hwang, S., Niwa, Y., Kobayaski, H., and Galbraith, D. W. 1995. Green-fluorescent protein as a new vital marker in plant cells. *Plant J.* **8**:777–784.

Spelling, T., A. Bottin, and R. Kahmann. 1996. Green fluorescent protein (GFP) as a new vital marker in the phytopathogenic fungus *Ustilago maydis*. *Mol. Gen. Genet.* **252**:503–509.

Stanghellini, M. E., and Rasmussen, S. L. 1989. Root prints: a technique for the determination of the in situ spatial distribution of bacteria on the rhizoplane of field-grown plants. *Phytopathology* **79**:1131–1134.

Sudarshana, P. 1995. Dynamics of plasmid transfer via conjugation and mobilization in the spermosphere and rhizosphere of pea. Ph.D. thesis, University of Idaho.

Tedla, T., and Stanghellini, M. E. 1992. Bacterial population dynamics and interactions with *Pythium aphanidermatum* in intact rhizosphere soil. *Phytopathology* **82**:652–656.

Thomas, D. D., and A. P. Peterson. 1990. Chemotactic auto-aggregation in the water mold *Achlya*. *J. Gen. Microbiol.* **52**:479–483.

Thrane, C., M. Lübeck, H. Green, Y. Degefu, S. Allerup, U. Thrane, and D. F. Jensen. 1995. A tool for monitoring *Trichoderma harzianum*. I. Transformation with the GUS gene by protoplast technology. *Phytopathology* **85**:1428–1435.

Trangmar, B. B., R. S. Yost, and G. Uehara, 1975, Application of geostatisitics to spatial studies of soil properties, in: *Advances in Agronomy*, Vol. 38, N. C. Brady, ed., Academic Press, New York, pp. 45–94.

Vanden Wymelenberg, A. J., D., Cullen, R. N. Spear, B. Schoenike, and J. H. Andrews, 1997. Expression of green fluorescent protein in *Aureobasidium pullulans* and quantification of the fungus on leaf surfaces. *Biotechniques* **23**:686–690.

Whipps, J. M., 1990, Carbon economy, in: *The Rhizosphere*, J. M. Lynch, ed., Wiley, Chichester, UK, pp. 59–97.

Zhou, T., and Paulitz, T. C. 1993. In vitro and in vivo effects of *Pseudomonas* spp. on *Pythium aphanidermatum*: zoospore behavior in exudates and on the rhizoplane of bacteria-treated cucumber roots. *Phytopathology* **83**:872–876.

Chapter 6

MICROBIAL DISTRIBUTIONS AND THEIR POTENTIAL CONTROLLING FACTORS IN TERRESTRIAL SUBSURFACE ENVIRONMENTS

R. Michael Lehman
US Department of Agriculture, North Central Agricultural Research Laboratory, Brookings, SD 57006, USA

Abstract: Terrestrial subsurface environments (below the plow layer) contain an enormous amount of the earth's biomass, yet are relatively undersampled compared to topsoil, aquatic, and marine environments. Depth emerges as a primary axis for relating distributions of microorganisms and the factors controlling their distribution. There is generally a sharp drop in microbial biomass, diversity, and activity as organic-rich topsoils deepen to mineral-dominated subsoils. Progressively deeper samples from the vadose zone to the capillary fringe and into saturated zones often reveal increases in biomass and changes in dominant microbial populations. Biomass appears to slowly decline with depth, and cell viability is limited by temperature between 4.5 and 6 km. In many subsurface environments, spatial distributions of microorganisms are extremely variable, frequently defying prediction. In a few highly structured saturated environments, such as confined or contaminated shallow aquifers, predominant terminal electron accepting activities are arranged in a spatially ordered manner that is consistent with selected geochemical measurements. Sampling issues specific to subsurface environments still require substantial added effort and expense to achieve a reasonable sample density in comparison to most other environments. Technological advances in microbial assay methodologies are easing some of the methodological boundaries that are often exceeded by subsurface samples.

Keywords: aquifer, distribution, microorganism, spatial, subsurface, vadose

1. INTRODUCTION

It has become apparent that terrestrial subsurface microorganisms are distributed in an extremely patchy manner and that the patches are quite variable in magnitude (Brockman and Murray, 1997). Neither the factors that control these distributions nor the distributions themselves have been rigorously quantified. To quantify relationships between microbial spatial distributions and their controlling factors, the relevant scales and most probable controlling factors need to be identified. To that end, published data on subsurface microbial properties from spatially arranged sample sets are reviewed and used to support conceptual models of microbial distributions in terrestial subsurface environments.

1.1. Historical expectations regarding subsurface microbial distributions

It has been estimated that the amount of microbial biomass in subsurface environments approximates the amount of *all* aboveground biomass (Gold, 1992; Onstott et al., 1999; Pedersen, 2000; Whitman et al., 1998), yet knowledge of subsurface microbial distributions rests on a sparse array of analyzed samples. Historical expectations were that biomass declined rapidly with depth and soil carbon content, and that the subsurface was largely devoid of life (Federle et al., 1986; Ghiorse and Wilson, 1988), despite occasional reports of microorganisms in produced waters from petroleum exploration and development (Bastin, 1926; Davis, 1967). Interest in subsurface microbiology grew in association with accelerated groundwater consumption and deterioration, and microbiologically oriented sampling of the subsurface largely commenced in the 1970s (Keswick, 1984). Based primarily on findings by the USEPA, USGS, and university researchers, it was generally acknowledged by the mid-1980s that shallow aquifer sediments were colonized by substantial numbers of microorganisms, largely bacteria (Ghiorse and Wilson, 1988). In the late 1980s, the United States Department of Energy (USDOE) initiated a systematic program to investigate the microbiology of deeper subsurface environments. Other countries, notably Canada, Sweden, Finland, Germany, England, Russia, and Japan, established similar programs, stimulated by groundwater contamination issues or by the challenge of isolating high-level radioactive waste over long duration (>10,000 years) in deep, geologic repositories. Overwhelming, subsurface biomass is prokaryotic (Ghiorse and Wilson, 1988; Sinclair et al., 1990; Whitman et al., 1998), and it appears that bacteria are resident in all areas of the terrestrial subsurface that have been sampled, including cores collected from 2,800 m depth

(Boone et al., 1995; Onstott et al., 1998), and groundwater thought to originate from at least 5,300 m depth (Szewzky et al., 1994).

2. BUILDING CONCEPTUAL MODELS OF DISTRIBUTION: DEPTH AS A PRIMARY AXIS

An inherent and dominating consideration of the terrestrial subsurface habitat is the depth axis. At any location, gradients in the subsurface habitat will be anchored by their depth from the surface, where organic matter primary production is sustained by light energy and dissolved or free gas concentrations approach equilibrium with atmospheric levels. Throughout the depth profile, the subsurface is strongly structured by the vertical (generally) arrangement of geologic units and their weathering profiles. Due to variations in soil profiles and underlying geology, some discussion arises as to what is "subsurface" as opposed to soils. The depth of soils is often considered synonymous with the rooting zone, although the depth of root penetration may vary greatly with vegetation, climate, etc. One convention is to define soil inclusive of the A, E, and B horizons, but not the weathered parent material or regolith composing the C horizon (Hunt, 1986). Alternative definitions of soils include the C horizon, respecting deep-rooted plants (Richter and Markewitz, 1995), or use a 2 m depth criterion (Soil Survey Staff, 1975). The "subsurface" includes everything below the "soils" or alternatively, everything below 1 m (Onstott et al., 1999) or 8 m (Whitman et al., 1998). For the current discussion of microbiology, "subsurface" will be defined as the region beneath the intensively studied plow layer soils that chiefly comprise the A and O horizons. The less frequently sampled "subsoils" (B horizon to bedrock) usually contrast sharply with the overlying topsoils in organic matter and mineral content.

2.1. Soil-unsaturated subsoil transition: the first few meters

Microbial biomass, diversity, and activity often exhibit parallel declines with depth during the transition from the organic-rich topsoils to the mineral-dominated subsoils , although at some point in the C horizon or bedrock, biomass levels become relatively constant with further depth increases as long as unsaturated conditions prevail (Fig. 6-1). A decline in biomass (by lipid analyses), activity (by FDA hydrolysis), and diversity (PLFA profile complexity) was observed >2 m deep profiles sampled within three unimproved pastures and a minimally managed soybean field in Alabama (Federle et al., 1986). Wood et al. (1993) observed 3–4 orders magnitude decrease in culturable aerobic heterotrophs >7 m depth profiles in silty loam soils at two sites in western North America. Bone and Balkwill (1988) reported decreasing values in total cells, culturable heterotrophs, and heterotrophic diversity over a 3 m depth profile containing loamy sand topsoil underlain by fine, silty clay at an Oklahoma site. Numerically dominant phenotypes differed between surface soils and subsoils. Dodds et al. (1996) showed that biomass, respiration, number of CTC-respiring cells, and numbers of culturable aerobic heterotrophs decreased >4 m of unsaturated clayey soils at a grassland site in northeast Kansas, mirroring the sharp drop in total soil organic matter. However, total numbers of cells remained relatively constant over the same interval, as did the amount of soluble organic carbon. In cropped soil at the same prairie site, microbial properties remained relatively constant over the same depth interval, suggesting that tilling may have resulted in increased homogenization, even below the actual depth tilled. In several coniferous forest soils in Finland, microbial biomass (total PLFA) was observed to decrease roughly tenfold over a 0.4 m depth profile (Fritze et al., 2000). These authors also found changes in community structure (PLFA profiles) between horizons such that $O \neq E \neq B = BC$ and that actinomycetes were relatively enriched with depth. Roughly 100-fold decreases in bacteria and fungal biomass were observed in forests soils at two sites in Denmark (Ekelund et al., 2001). Over a 1.5 m depth profile in two Indiana agricultural soils, Blume et al. (2002) observed sharp decreases in bacterial biomass and found that the community structure (PLFA) in the subsurface soils was relatively enriched in gram-positive bacteria compared to the topsoils. In 4 m depth profiles from two mid-western US agricultural soils of differing textures, Taylor et al. (2002) found sharp decreases in microbial abundance (various measures) and activity (enzymes) with depth that were well-correlated with declines in soil organic matter content. At two California loamy grassland sites, Fierer et al. (2003), found that microbial biomass and diversity declined, and community structure was relatively enriched in both actinomycetes and gram-positive bacteria with increasing depth over a 2 m vertical interval.

An exception to the general decline in microbiological parameters with depth in subsoils may be in deep, organic-rich soils that have been recently drained, such as observed in Florida Histosols, where oxygen concentrations regulate activity (Tate, 1979). Similarly, in the partially waterlogged soils of a Danish bog, higher microbial numbers observed at the depth have been attributed to preservation of organic matter and exclusion of protozoan predators by the low oxygen tensions (Ekelund et al., 2001).

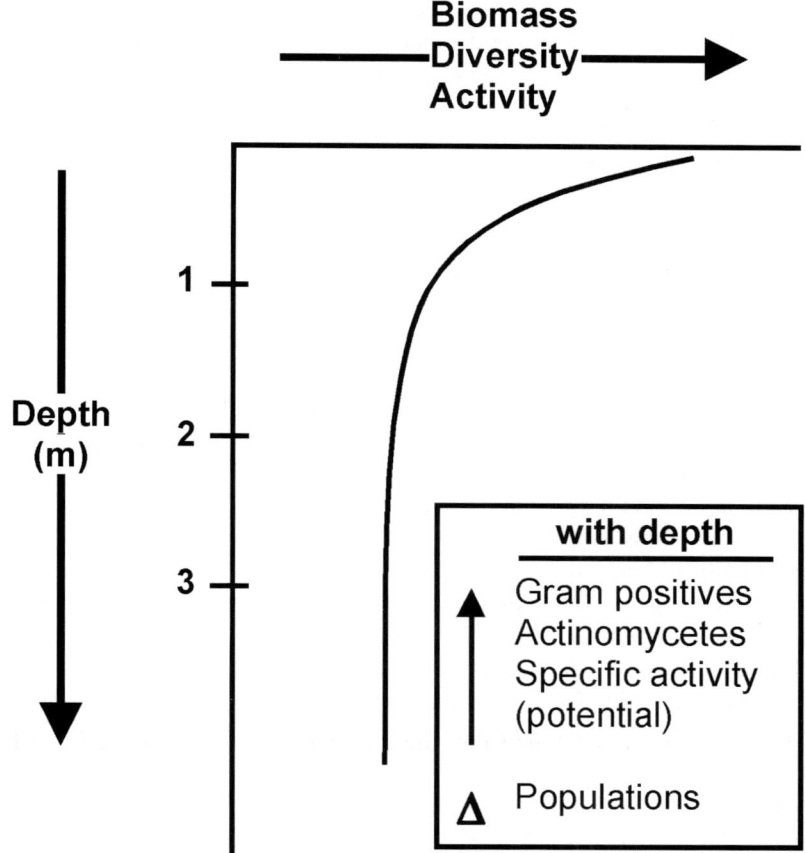

Figure 6-1. Generalized depth distribution of microbial properties from the surface through the first several meters of unsaturated subsoil.

2.2. Unsaturated–saturated transition in the shallow subsurface: meters to dekameters

As the depth of sampling increases in the unsaturated or vadose zone past several meters, and often into a "C" horizon, microbial properties tend to level off, without further predictable decreases with depth through the "shallow" subsurface. Several scenarios generally present themselves: (i) a "shallow" aquifer is encountered, or (ii) a large vadose zone overlies a "deep" water table aquifer; (iii) less permeable layers overlay a shallow or deep confined aquifer. Discussion again arises as to what is "shallow" and what is "deep" subsurface. The "deep subsurface" has been arbitrarily defined by depths >10–20 m (Fliermans and Balkwill, 1989), >50 m (Sinclair and Ghiorse, 1989), >50–100 m (Pedersen, 1993), or other depths within these ranges. Often, an aquifer is the target of subsurface microbiological investigations. Although it may be convenient to consider water table aquifers shallow, in areas of the arid western USA, the water table may lie at depths exceeding 200 m. Due to regional geologic distinctions, it has been proposed that the classification of the subsurface be based on aquifer recharge characteristics and that the term "deep subsurface" refer to intermediate and deep aquifer systems, but not local aquifers (Lovley and Chapelle, 1995). However, not all relatively deep locations that have been sampled are aquifers. For the current purposes, the shallow subsurface will be loosely defined as within 20 m of the surface, generally accessible by direct push drilling technologies, and if an aquifer is present, it will be a water table aquifer usually consisting of weakly or unconsolidated clastic material.

Depth-related profiles of solid-associated microbial properties as a shallow aquifer is encountered have been largely performed in sedimentary systems where there is a capillary fringe. Wilson et al. (1983) reported that bacterial cells and culturable heterotrophs decreased from 1.2 (B horizon) to 3.0 m (C horizon) and then increased or remained constant at 5.0 m in the saturated zone of a sandy Oklahoma aquifer. A second sampling at this Lula, Oklahoma location also demonstrated an increase in culturable aerobic heterotrophs at the capillary fringe and then relatively constant numbers through the saturated zone to 6 m (Balkwill and Ghiorse, 1985). Other reports from Lula with more intensive depth-stratified sampling demonstrated the progressive decrease in culturable numbers and diversity from topsoil through 3 m of subsoil and then an increase to relatively constant numbers through 3 m of the saturated zone (Beloin et al., 1988; Bone and Balkwill, 1988). Subsurface samples collected at this location contained a predominance of gram-positive cells and small cellular morphologies (Balkwill and Ghiorse, 1985; Bone and Balkwill, 1988; Wilson et al., 1983), although deeper (7.5 m) aquifer samples produced relatively more gram-negative bacteria (Bone and Balkwill, 1988). Lula subsurface isolates tended to be nutritionally flexible with respect to substrate concentrations and were often affiliated with taxonomic groups with broad substrate ranges (Balkwill and Ghiorse, 1985; Bone and Balkwill, 1988). Predominant populations varied with depth and were distinct from surface populations (Bone and Balkwill, 1988).

At two sites in western North America, Wood et al. (1993) presented data that suggested that microbes were most active, producing CO_2, at the interface between the unsaturated zone and a sandy aquifer. At both the cropped and grassland Kansas prairie soil profiles studied by Dodds et al.

(1996), microbial parameters indicating biomass and activity exhibited transient increases as the capillary fringe was encountered and then leveled off or rose though the saturated zone to 10 m depth. At several sites in agricultural fields overlying a sandy aquifer on Virginia's Eastern Shore, Zhang et al. (1997), witnessed a 100-fold decrease in culturable aerobic heterotrophs with depth over about 3 m, and then numbers increased as the water table was encountered. At a nearby location on the Virginia Eastern Shore, Zhou et al. (2004) produced 16S rDNA clone libraries for three depth intervals, the deepest was saturated sands at 4 m. These authors found the subsurface communities to be less diverse and harboring populations distinct from the surface. Moreover, the composition of subsurface communities varied dramatically (>90% difference in operational taxonomic units (OTUs)) between samples vertically separated by meters in an apparently homogeneous sandy subsoil. In a related report that included samples from this Virginia site, 16S rDNA clone libraries from several saturated subsurface samples had fewer OTUs, but more dominant (high frequency) OTUs compared to corresponding low carbon surface samples. Unsaturated subsoils exhibited intermediate OTU richness and evenness between the surface and saturated subsurface samples. At sites where surface carbon levels were higher, surface soil communities were still very diverse, but exhibited more dominance, similar to the subsoils. These authors hypothesized that increased access to carbon, either by higher concentrations in the unsaturated surface soils or by increasing homogenization in saturated aquifer sediments, leads to decreased competetion and decreased evenness in population distributions.

So, after the first decline in microbial numbers and diversity from topsoil to subsoil, there appears to be no further decrease with depth in the unsaturated zone. Further, once saturated systems are encountered in shallow systems, there is a generally a rise in solid-associated microbial activities and numbers and no further depth-related decreases as depth in the saturated zone is increased (Fig. 6-2). Several additional themes emerge from the current body of work that may apply to shallow subsurface environments: (i) subsurface communities are less diverse than surface communities; (ii) subsurface communities appear to have different populations than surface communities, (iii) gram-positive bacteria and actinomycetes become relatively more abundant in unsaturated subsoils with increasing depth; (iv) gram-negative bacteria become relatively more abundant in the saturated zone; (v) community composition varies considerably among subsurface samples collected from the same site; and, (vi) saturated subsurface communities tend to have dominant populations.

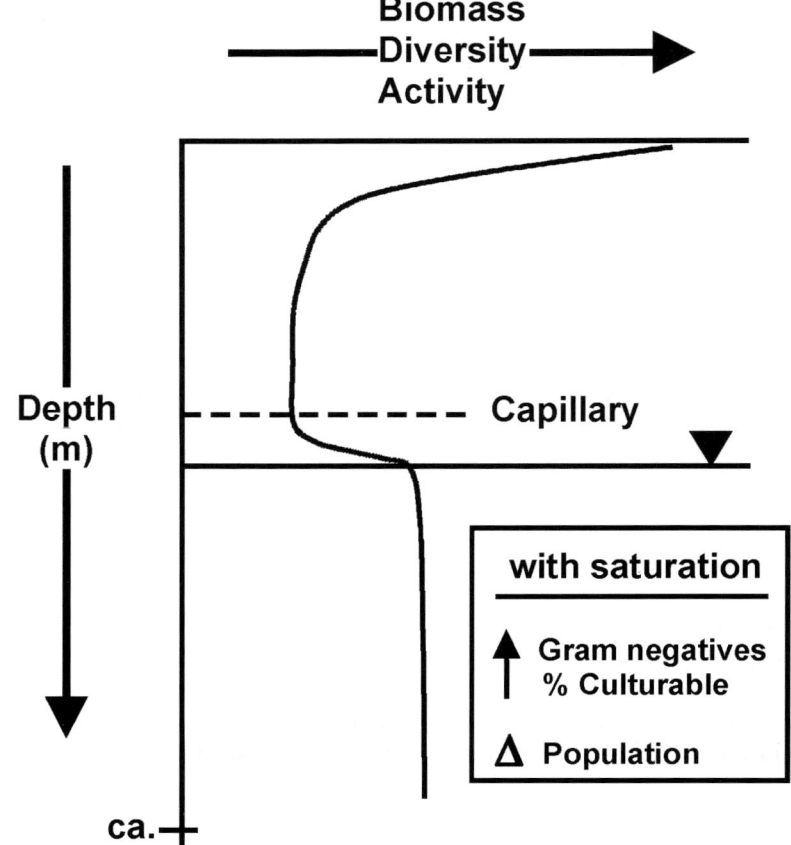

Figure 6-2. Generalized depth distribution of microbial properties though deepening unsaturated subsoils and through an encounter with a shallow water table aquifer.

2.3. Deeper depth distribution: dekameters to kilometers

The initial decline in microbial parameter from the surface to the subsurface followed by little or no discernable depth-related trend has also been witnessed in sediment or rock samples collected over wider depth intervals from deeper profiles. Sampling campaigns comparing surface samples to depth-stratified samples from deeper regions have been collected from (a) deeper sedimentary aquifers, (b) cross sections that contain both aquifer and aquitard formations, (c) sedimentary vadose zones of the western USA, and (d) fractured rock aquifers and highly consolidated deeper formations.

2.3.1. Deeper sedimentary aquifers

In an extensive study of isolates cultured from five coreholes penetrating saturated sedimentary formations in the Germany (Kolbel-Boelke et al., 1988), the number of viable cells in sediment samples was 10- to 1,000-fold lower than surface sample, yet there was no systematic decrease with depth >40 m. The numbers and types of culturable organisms strongly differed among samples of different depths. Coefficients of variation (CV) for aerobic colony forming unit (CFU) ranged from 123% to 226% in the five transects. Within a sand–gravel aquifer in northeastern Kansas, total cell numbers and cultureable heterotrophs in sediment samples were 100- to 1,000-fold lower than surface samples, but did not decline even up to a depth of 90 m (Sinclair et al., 1990). Microbiological properties of a 30 m sequence of sedimentary formations of varying lithology that contained two aquifers was studied in southeast Texas (Martino et al., 1998; Ulrich et al., 1998). The shallow aquifer was primarily oxidizing and the deeper primarily reducing. There was an initial 2–3 log decrease from surface values of total cells and viable aerobic heterotrophs in vadose zone sediments, and then values did not decline further. Anaerobic heterotrophs and sulfate-reducing bacteria (SRB) decreased from surface values of 10^4–10^5 cells per gram to zero in unsaturated zone sediments. Once the saturated zone was encountered (8 m), total cells and viable aerobes remained relatively constant for the remainder of the intervals, while viable anaerobes and SRB increased to 10^2 and peaked at 10^4–10^5 in the lower, reducing aquifer (Martino et al., 1998). No systemic decrease in SRB or iron- and sulfur-oxidizing bacteria with depth was observed across the entire profile (Ulrich et al., 1998).

2.3.2. Cross sections that contain both aquifer and aquitard formations

In samples collected across multiple horizons of eastern US coastal plain saturated sediments (Maryland), the number of cells and viable bacteria initially declined from the uppermost 21 m deep sample to about 100 m deep, but then no further declines were observed to a total depth of 168 m (Chapelle et al., 1987). In the late 1980s, several 250 m deep profiles of microbiological properties were established across at least seven geologic formations, including several aquifers, in southeastern US coastal plain sediments (South Carolina). Microbial populations generally decreased from surface values through the vadose zone, and then increased under saturated conditions. While numbers of culturable aerobic heterotrophs were about 100-fold less numerous and species richness was less in subsurface samples compared to surface samples at this site, no depth-related decreases were observed in numbers or diversity once saturated sediments were encountered (Sinclair and Ghiorse, 1989). Even at depths to 265 m, about 10^7 CFU g^{-1} were recovered and relatively high numbers of other oxygen-utilizing organisms were maintained (Fredrickson et al., 1989); other methods for enumerating aerobic heterotrophs yield similar results (Balkwill, 1989). Gram-positive bacteria tended to have highest relative abundances in the vadose zone, while gram-negative bacteria increased in the saturated zone. From sample to sample throughout the depth profile, bacterial populations varied considerably evidenced by both physiology (Balkwill et al., 1989; Fredrickson et al., 1991) and colony morphology (Balkwill, 1989). Aerobic CFU had CVs of 128–178% in each of the vertical transects. Morphological, physiological, and molecular evidence suggests that the aerobic heterotrophic isolates from the subsurface were primarily different populations from those isolated from surface soils at this site (Balkwill et al., 1989; Jimenez, 1990). Several studies indicated that subsurface isolates tended to be capable of existing under low-nutrient conditions and were affiliated with taxonomic groups that are nutritionally versatile and often capable of degrading complex compounds including aromatics (Balkwill et al., 1989; Jimenez, 1990). Frederickson et al. (1988) reported that isolates from progressively deeper horizons had a higher occurrence of plasmids and larger plasmids.

2.3.3. Sedimentary vadose zones of the western USA

In areas of the Western USA, where the water table may be at great depths, several studies have examined microbiological properties in unsaturated sediments over vertical transects. Deep vadose zones in arid climates tend to have very low levels of biomass compared to surface soils and extremely patchy distributions (Colwell, 1989; Hersman, 1997; Kieft et al., 1993). At five sites with contrasting recharge characteristics in Washington, Balkwill et al. (1998) observed that the numbers of culturable organisms and biomass (PLFA) declined sharply in the first 1–2 m and then remained relatively constant, but often very low, to total depths examined between 10 and 15 m. At several of the five sites, the percentage of actinomycetes increased with depth. Kieft et al. (1998) reported on microbial depth distributions across chronosequences of unsaturated buried loess at two sites in Washington. The authors found that at both sites biomass (PLFA), total and viable cells (plate counts), and activities (various methods) declined sharply in the first several meters, but then remained relatively constant from 10 to >30 m total depth. Community composition (PLFA profiles) shifted and appeared to become less complex with depth. At an uncontaminated location in south central

Washington, viable counts and activity (heterotrophic potential) in vadose sediments from a depth range of 30–90 m were compared with sandy surface soils (Fredrickson et al., 1993). At this site, viable counts and activity were much lower in the subsurface (frequently at or below detection) than the surface, although the highest subsurface values occurred in the deepest samples collected.

2.3.4. Fractured rock aquifers and highly consolidated deeper formations

Microbial properties have been determined from sediment and rock samples collected from additional sites at varying depths. Often, vertical sampling transects were concentrated at certain depths and/or comparative analyses were not performed on surface samples. Nonetheless, total cell counts on subsurface sample from multiple studies, including those with limited vertical distributions, can be plotted on a depth axis to provide a rough view of cell number with depth (Fig. 6-3). Although Fig. 6-3 does present total cells (for which there is the most data), the majority of samples where cells were observed also yielded evidence of cell viability. The primary exceptions are some deep, dry vadose zone sediments (Colwell, 1989; Fredrickson et al., 1993; Kieft et al., 1993), igneous rock samples (massive basalt) (Kieft et al., 1993; Lehman et al., 2004), and saturated sediments of low permeability (Colwell et al., 1997; Fredrickson et al., 1997; Kieft et al., 1995) including massive clays (Boivin-Jahns et al., 1996; Lawrence et al., 2000), where the limited number of tests performed did not reveal cell viability. Over a depth range of several kilometers, cell numbers are lower in the subsurface than surface soils (~10^8–10^9 cells g^{-1}), however, there remains great variability in cell numbers that does not appear associated with depth. So, what are the ultimate limiting factors for microbes in the subsurface? Is there an ultimate depth past which they will not be found? If 110°C is assumed to be the maximum for survival and the average geothermal gradient is approximately 25°C km^{-1}, then about 4.5 km may be an absolute depth for microbial life to persist. The estimated *in situ* temperature at 2,800 m where viable organisms were cultivated was approximately 75°C (Onstott et al., 1998). Alternatively, if 150°C (with pressure) is assumed to be the limit of microbial survival, then the absolute depth would be closer to 6 km. Other potentially limiting factors such as the presence of toxic substances, extreme pH, or radioactive fields will only exclude microorganisms from very limited portions of the subsurface. Ionic strength is not likely to be an absolute limitation as microorganisms have been shown to tolerate saturated (32% NaCl) solutions, and while pressure will rise about 100 atm per 1,000 m, many microbes can endure 100s of atm and some can grow at 600–1,000 atm (Madigan et al., 1997). At deeper depths, physical space may become a factor – if pore volumes decrease to <0.1 μm^3 – then there will simply be no room for microbial cells. Oxygen generally decreases with depth from its source at the surface as it is consumed during organic matter oxidation, although many relatively deep aquifers remain well-oxygenated (Winograd and Robertson, 1982). The presence and concentrations of dissolved or free gases will vary with depth, distance from recharge, nature of the overlying vadose zone, and other location-specific characteristics, yet the most likely impact of local atmosphere will be a selection for certain physiologies, not uniform elimination. The availability of water or water activity will influence the distribution and activities of microorganisms in the unsaturated subsurface, but will probably not decrease to levels that cause cell dessication and death (Kieft et al., 1993).

Microorganisms need to conduct energy yielding reactions and assimilate carbon to maintain cell integrity. The distribution of physiological types and the magnitude of activity in subsurface environments will reflect amounts, and perhaps quality of available carbon, as well as terminal electron acceptors and other nutrients that are sustained by advective and diffusional fluxes at a given location. These geochemical parameters will be constrained on a larger scale by the broader geological history and regional climate of each site, but will also be expressed on the centimeter–meter scale by the immediate geochemical and hydrological conditions imposed by the specific location.

3. PATTERNS AMONG SPECIFIC GEOLOGIC FORMATIONS AND DEPOSITIONAL ENVIRONMENTS

Up to this point, the microbiological attributes of core samples, sediment or rock, have been discussed with respect to depth. Obviously, a lot of knowledge concerning terrestrial subsurface microbiology has been gathered by analysis of groundwater samples. At some sites, it has been shown that free-living groundwater communities may differ significantly in structure and potential function from those attached to the geologic media (Bekins et al., 1999; Godsy et al., 1992; Kolbel-Boelke et al., 1988; Lehman et al., 2004; Lehman et al., 2001) and this represents another dimension of spatial variability. However, it is important to recognize the limitations inherent in spatially locating the origin of a particular sample of groundwater. Many groundwater samples have been collected in bulk from wells that are completely open,

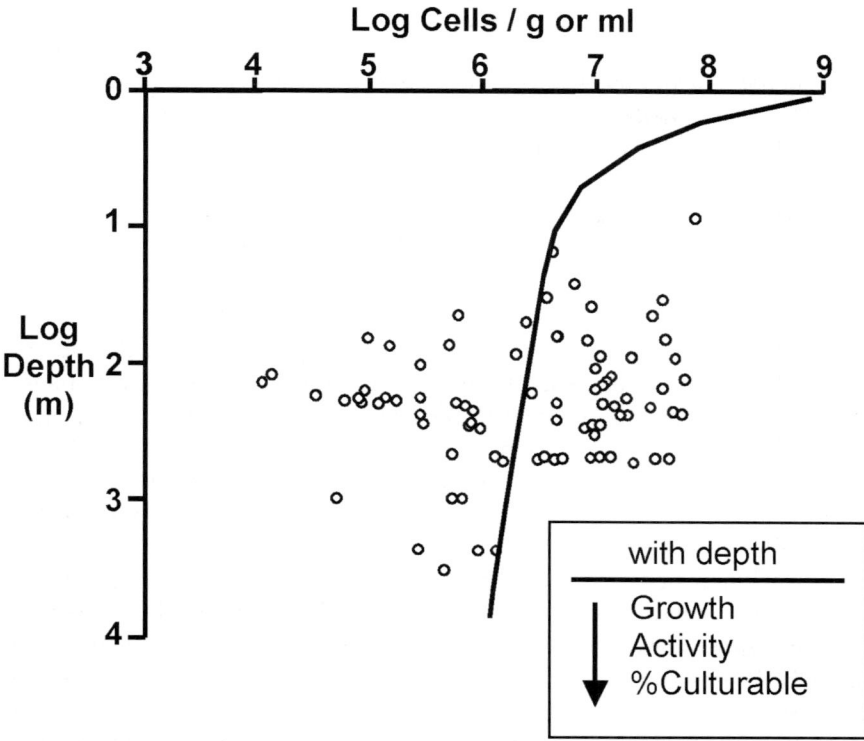

Figure 6-3. Microbial cell numbers over a >3 km depth in terrestrial subsurface environments. (Adapted from Onstott et al., 1999; Courtesy R. Colwell.) Hand-drawn trend line superimposed on data to suggest a conceptual model for cell distributions across this broadscale.

open at the bottom, or screened across some length. In other cases, multi-level wells have been installed wherein individual inlet ports occur at certain intervals (Smith et al., 1991). Straddle-packers have been used to isolate sections of an open hole or screened interval to pump water specific for a given interval (Lehman et al., 2001). Lastly, diffusion cells separated by baffles have been used to collect depth-discrete water samples (Lehman et al., 2004; Takai et al., 2003). In many of the studies of vertically stratified groundwater samples, decimeter–meter variations in microbial properties have been observed. But, regardless of the methods used to acquire the groundwater samples, there is uncertainty regarding the water's previous location. Even when using multilevel wells, straddle-packers, or diffusion cells separated by baffles, there is the chance that the groundwater originated from locations above or below the desired sampling point by traveling vertically through the formation by natural or induced gradients. The well itself presents an intrusion that may lead to short-term artifacts associated with well construction, or long-term artifacts due to colonization of the foreign well environment. For those reasons, only selected examples of groundwater analyses that were intentionally conducted on a number of vertically or horizontally arrayed samples will be included in the following discussion.

The microbiology of a variety of sedimentary sequences has been studied. The individual sedimentary units differ in their lithology, exhibiting characteristic mineralogy, grain sizes, carbon content, etc. according to the origin of the particles, their original depositional environment and subsequent history, including metamorphosis. If saturated, the microbiology in a given horizon will be influenced by the local geochemistry of the water. Consideration of subsurface sedimentary formations as habitats compels us to assess how the subsurface environment differs from that observed at the surface. For instance, some locations were originally the sediments of an inland sea, others were the floodplain of a meandering river, and some were soils in their own right, at various stages of development. All were initially subjected to processes that occur at or near the earth's surface. Yet, over staggering time spans, these sediments have been relocated and physically transformed so that current conditions do not resemble historic conditions. In comparing one sedimentary unit with another, it must be recognized that these units also span large time intervals and that deeper is older (usually). The timescale becomes important when considering spatial scales of microbes in the subsurface. Current evidence suggest that bacteria in some sedimentary rocks may be progeny of those that were associated with original deposition, while others were subsequently transported to that location at uncertain intervals over a long time period (Fredrickson et al., 1995, 1997; Kieft et al., 1998). This evidence is produced by examining the ages of the sediments or rock, the pore water or groundwater ages, the pore structure of the formation and neighboring formations, the overall history of geologic formation (e.g., temperature history), current and historical recharge, and other hydrogeological conditions. Therefore, when trying to ascertain distributions, do we expect organisms that might have thrived in the originally depositional environment? Or, do we confine the search to organisms that have physiologies consistent with the current circumstances? There is fairly strong evidence that microorganisms can persist isolated in geologic formations for very long periods of time (Fredrickson et al., 1995; Lawrence et al., 2000; Onstott et al., 1998), with estimated growth rates that appear static (Chapelle and Lovley, 1990; Lawrence et al., 2000; Onstott et al., 1998, 1999; Phelps et al., 1994). If that is the case, then inactive, but viable cells (original progeny or transported) may compose much of the biomass and account for

Distribution in Subsurface Environments 149

the low percentage of cultivable cells observed in many deeper locations. Because of these issues that are unique to or exaggerated in subsurface environments and the necessarily selected (and selective) methods used to analyze the samples, it might be expected that relatively few of the inhabitants have actually been evaluated. The origin of subsurface microbes in a particular formation, their degree of adaptation to the current conditions, and their ability to respond to given assays may have strong consequences for our perception of subsurface microbial distributions beyond total cells.

3.1. Vertically distributed patterns within the subsurface: centimeters, meters, dekameters

Microbiological analyses of depth transects through stacked sedimentary layers of differing lithologies have produced several trends concerning the distribution of microorganisms and their activities in relation to their physicochemical location. These trends can be broken into two groups: (i) those that relate to the characteristics of the individual sedimentary unit examined; and (ii) those that relate to the juxtaposition of two units with differing characteristics.

3.1.1. Vertical trends according to lithology, meters to dekameters

A primary trend in the analysis of unconsolidated sedimentary units is a relationship between the microbiology and sediment texture. Sediment texture directly influences key characteristics including porosity, permeability, water potential, and frequently is related to organic carbon and particle mineralogy. Chappelle et al. (1987) observed increased numbers of total cells in clay layers compared to coarser-grained layers, but the number of viable cells was not correlated with sediment texture in their study of Miocene-Lower Cretaceous age northern Atlantic coastal plain sediments. Albrechtson and Winding (1992) found that sediment texture was correlated with microbiological properties in a series of Quarternary glacio-fluvial sediments in Denmark. They reported that clay content was associated with higher numbers of total cells, although there was a high degree of correlation between clay content and organic matter content in these samples. Conversely, activity (aerobic C-14 substrate mineralization) was associated with sediment sand content and higher permeability (calculated from grain size distributions). In size-fractionated aquifer sediments, Albrechtsen (1994) found that the number of viable cells and the amount of activity (aerobic C-14 substrate mineralization) was highest in the silt-sized particle fraction. In an earlier study of size-fractionated sediments collected from a coastal aquifer, Harvey et al. (1984) reported that the majority of bacterial cells were associated with fine-grained (<20 μm) sediments. In analyses of samples collected from four coreholes penetrating a glacial till aquifer (Kansas), total cells and viable cells were negatively correlated with sediment clay content, and positively correlated with sand content (Sinclair et al., 1990). A number of papers on samples collected from Tertiary and Cretaceous age coastal plain sediments of the southeastern USA indicate that sediment clay content was negatively correlated (and sand content positively correlated) with total cell numbers (Sinclair and Ghiorse, 1989), culturable aerobic bacteria (Balkwill, 1989; Fredrickson et al., 1989; Phelps et al., 1994; Sinclair and Ghiorse, 1989) and anaerobic bacteria (Jones et al., 1989; Phelps et al., 1989, 1994), eukaryotes (Sinclair and Ghiorse, 1989), chemoheterotrophic diversity (Balkwill et al., 1989), denitrifying activity (Francis et al., 1989), aerobic activity (C-14 substrate mineralization) (Hicks and Fredrickson, 1989; Phelps et al., 1994), and anaerobic activity (C-14 substrate mineralization) (Hicks and Fredrickson, 1989; Phelps et al., 1994, 1989), and growth (C-14 acetate incorporation into lipids) (Phelps et al., 1994). Samples with high clay content were associated with relatively higher proportions of gram-positive bacteria (Balkwill, 1989; Sinclair and Ghiorse, 1989). Despite the overall relationships of microbial biomass and activities with sediment texture, there appeared to be considerable variation in chemoheterotrophic community composition in samples within a single formation and from different formations (Balkwill, 1989; Fredrickson et al., 1991). Even given the distance from the surface (>100 m) and the apparent ages of the groundwater (up to 1,000s of years in some samples (Phelps et al., 1994)), the higher nutrient fluxes enabled by the relative high hydraulic conductivities (20–60 kD) of the sandy formations should support higher microbial activities in the sandy aquifer layers than the clay aquitards (Phelps et al., 1994). In support of this conclusion, the addition of water to clay sediments dramatically increased their activity, presumably by increasing the availability of nutrients in the low-water potential clays (Phelps et al., 1994).

A contrasting relationship of microbial properties with sediment texture was observed during analysis of depth-stratified saturated samples (~1 m intervals >172–197 m depth) from a Miocene age profile consisting of fine-grained lacustrine sediments, a paleosol sequence, and coarse-grained fluvial sediments located in south-central Washington. Biomass (PLFA) (Fredrickson et al., 1995), number of bacterial cells, aerobic basal respiration (CO_2 production), and aerobic activity (C-14 glucose and acetate substrate mineralization) were significantly higher in the fine-grained lacustrine sediments compared to the other layers (Kieft et al., 1995). Very little anaerobic activity (C-14 substrate mineralization) was apparent in any of the samples, although multiple lines of evidence suggest that the presence and activity of dissimilatory iron reducing bacteria and SRB were also highest in the lacustrine sediments (Fredrickson et al., 1995; McKinley et al., 1997). There was a high degree of correlation of total cells and aerobic activities (basal respiration and glucose mineralization) with total organic carbon (TOC) content of the sediments, which was markedly high in the lacustrine sediments (~1%). Unexpectedly, the paleosol sequence did not have high amounts of organic carbon (~0.1%), although the fluvial sands were even lower (~0.03%). Glucose mineralization exhibited an 85% CV over the vertical series of samples. It was also noted that the cell numbers observed in these saturated zone samples were lower than cell numbers previously reported from the shallow unsaturated zone samples collected from the same borehole (Kieft et al., 1993). This arid site has much lower rates of surface primary production and almost negligible annual recharge rates compared to the southeastern coastal plain site. Groundwater ages are estimated to exceed 13,000 years and permeabilities of all sediments ranged from 1 to $<10^{-3}$ mD. Thus, the fluvial sediments also had low permeability, although relatively higher than the lacustrine sediments, and may not have received sufficient fluxes of exogenous carbon to maintain any level of activity. Therefore, it may be the higher amounts of carbon originally deposited in the sediment that provides for the higher microbial numbers and activities in the lacustrine sediments, although their *in situ* activities may be severely limited by electron acceptor flux (Fredrickson et al., 1995; Kieft et al., 1995).

3.1.2. Juxtaposed sedimentary formations: centimeters to meters

A second recognizable trend in the distributions of microbes and activities in stacked sedimentary layers has also been observed at several sites. When a low-permeability, higher carbon, fine-grained sedimentary formation lies adjacent to a more permeable, coarser-grained formation, diffusion of organic substrates from higher concentrations in the low permeability sediments into the higher permeability formation where electron acceptor flux may be higher, results in elevated respiratory activities (McMahon and Chapelle, 1991) (Fig. 6-4). The organic substrates can be solubilized detrital carbon or fermentation products thereof. Respiration in the coarse grained sediments is limited by the low flux of exogenous carbon in the groundwater. This phenomenon was initially proposed following study of pore water organic acid concentrations in a clay-rich aquitard overlaying a sandy aquifer in the southeastern USA (McMahon and Chapelle, 1991). It was concluded that fermentation predominated in the organic-rich aquitard producing low-molecular weight fatty acids that diffused into the aquifer where they supported anaerobic respiration of advectively supplied electron acceptors. Peaks in respiratory activity in the coarser-grained sediments may be particularly pronounced at the interface between two layers. Such a pattern was reported in a Cretaceous sandstone–shale sequence cored to a total depth of 230 m in northwestern New Mexico. In these anaerobic consolidated sediments, peaks in sulfate-reducing activity potentials were observed in sandstones adjacent to the interface with high-carbon shales (~1% TOC) (Krumholz et al., 1997). The autoradiograph technique used indicated that SRB activities varied considerably within samples at millimeter scales. No evidence was found of active fermentation in the shales, but it was proposed that acetogens co-occurring at the interface converted the diffusing detrital carbon to fuel the SRBs (Krumholz et al., 1999). General anaerobic activities (C-14 substrate mineralization) were recorded in sandstones with pore throat diameters >0.2 μm, but not shales with pore-throat diameters <0.2 μm, although overall detectable microbes and activities were extremely patchy thoughout the profile (Fredrickson et al., 1997). Over the entire vertical sequence, microbial biomass (PLFA) exhibited 185% CV. Permeability is directly related to pore throat diameter, and therefore the shales would experience low groundwater fluxes and are probably electron acceptor limited. Small pore diameters (<0.2 μm) that reduce diffusive fluxes (less connectivity, more tortuosity) and represent a critical boundary for cell movement may explain the carbon preserved in the shales. A later study (Walvoord et al., 1999) indicated an upwards groundwater gradient at this location and that microfractures in the shale may have also contributed to the observed microbial distributions. Vertical distributions of groundwater Archaea were studied at this site using dialysis chambers incubated in the open corehole over an 8 m interval spanning a shale–sandstone interface (Takai et al., 2003). rDNA profiles (tRFLP) of Archaeal communities shifted considerably at scales of 10s of cm across the interface of the two formations. In conjunction with groundwater geochemistry, methanogens were found to be present and active in these groundwater samples collected adjacent to the shale. No evidence of methanogens or their activity had been previously observed during study of the core samples (Fredrickson et al., 1997; Krumholz et al., 1997).

Conversely, while fermentation may be occurring in the low permeability, higher carbon sediments, respiration in the same would be controlled by diffusion of electron acceptors from the higher permeability sediments. This situation was hypothesized to occur in the Miocene lacustrine sediments by Fredrickson et al. (1995) based on dissolved pore-water sulfate profiles and by McKinley et al. (1997) with additional data on fermentor and SRB distributions surrounding the contact interval. It appeared as if fermentation was ongoing within the low permeability lacustrine sediments and

SRB activity peaked near the edges of the fine-grained sediments where SO_4^{2-} was supplied. So, peaks in respiratory activity should occur at the interface, but could exist on either side. Samples collected from 30 m of unconsolidated Eocene sediments in east central Texas are another location where the diffusion of organic acids (presumably fermentation products) from lower permeability high-carbon clays or lignites stimulated SRB activity in the adjacent higher permeability sands, particularly at the interface (Ulrich et al., 1998). Sulfur and iron-oxidizing bacteria were also preferentially present in the sandy sediments at these interfaces, even under reducing conditions. Similarly, SRB activity was observed in areas that were considered oxidizing. Over the 30 m profile, SRB activity varied several orders of magnitude, even when samples were taken <10 cm apart. Finer scale studies performed with an autoradiograph technique demonstrated that SRB activity in cores also varied at the millimeter scale. It was proposed that pyrite oxidation preferentially occurring in the more shallow areas supplied SO_4^{2-} for SRB activity in the deeper aquifer.

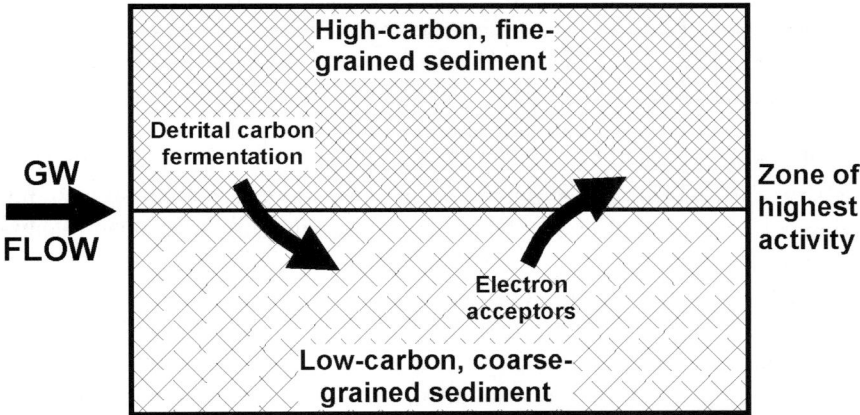

Figure 6-4. Diffusion gradients of organic matter from high-carbon, fine-grained sediments encounter higher electron acceptor concentrations in low-carbon, coarser-grained sediments, and vice versa.

3.2. Trends within shallow sandy formations: centimeters to meters

Barbaro et al. (1994) examined microbial properties at 10 cm intervals over nine 1.5 m vertical transects though a shallow sandy aquifer near Borden, Ontario. At these sites dissolved oxygen declined sharply over this depth interval from values of 5–6 mg L^{-1} near the top of the aquifer to 1–3 mg L^{-1} near the bottom. They found that aerobic numbers and activities (ETS) generally declined in parallel with depth and oxygen, although occasional peaks were exhibited. No assessment of anaerobic physiologies was performed. Samples collected at 10 cm intervals from three 1.5 m vertical transects in unsaturated sandy sediments from Virginia's Eastern Shore were analyzed for a variety of microbial properties (Musslewhite et al., 2003). The transects were established at 1–3 m depth and were confined to a sedimentologically homogenous stratum. No evidence was provided that indicated redox changes over the profile. Microbial properties (biomass, aerobic and anaerobic H_2 oxidation) varied little over the vertical direction, with slight increases near the bottom of the profile where moisture was higher (possibly capillary fringe, no groundwater table depth given).

3.3. Deep vadose zones in arid regions: meters

In the deep vadose zones of arid regions, numbers and activities are very low and very sparsely distributed. It appears that water potentials exert an overriding control on the activity, and perhaps persistence of microbes in these dry sediments. Even at the low (−0.1 to −1.0 MPa) matric water potentials commonly observed in arid vadose zones sediments; however, it is unlikely that that microbes will be killed by dessication (Kieft et al., 1993). In comparison, surface soil water potentials in these same arid regions can fall below −10 MPa (Kieft et al., 1993), and some microorganisms can tolerate water potentials as low as −400 MPa (Potts, 1994). The effect of low water potential is likely to be indirect (Fredrickson et al., 1993; Kieft et al., 1993). There are infrequent (if any) saturated episodes to advectively supply or redistribute nutrients. Decreasing water potentials (<−0.1 MPa) in the unsaturated sediments further restrict diffusional fluxes of carbon and nutrients required for maintenance activities or growth. Further, the movement of bacteria would be expected to decrease as water films become thinner than 1 μm below −0.1 MPa (Kieft et al., 1993). Kieft et al. (1998) assessed the microbiological properties of two vertical vadose zone sediment profiles with contrasting porewater ages. Samples from profiles were composed of relatively uniform silty-fine sand sediments that primarily differed with age as a function of depth (chronosequence). A unified and steady decrease with depth (15 and 35 m) was observed in all microbiological parameters measured on samples collected at approximately 1–2 m intervals. In a similar study of arid vadose zone sediments in southeastern Washington, Ballkwill et al. (1998) analyzed samples from ~7 m depth profiles from five sites that differed with respect to recharge characteristics. The authors concluded that at the higher recharge sites, episodes of preferential flow (at or near saturating conditions) via vertical fissures allow for periodic recolonization of spore-forming bacteria (i.e., Streptomyces) from the surface. The result of more frequent recolonization is a more uniform depth profile of populations prone to transport that can also thrive under the subsurface conditions.

3.4. Deeper, low permeability saturated formations

In samples collected via mining from a massive clay at 224 m depth, very low activities and viable cells (aerobic and anaerobic), frequently below detection, have been reported (Boivin-Jahns et al., 1996). These authors conclude that the small pore throat dimensions (<0.1 µm) of this formation preclude sufficient fluxes to sustain microorganisms, despite the high carbon content (3%) of the clay. In a series of massive clay aquitard and surrounding clay-rich sediments, Lawrence et al. (2000) used a variety of culture-dependent and -independent methods to characterize the microbial communities to a total depth of 122 m. In the upper range of these sequences, numbers and diversity of physiological types were high in the less dense, more weathered saturated sediments overlying the aquitard. Progressing downward through the profile into a more undisturbed, massive clay aquitard, distributions of microorganisms became extremely patchy, with occasional peaks of uncertain trend. These authors conclude that the low permeability ($<10^{-3}$ mD) generally limits fluxes of nutrients, and only occasional localized areas (less than millimeter scale) possess favorable conditions.

Core segments from deep (~860, 2,000, and 2,100 m) Cretaceous and Tertiary sandstones interbedded with shales were analyzed for a variety of microbiological properties with considerable emphasis on anaerobic bacteria (Colwell et al., 1997). These strata had previously experienced sterilizing temperatures (>120°C) and present temperatures ranged from 43°C in the 860 m samples to 85°C in the 2,100 m samples. Primarily fermentors and iron-reducing bacteria were recovered from some of the 860 m samples, while the deeper samples yield little evidence of microorganisms. The positive enrichments were from 860 m deep sandstone cores that had relatively higher permeability (up to 1 mD), higher porosity (up to 12%), and higher average pore-throat dimensions (up to 1 µm) compared to adjacent shale cores with low permeabilities ($<10^{-3}$ mD), lower porosity (<5%), and smaller pore-throat diameters (<0.2 µm). According to this report, recolonization of these sandstones was hypothesized to occur via vertical fractures due to the relatively young groundwater age. The absence of organisms in the deeper cores containing more ancient groundwater was attributed to those formations being hydrologically closed since the period of sterilization.

3.5. Fractured, crystalline rock

Fractured igneous rocks such as granite and basalt have also been sampled intensively at a limited number of locations to characterize their microbiology. The origin of these habitats strongly contrasts with sedimentary systems as they were formed en mass, either as a plug emplaced from below, or as flow extruded onto the earth's surface. Igneous rocks exposed to the earth's surface will have encountered various periods of weathering, sometimes being covered with sedimentary deposition. Intrusives may undergo further changes via metamorphosis or secondary mineralization. Regardless, the temperature of the originally molten rock would have rendered them initially sterile.

The microbiological distributions within the ash-fall volcanic tuff of the Ranier Mesa (Nevada Test Site) near the proposed Yucca Mountain (Nevada) geologic repository for high-level nuclear waste have been studied in several papers. Detectable numbers of cells, biomass (PLFA) viable heterotrophs, and diversity of heterotrophs were observed in seven samples arrayed over a 50–450 m depth range from three locations (within 10 km) in unsaturated tuff (Haldeman and Amy, 1993). Microbiological properties strongly varied and samples separated by several meters were as different as those separated by >1 km, despite the geophysical similarity of the tuffs. There were no reported correlations between microbiological and physical properties (including moisture and temperature) of the rock samples. An intensive study of the heterogeneity of chemoheterophic bacteria was conducted on a three-dimensional array of samples collected from a 21 m^3 section of the same tuff (Haldeman et al., 1993). Again, a great deal of variation was observed in the numbers and populations of isolates among the rock samples, and there were no apparent correlations with rock physical properties. A thorough statistical analysis was performed on the relationships between biological and abiological properties of these same tuff samples (Russell et al., 1994). The magnitude of culturable counts exhibited a coefficient of variation of 276% (3D) and it was concluded that the chemoheterotrophic populations were distributed randomly thoughout the 21 m^3 tuff section.

Vertical transects (63–134 m) of microbiological properties were studied in two sets of saturated cores collected from the layered basalt flows of the Snake River Plain Aquifer in southeast Idaho (Lehman et al., 2004). Twenty-three basalt cores were analyzed from a well (TAN-37) that was influenced by an organic-rich, mixed waste plume and 19 cores from another well (TAN-33) that was minimally influenced by this plume. The basalt core samples were compositionally similar; however, the physical structure of the individual basalt cores at the centimeter–meter scale varied from extremely dense to highly vesicular and each core possessed varying occurrences of small fractures (generally infilled with clays or calcite). At the larger scale spanning meters to 10s of meters, the core samples varied with respect to their proximity to vertically zoned areas of high-lateral hydraulic conductivity. Although groundwater is contained in the basalt matrix, groundwater flow primarily occurs via formation level features such as large, open fractures and weathered, rubble zones existing at basalt flow interfaces. The basalt is considered a dual permeability geologic media wherein high formation permeabilities result from the large fractures and rubble zones, while the bulk of the basalt matrix possesses low permeability (ca. <1 mD). There may be relatively high pore space (~15%) represented by vesicles in the basalt matrix, yet these vesicles may be only minimally connected by small apertures, microfractures, or mineral grain boundaries that greatly restrict flow. It was originally hypothesized that the occurrence of the highly conductive fracture and rubble zones would control microbiological properties through the vertical transect. In basalt cores collected from the relatively pristine TAN-33 location where oxic conditions predominate, over a dozen assays (total cells, viable heterotrophs, enumerations of various aerobic and anaerobic physiologies, aerobic and anaerobic C-14 substrate mineralization) demonstrated little evidence of viable or active microorganisms (Lehman et al., 2004). For the limited number of assays that had a positive response, CVs ranged from 100% to 200% over the entire vertical transect (Lehman et al., 1999). At the suboxic, contaminated TAN-37 site, much higher values were obtained from all assays, however individual parameter magnitudes varied considerably over the vertical transect, often exceeding 200% CV. No significant positive correlations were observed between measurable microbiological properties and the occurrence of highly conductive flow zones (identified by multiple geophysical measures) in either well.

At the Mineral Park Mine (Arizona), 16 cores collected at regular intervals between 11 and 103 m deep within an igneous intrusive (biotite-quartz-monzonite porphyry) were subjected to intensive analysis of microbiological properties (Lehman et al., 2001). The rock itself had very low permeability and porosity, and groundwater was largely conducted via larger, open macrofractures. Groundwater in the upper 25 m had detectable dissolved oxygen (1–3 mg L^{-1}) compared to the lower anoxic portion of the well. Cores often contained smaller fractures of various sizes which were heavily mineralized with quartz, pyrite, chalcopyrite, and molybdenite. No neutrophilic aerobic or anerobic chemoheterotrophic bacteria were cultured from any of the cores. Using solid media plating approaches, acidophilic chemolithotrophic bacteria were also universally absent, while acidophilic chemoheterotrophs were enumerated in about half the cores spanning the entire depth. No fracture surfaces or fracture infills were examined as these were presumed to be contaminated during sample collection. Liquid enrichments using various donors for iron- and sulfur-oxidizing bacteria were largely negative, except $FeSO_4$ enrichments on samples from the lower half of the vertical sequence. Considering the data in sum, there was little evidence of vertical stratification of microbial properties in this relatively uniform crystalline rock.

The microbiology of deep granitic aquifers in Finland and Sweden have been the focus of numerous studies (Pedersen, 1997). This work has been primarily conducted on anoxic groundwater collected from boreholes and excavations or on biofilms from substrata incubated in the flowing groundwater. Total cells range from 10^3 to 10^7 mL^{-1} and a variety of physiologies have been documented including facultative anaerobic chemoheterotrophs, DIRB, SRB, autotrophic and heterotrophic acetogens, and methanogens. Analysis of 16 groundwater samples collected from a 65 to 350 m range of depths from multiple wells showed total cells did not decrease with depth. Multivariate statistical analyses showed that the distributions of physiologically distinct populations were largely independent of a range of groundwater geochemistries (Haveman and Pedersen, 2002). One exception is that DIRB and SRB were not cultured from deeper, older, more saline groundwater. Anaerobic physiologically distinct populations were enumerated from groundwaters spanning 200–950 m depth across four sites (Haveman et al., 1999). Total cells did not decrease with depth. This study concluded that the nature of fracture infills in the vicinity of the groundwaters predicted the relative population distributions. Specifically SRB numbers were highest where iron sulfide minerals filled fractures and DIRB were more abundant where fractures were filled with iron hydroxides. In contrast, groundwater geochemistry did not predict these anaerobic populations. In a previous study of granitic groundwaters, it was found that groundwater TOC correlated with total cell numbers (Pedersen and Ekendahl, 1990). A similar correlation was observed between TOC and total cell numbers in groundwaters collected from up to 105 m deep from sedimentary formations in Gabon, Africa (Pedersen et al., 1996).

3.6. Hot spots – vertical gradients stimulated from below

The possibility exists for deep subsurface communities to exist independent of photosynthetically fixed carbon. It has been proposed that autotrophic microorganisms form the base of these ecosystems and that they are fueled by either hydrogen generated by low-temperature water–rock interactions (Stevens and McKinley, 1995) or by H_2 and CO_2 generated from thermal processes in deeper, hotter regions in the crust or mantle (Chapelle et al., 2002; Pedersen, 2000). Alternatively, inverted vertical gradients of microorganisms may be sustained by thermal decomposition of buried, photosynthetically fixed carbon that produces hydrogen, methane, and short chain hydrocarbons. In either case, microbial populations would be distributed according to the presence of these abiotic reactions that supply the electron donors and the availability of CO_2.

4. HORIZONTALLY DISTRIBUTED MICROBIAL PATTERNS IN THE SUBSURFACE

There are few studies that report on horizontally distributed samples of subsurface sediments or rocks, primarily due to the expense and effort in coring at multiple locations. In order to interpret lateral continuity in microbiological properties, not only would depth have to be comparative, but also relative sample position within a given geologic formation. Shallow unconsolidated sediments that are accessible by push probe technology could be sampled at reasonable expense and effort, but there are few instances where such data has been reported. These few exceptions concerning samples within a single formation will be discussed below. Several studies report on samples that progress laterally from a contaminant point source that strongly influences and structures the subsurface communities. Other studies have examined groundwater microbiology in association with spatial distributed differences in groundwater geochemistry, frequently due to the occurrence of contamination. Some of these will be included because of the lack of appropriate comparisons for core samples.

4.1. Horizontally distributed patterns within a sedimentary unit

Information on horizontal distributions can be gained from study of cores collected from different locations within laterally contiguous formations. Mineralization patterns (aerobic C-14 substrates) (Hicks and Fredrickson, 1989) and cell growth (acetate and thymidine incorporation) (Phelps et al., 1989) according to sediment textural characteristics were observed within vertical transects through Tertiary and Cretaceous formations cored in the southeastern coastal plain. These same mineralization patterns were observed in all three boreholes separated by distances up to 16 km. The distribution of physiological distinct chemoheterotrophs isolated showed less variability in samples collected from the same formation in different boreholes than from samples collected at different depths or formations within a single borehole (Balkwill et al., 1989). These results indicate that where formations exhibit sharp contrast, there is a degree of lateral continuity in these associated microbiological properties with the formation. Although at this same site, even closely spaced vertical samples confined to a single formation showed substantial heterogeneity in their chemoheterotrophic populations (Fredrickson et al., 1991). Together, these observations indicate that the observed lateral continuity in formation microbial properties was a relative consequence of the coarse differences between formations.

In the study of fluvial sediments of varying texture from eight boreholes (20–30 m deep) within a ~6 km^2 (Germany) area, community structure (based on phenotypic testing of chemoheterotrophic isolates) varied greatly among the boreholes (Kolbel-Boelke et al., 1988). In nine cores collected within an 20 m^2 area from a shallow (<3 m), sandy aquifer (Borden Ontario), depth profiles of activity (ETS) and viable cells were consistent in five of the nine boreholes, but exhibited various departures from this pattern in the other four boreholes (Barbaro et al., 1994). The rather low numbers and activities observed in these sediments with relatively high dissolved organic carbon (3–29 mg L^{-1}) led investigators to propose that microbial activity may be regulated by the quality of available carbon.

In a study of excavated samples from three vertical transects each separated by 5.5 m in a shallow (2 m), sandy, vadose zone on Virginia's Eastern Shore, depth profiles of microbial properties were strongly consistent (Musslewhite et al., 2003). In this sedimentological uniform subsurface environment, little lateral variability in microbiological properties was thus observed, although most values were generally low. In nearby shallow (<7 m), sandy aquifer, Franklin et al. (2000), examined the distribution of microbial community profiles (DNA–RAPD) in groundwater samples with respect to groundwater geochemistry as it varied in distinct oxic and anoxic zones occurring within a 0.02 km^2 area. Groundwater communities from oxic and anoxic locations were easily distinguished, but geostatistical analysis demonstrated no relationship between well location and community profiles within each location (oxic or anoxic). The results suggest that scales of spatial correlation in groundwater communities in this relatively homogeneous sandy aquifer must be smaller than the distance between wells in this study (10 m).

4.2. Horizontally distributed patterns within a fractured rocks

Within uncontaminated, unsaturated volcanic tuffs of the Nevada Test Site, there was no evidence of lateral continuity in the microbiology of geophysically similar tuff samples within a 21 m^3 volume (Haldeman et al., 1993; Russell et al., 1994).

In basalt flows of the Snake River Plain Aquifer (southeastern Idaho), microbiological properties (inclusive of aerobes and anaerobes) were measured in cores from three boreholes arranged along the local hydrological gradient at varying distance from a point source of a mixed-waste contaminant plume (injection well). Mean property values for vertically distributed cores were plotted against distance from the injection well (Fig. 6-5). In this case, the eutrophication of an oligotrophic aquifer can easily be detected. While there is substantial variation in the magnitude of microbial properties in the lateral direction, this variation was associated with contaminant plume as all of the detected quantities had correlations >0.93 with TCE concentration (used as a relatively conservative tracer of the plume).

In a spatially broader study of Snake River Plain Aquifer microbial communities, groundwater samples were analyzed from 85 wells variably distributed over a roughly 1,500 km^2 section of the aquifer. Groundwater was pumped from the layered basalt aquifer from depths of 60–250 m (depth to water table increases in a southerly direction across the region). A total of 132 groundwater samples were analyzed in triplicate by community-level physiological profiling with Biolog plates (O'Connell and Lehman, unpublished data) in an effort to determine a baseline against which perturbations could be compared. The resulting responses were so variable that multivariate comparisons were not required, even when comparing samples collected from a single well at different dates. Using the number (of 95) carbon sources respired as the response, numbers ranged from 2 to 95 (mean = 56) positive carbon source utilization tests by the groundwater communities. The CV for the number of carbon sources (a very coarse measure) used by these 132 samples was 466%. Despite the physical heterogeneity of the basalt flows that comprise this aquifer, the mineralogical make-up of the basalt is very uniform and with very few exceptions, the groundwater is oligotrophic with DOC <1 mg L^{-1}. The results indicate a very unpredictable pattern in the distribution of groundwater microbial communities in the absence of a strong selective gradient (e.g., contaminant plume) or sharply contrasting lithological features.

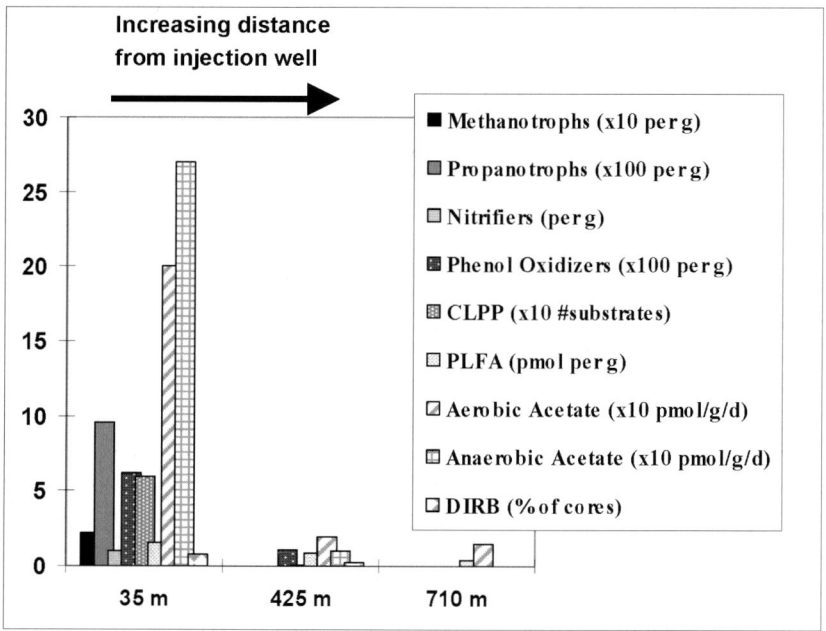

Figure 6-5. Distribution of mean values (all cores) of microbiological properties from three boreholes in the fractured basalt Snake River Plain Aquifer (SE Idaho) arranged with increasing distance (TAN-37, 35 m; TAN-33, 425 m, TAN-48, 710 m) from an injection well used for the disposal of sanitary and mixed industrial wastes. (Modified from Lehman et al., 1999.)

4.3. Horizontally distributed patterns along a groundwater flowpath

A trend that has been observed in the microbiology of aquifers relates to the distribution of predominant terminal electron accepting processes (TEAP) along a groundwater flow path. Redox potentials in systems with restricted oxygen resupply are driven progressively downwards with the oxidation of organic matter. Most subsurface systems have restricted access to atmospheric oxygen, either dependent on free gas diffusing though the overlying vadose zone or by dissolved gas in water infiltrating from the surface or from some horizontally located recharge area. Some of the oxygen is consumed by organic oxidation occurring during recharge/infiltration and the remainder is consumed within the aquifer during oxidation of dissolved carbon entrained in the groundwaters during its passage through soils. In some instances, detrital carbon deposited in deeper sediments may support respiratory activities, although this is likely to be at slow activity rates. Once oxygen is consumed, a predictable sequence of TEAPs will be observed based on the relative energy yield of these reactions and the availability of such acceptors either dissolved in the incoming water or present in the rocks (Fig. 6-6) (Lovley and Chapelle, 1995). Oxidized iron may be particularly important in the subsurface due to its abundance in solid forms (Lovley, 1991). This progressive sequence with distance from a recharge site, that is similar to that observed with depth in aquatic and marine sediments, will result in physiologically distinct populations being more numerous and active in horizontally distributed aquifer locations. Exceptions to this pattern occur when an aquifer is resupplied with oxygen from the atmosphere, or there is little dissolved carbon entering the aquifer in oxygenated recharge waters. The evolution of the organic carbon, both in quantity and quality, is suspected to govern overall activity rates and the distributions of populations associated with various TEAPs (Murphy et al., 1992). In aquifers that receive little infiltration from meteoric waters, the evolution of dissolved carbon and microbial activities will be associated with progressive age of the groundwater (Murphy et al., 1992). Unfortunately, there are few data concerning the composition or quality of organic carbon in aquifers and relationships with microorganisms.

In samples collected at varying distance from a contaminant point source, a continuum of elevated biomass can be observed and/or a continuum of predominant physiologies consistent with progressive changes in the local redox state (Lovley, 1991). The resulting distributions of biomass are not unlike the increase of biomass in a stream or lake near a sewage outfall and the sequence of predominant TEAP along the flow path is the reverse of uncontaminated aquifers. In this analogous case, it is the overall BOD that decreases with distance from the source. Once the added carbon is consumed, the aquifer microbial communities will resume the pattern described in the previous paragraph. Contaminated sites tend to receive the most study, and this pattern has been commonly observed in groundwater samples.

One of the most thoroughly documented examples of sediment-associated microbial community changes associated with distance from a carbon source is a study of core samples collected from a shallow, sandy aquifer along the flow path at increasing distances from a landfill (Denmark) (Ludvigsen et al., 1999). The top of the aquifer is about 3 m deep and the aquifer is composed of a heterogeneous assortment of Quaternary- to Tertiary-aged sediments of varying texture and origin. Sediment samples from depth profiles of nine boreholes arranged along a horizontal transect, 0 to 305 m from the landfill. Biomass (PLFA) generally decreased with distance

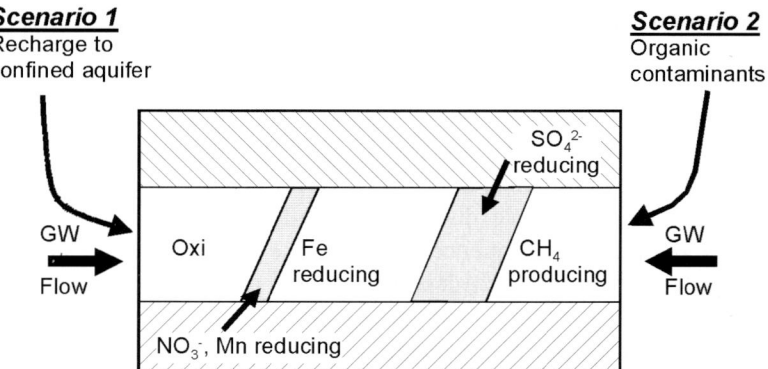

Figure 6-6. Distribution of predominant TEAPs: Scenario 1 – oxidation of groundwater organic matter in a confined or semi-confined aquifer results in decreasing redox potential with distance from recharge; Scenario 2 – high concentrations of organic contaminants drive redox potential to low levels that subsequently increase with distance from the contaminant source as the introduced organic matter is oxidized and dispersed to ambient levels. (Adapted from Lovely and Chapelle, 1995.)

from the landfill while total cells (AODC) show little discernable pattern. With increasing distance from the landfill, numbers (MPN and PLFA) of the following groups reached their peak and then declined in spatially distributed sequences: methanogens, SRB, DIRB, Mn-reducers, and nitrate-reducers.

4.4. Other examples of spatial distributions at contaminated sites

In addition to the study of Franklin et al. (2000) described above, several other studies have examined subsurface microbial communities or activities using a collection of spatially distributed wells. Although these studies do not examine core samples, they are included because they specifically illustrate horizontal distributions of subsurface microbial properties. Adrian et al. (1994) examined methane production rate from well clusters at two geochemically distinct locations impacted by landfill leachate in an otherwise homogeneous, shallow (<2.5 m), sandy aquifer. Over a 9-month period, the headspace methane concentrations at the two sites were measured in sixteen wells arranged in a grid covering 35 m². Methane production rates over the entire period at both sites were extremely variable, with CVs exceeding 340%. When individual seasons were examined, CVs of methane production in each of the two sites still ranged between 200% and 300%. It was concluded that the magnitudes of methane production in the aquifer were log-normally distributed.

The association of the geochemical and microbiological characteristics of groundwater samples has rarely been quantified. Groundwater samples from multiple depth intervals from a series of wells located upstream and at regular intervals downstream from a landfill were used to relate microbial community structure (16S rDNA DGGE profile) to the groundwater geochemistry (Roling et al., 2001). Overall, large variations in very complex community profiles among the groundwater samples were observed. Multivariate statistical analysis of community profiles was able to distinguish groundwater samples from polluted and unpolluted zones and to correlate these communities and their members with contaminant concentrations. This data set was used in subsequent geostatistical analyses (Mouser et al., 2005) that found horizontal correlation distances of 40–50 m in the community profiles. It must be emphasized, that these groundwater samples were taken from a location that was highly structured (horizontally) by the landfill leachate. Nonetheless, the pattern could not have been quantified without the regular intervals of access, horizontal and vertical, that the multiple wells provided in a shallow (<8 m), unconsolidated sediments. Some sediment samples were also studied at this location, but the geochemistry of the groundwater was not reflected in the sediment communities. In another example linking microbiology to geochemistry in wells that were not spatially structured, but varied with respect to the occurrence of mixed contaminants, distributions of genes involved in nitrate and sulfate reduction were studied in groundwater samples (Palumbo et al., 2004). Artificial neural network models were used to establish relationships between the occurrence of two of five groups of dsrAB-related genes with uranium and sulfate concentrations and the occurrences of the other three groups with pH, nickel, and organic carbon concentrations.

4.5. Horizontally distributed patterns across regions

At most study locations, there are insufficient spatially arranged core samples to determine trends, let alone quantify these trends. However, the distribution of the study sites themselves may allow additional trends to be identified. Superimposed on the mosaic of historically evolved and complex geologic formations that comprise the subsurface environments will be patterns associated with modern-day climates. While climate may influence subsurface microorganisms, its influence will be largely a function of annual means (i.e., temperature and precipitation), rather than the diurnal and seasonal fluctuations that may strongly influence terrestrial surface microbes. Regional variations in climate will influence the amount and type of surface primary production, land use patterns, and water. So, to the extent that these regional trends may result in difference in the delivery of either carbon or water, it would be expected that subsurface communities will respond. An example might be the deep vadose zone of arid regions, where microbial population seem to be considerably less numerous, less diverse, and less active than their counterparts in more humid environments.

5. LIMITATIONS TO UNDERSTANDING TERRESTRIAL SUBSURFACE MICROBIAL DISTRIBUTIONS

There are a number of considerations that are either unique to subsurface environments or elevated in comparison to surface environments. These considerations may significantly alter conclusions regarding subsurface microorganisms.

5.1. Sampling constraints

Spatial distributions in subsurface environments have generally resisted quantification because subsurface environments are: (i) highly complex, (ii) not directly observable, (iii) technically and economically difficult to sample, and (iv) not intensively studied over longer periods (>10 year). As a result, few (if any) subsurface environments have been representatively sampled. The overall density of samples is often too low for geostatistical comparisons. Sample replication is not usually achieved, and estimates of variance for replicate samples attempting to characterize a limited volume are rare. At many locations, only a single borehole is investigated, so often $n = 1$. The scale of variability is invariably smaller than the sampling scale, which is constrained by the required material for analyses. In a study to determine the representative elemental volume (REV) for culturable heterotrophs in vadose zone sediments, investigators found that large REVs would be required to account for the extreme patchiness of microbial distributions observed in these samples (and other subsurface samples) (Stevens and Holbert, 1995). Further, a good estimate of REV was not achieved for these sediments due to density-dependent growth during dilution plating, a phenomenon that was observed in other subsurface studies. Even simple correlations between biotic and abiotic variables are difficult as measurements often cannot be made on the same material, especially if a large REV is necessary for microbiological analyses. Given multiple interacting factors controlling microbial distributions, it is certain that the distributions of these factors will be on differing scales. The exception is when there is a strong overriding selective factor such as organic contamination.

5.2. Methodological constraints

In addition to the patchy distributions of subsurface microorganism, several other characteristics of microorganisms present exaggerated difficulties when subsurface samples are analyzed.

1. The percentage of total cells that are culturable is generally orders of magnitude lower than surface environments.
    a) Enumeration detection limits are confronted.
    b) A small percentage of the community is evaluated.
2. Cells are frequently small and often indistinguishable from sediment particles by direct observation.
3. Activities are generally very low.
    a) Assay detection limits are confronted.
    b) Longer assay durations encourage selection.
    c) Laboratory conditions sharply contrast with *in situ* conditions.
4. *In situ* activities are technically difficult to measure.

The high disparity between laboratory estimates of subsurface microbial activities and their apparent *in situ* activities has been thoroughly documented (Chapelle and Lovley, 1990; Phelps et al., 1994).

6. TENDENCIES AND TRENDS IN TERRESTRIAL SUBSURFACE MICROBIAL DISTRIBUTIONS

Like other environments, microorganisms exhibit heterogeneous distributions at a number of different scales. At the micron scale, microorganisms experience the local geochemical gradients that regulate their metabolism, although this scale remains a challenge. Minerals are often distributed in rocks at the micron and tens of microns scale. Some work has revealed these fine-scale distributions of microorganisms with respect to mineral phases and their boundaries (Edwards et al., 1998; Lawrence et al., 1997; Taunton et al., 2000). At the sediment particle or aggregate scale (microns to millimeters), work has shown substantial gradients in populations, both in number and type (Albrechtsen, 1994; Murphy et al., 1992). A few studies have specifically addressed variations of subsurface microbial properties at the centimeter scale, particularly with respect to REV (Brockman and Murray, 1997; Stevens and Holbert, 1995). Analyses of frequency distributions of subsurface microbial properties (their magnitude) suggest that they are log-normally distributed (Brockman and Murray, 1997; Zhang et al., 1997), like many other environments. Large CVs for microbiological parameters measured on spatially distributed samples are consistent with log-normally distributions. Conventionally, sample sizes are approximately 100 g (~35 cm^3), which may be a subsample from a larger (~1–2 kg homogenized sample). Yet, the frequency distributions must depend on the sample size in relation to the patch size, which is undetermined for most environments. Subsurface microbial properties clearly vary at the meter scale, both within and between formations, and all higher spatial scales. Considering that the spatial variations in populations and the magnitude of their activities may be as great over 1 mm as they are in 1 km, the scale of interest must be determined apriori, and then the investigation can turn to detecting the scales of the controlling factors. Few reports have even attempted geostatistical analyses of subsurface microbial distributions. Unpublished geostatistical analyses of the centimeter-spaced samples from relatively homogenous, bedded vadose sediments indicated a vertical variogram range of 15 cm and a horizontal range of 250 cm at one site and a vertical range of 1 m at another site for aerobic mineralization (Brockman and Murray, 1997). The other two studies analyzed groundwater samples. The first (Franklin et al., 2000) could not detect a correlated lateral range of microbial community composition at the minimum distance sampled (10 m). The second (Mouser et al., 2005) found a lateral range of 40–50 m at a site strongly structured by landfill contamination.

6.1. Vadose versus saturated zones

The trends identified in Figs. 6-1 and 6-2 appear to be most reproducible – increasing depth beyond some point is not accompanied by decreases in microbial biomass. The initial decrease in biomass from the subsurface is strongly related to changes in TOC. In the subsoil region, depth appears to be associated with a greater predominance of gram-positive bacteria and Actinomycetes. Some observations indicate that cell-specific activities may be higher in subsoils compared to surface soils (Blume et al., 2002; Federle et al., 1986; Fierer et al., 2003), suggesting these organisms may be opportunistic, poised for a fresh supply of carbon. In deeper vadose zones, particularly in arid regions, water seems to be the controlling factor on populations and activities. Distributions are frequently extremely patchy and a low fraction of the total biomass is culturable. The very low activities respond to water as it delivers a flux of nutrients (Fredrickson et al., 1993;

Kieft et al., 1993). Even under the driest subsurface conditions, the lack of water-induced nutrient fluxes will not ultimately limit survival of all bacteria, especially when considering the viability of cells that have been isolated from nutrient fluxes in amber (Cano and Borucki, 1995), fluid inclusions (Vreeland et al., 2000), permafrost (Shi et al., 1997), and highly impermeable formations (Fredrickson et al., 1995; Lawrence et al., 2000; Onstott et al., 1998) for millions of years.

Biomass generally increases when sediments are saturated and it has been frequently reported that gram-negative bacteria become increasingly predominant. The more uniform distribution of nutrients may produce a more dominant community structure due to increased competitive opportunities. Redox potential in the saturated zone will be a function of its isolation from atmosphere, the amount of carbon oxidized near the surface during infiltration or recharge, the amount of carbon that reaches the aquifer (including anthropogenic), and the distance from the recharge source. The ambient redox potential should predict the predominance of physiologically distinct populations and their activities (Fig. 6-6). There are two primary habitats in the saturated, groundwater and sediments (or rock), and different microbes may inhabit these environments and respond to perturbations in different ways.

6.2. Shallow versus deep saturated sediments

The microbiological activity of shallow aquifers is more likely to be controlled by quantity of carbon reaching the aquifer from the surface. In comparison, activities in deeper sedimentary aquifers are more likely to be related to detrital carbon sources. Shallow aquifers are usually unconsolidated sediments and fluxes of electron acceptors should be sufficient, although ambient redox conditions will modulate dominant populations and activities. In deeper aquifers, groundwater flow is generally slower and the flux of electron acceptors may become as important as the flux of carbon in controlling overall levels of activity. Deeper formations have lower porosity and smaller pore-throat dimensions that restrict movement of cells and the diffusion of electron donors and acceptors. Some evidence suggests that bacteria cells in shallow aquifers (Balkwill and Ghiorse, 1985) or deeper transmissive formation (Brockman and Murray, 1997; Sinclair and Ghiorse, 1989) may have relatively high culturability and potential for activity, while most evidence indicates very low culturability in deep, lower permeability and unsaturated formations (Brockman and Murray, 1997; Onstott et al., 1999).

It appears that in many deeper subsurface environments, it is the flux of nutrients (especially electron donors and acceptors) that limits and distributes microbial populations and their activities. Therefore, the factors that control the hydrology seem paramount. Carbon must be present either associated with sediments at deposition or remaining in the groundwater after its travel from the recharge or infiltration site. In deeper areas with older groundwater, the amount and quality of carbon supplied by groundwater fluxes may be insufficient to support much activity. If detrital carbon in sediments is sustaining microbial activity, it must be physically accessible to cells via sufficiently size pore throats (>0.2 µm) or there must be sufficient water in the small pores to sustain diffusional fluxes to areas where cells exist. Even given the existence of sufficient carbon, fluxes of electron acceptors may limit respiratory activity to higher permeability regions where groundwater provides a supply of electron acceptors (Fredrickson et al., 1995). Ultimately survival will be limited by temperature in formations deeper than 4–6 km.

6.3. Vertical versus horizontal distributions

Microbial biomass and activity initially declines with depth, then remains fairly uniform or slightly decreases over increasing depth (Fig. 6-3). Many potential controlling variables (T, TOC, porosity) covary with depth over varying intervals. The occurrence of different formations varies according to changing depth intervals within a single site and between sites. Subsurface microbial properties in a single borehole have been observed to vary considerably within formations and between formations with CVs of measured properties often exceeding 200%. The microbiology of sedimentary formations often correlates with textural characteristics. The number of cells in fine-grained sediments may be equal or greater than in coarse-grained sediments, but the viable cells and activity are usually higher in the coarse-grained materials. The amount of carbon associated with sediments and the relatively permeability of these formations may alter this relationship (Kieft et al., 1995). Where high-carbon, fine-grained sediments abut lower carbon coarse-grained sediments, areas of higher activity have been observed near the interface (Fig. 6-4). The most common explanation is that organic carbon or fermentation products thereof diffuse from the fine-grained sediments into the coarser-grained sediments where higher fluxes of electron acceptors are present.

Data on horizontal distributions of sediment- (or rock-) associated microorganisms are limited. Reports on US southeast coastal plain sediments indicate a degree of lateral continuity in microbiological properties within a formation (Balkwill et al., 1989; Phelps et al., 1989), although this continuity may be relative to high contrasts observed between formations at this location. Studies of shallow, uniform, unsaturated sands indicate a high degree of lateral continuity, although little variation was observed vertically at this site, and the magnitude of measured properties was very low overall (Musslewhite et al., 2003). In other instances discussed above, there appears to be high variation (high CVs) in microbial properties measured at laterally separated locations. High horizontal and vertical variation in microbial properties observed at most sites shed doubt on the reliability of using a single control well or borehole for comparative analyses. In locations where there is a strong selective pressure such as organic contamination, the ambient community may react strongly and predictably (higher biomass, TEAP sequence) to the structure imposed on the system. The magnitude of the response of the microbial community under conditions of organic enrichment will make ambient variations appear very small in comparison.

6.4. Sedimentary versus crystalline formations

The microbiology of fractured, crystalline rock environments contrasts strongly with that of sedimentary environments. Fractured, crystalline rock will not possess any detrital carbon and the groundwater in those environments that have been studied is usually low in dissolved carbon. The seemingly amorphous rock often defies attempts to find any horizontal or vertical trends in microbiological properties. Populations and activities are often low or undetectable in the low permeability rock matrix. Fluxes of nutrients will control populations and activities in fractured rock; however, fracture surfaces have not been well-studied because they are presumed contaminated during the sampling process. Groundwater samples often produce much higher populations and activities than samples of the rock. Given the low organic carbon in igneous rock aquifers, their frequent proximity to thermally active regions, and the occurrence of reduced inorganic energy sources (e.g., sulfides), there may be a higher probability of autotrophic populations and processes dominating these regions than many sedimentary formations. This has been observed in deep granite (Pedersen, 2000) and basalt aquifers (Chapelle et al., 2002; Stevens and McKinley, 1995).

REFERENCES

Adrian, N. R., J. A. Robinson, and J. M. Suflita, 1994, Spatial variability in biodegradation rates as evidenced by methane production from an aquifer, *Appl. Environ. Microbiol.* **60:**3632–3639.

Albrechtsen, H. -J., 1994, Distribution of bacteria, estimated by a viable count method, and heterotrophic activity in different size fractions of aquifer sediment, *Geomicrobiol. J.* **12**:253–264.

Albrechtsen, H. -J., and A. Winding, 1992, Microbial biomass and activity in subsurface sediments from Vejen, Denmark, *Microb. Ecol.* **23**:303–317.

Balkwill, D. L., 1989, Numbers, diversity, and morphological characteristics of aerobic, chemoheterotrophic bacteria in deep subsurface sediments from a site in South Carolina, *Geomicrobiol. J.* **7**:33–52.

Balkwill, D. L., and W. C. Ghiorse, 1985, Characterization of subsurface bacteria associated with two shallow aquifers in Oklahoma, *Appl. Environ. Microbiol.* **50**:580–588.

Balkwill, D. L., J. K. Fredrickson, and J. M. Thomas, 1989, Vertical and horizontal variations in the physiological diversity of the aerobic chemoheterotrophic bacterial microflora in deep southeast coastal plain subsurface sediments, *Appl. Environ. Microbiol.* **55**:1058–1065.

Balkwill, D. L., E. M. Murphy, D. M. Fair, D. B. Ringelberg, and D. C. White, 1998, Microbial communities in high and low recharge environments: implications for microbial transport in the vadose zone, *Microb. Ecol.* **35**:156–171.

Barbaro, S. E., H. -J. Albrechtsen, B. K. Jensen, C. I. Mayfield, and J. F. Barker, 1994, Relationships between aquifer properties and microbial populations in the Borden aquifer, *Geomicrobiol. J.* **12**:203–219.

Bastin, E. S., 1926, The presence of sulphate-reducing bacteria in oil-field waters, *Science* **63**:21–24.

Bekins, B. A., E. M. Godsy, and E. Warren, 1999, Distribution of microbial physiologic types in an aquifer contaminated by crude oil, *Microb. Ecol.* **37**:263–275.

Beloin, R. M., J. L. Sinclair, and W. C. Ghiorse, 1988, Distribution and activity of microorganisms in subsurface sediments of a pristine study site in Oklahoma, *Microb. Ecol.* **16**:85–97.

Blume, E., M. Bischoff, J. M. Reichert, T. Moorman, A. Konopka, and R. F. Turco, Jr., 2002, Surface and subsurface microbial biomass, community structure and metabolic activity as a function of soil depth and season, *Appl. Soil Ecol.* **20**:171–181.

Boivin-Jahns, V., R. Ruimy, A. Bianchi, S. Daumas, and R. Christen, 1996, Bacterial diversity in a deep-subsurface clay environment, *Appl. Environ. Microbiol.* **62**:3405–3412.

Bone, T. L., and D. L. Balkwill, 1988, Morphological and cultural comparison of microorganisms in surface soil and subsurface sediments at a pristine study site in Oklahoma, *Microb. Ecol.* **16**:49–64.

Boone, D., Y. Liu, Z. Zhao, D. Balkwill, G. Drake, T. Stevens, and H. Aldrich, 1995, *Bacillus infernus* sp. nov., an Fe(III)- and Mn(IV)-reducing anaerobe from the deep terrestrial subsurface, *Int. J. Syst. Bacteriol.* **45**:441–448.

Brockman, F. J., and C. J. Murray, 1997, Microbiological heterogeneity in the terrestrial subsurface and approaches for its description, in: *The Microbiology of the Terrestrial Deep Subsurface*, P. S. Amy and D. L. Haldeman, eds., CRC Press, Boca Raton, pp. 75–102.

Brockman, F. J., and C. J. Murray, 1997, Subsurface microbiological heterogeneity: current knowledge, descriptive approaches and applications, *FEMS Microbiol. Rev.* **20**:231–247.

Cano, R. J., and M. K. Borucki, 1995, Revival and identification of bacterial spores in 25- to 40-million-year-Old Dominican amber, *Science* **268**:1060–1063.

Chapelle, F. H., and D. R. Lovley, 1990, Rates of microbial metabolism in deep coastal plain aquifers, *Appl. Environ. Microbiol.* **56**:1865–1874.

Chapelle, F. H., J. Zelibor, J.L., D. J. Grimes, and L. L. Knobel, 1987, Bacteria in deep coastal plain sediments of Maryland: a possible source of CO_2 to groundwater, *Water Resour. Res.* **23**:1625–1632.
Chapelle, F. H., K. O'Neill, P. M. Bradley, B. A. Methe, S. A. Ciufo, L. L. Knobel, and D. R. Lovley, 2002, A hydrogen-based subsurface microbial community dominated by methanogens, *Nature* **415**:312–315.
Colwell, F. S., 1989, Microbiological comparison of surface soil and unsaturated subsurface soil from a semiarid high desert, *Appl. Environ. Microbiol.* **55**:2420–2423.
Colwell, F. S., T. C. Onstott, M. E. Delwiche, D. Chandler, J. K. Fredrickson, Q. -J. Yao, J. P. McKinley, D. R. Boone, R. P. Griffiths, T. J. Phelps, D. B. Ringelberg, D. C. White, L. Lafreniere, D. L. Balkwill, R. M. Lehman, J. Konisky, and P. E. Long, 1997, Microorganisms from deep, high-temperature sandstones: constraints on microbial colonization, *FEMS Microbiol. Rev.* **20**:425–435.
Davis, J. P., 1967, *Petroleum Microbiology*, Elsevier, New York.
Dodds, W. K., M. K. Banks, C. S. Clenan, C. W. Rice, D. Sotomayor, E. A. Strauss, and W. Yu, 1996, Biological properties of soil and subsurface sediments under abandoned pasture and cropland, *Soil Biol. Biochem.* **28**:837–846.
Edwards, K. J., M. O. Schrenk, R. J. Hamers, and J. F. Banfield, 1998, Microbial oxidation of pyrite: experiments using microorganisms from an extreme acidic environment, *Am. Mineralogist.* **83**:1444–1453.
Ekelund, F., R. Ronn, and S. Christensen, 2001, Distribution with depth of protozoa, bacteria and fungi in soil profiles from three Danish forest sites, *Soil Biol. Biochem.* **33**:475–481.
Federle, T., D. Dobbins, J. Thornton-Manning, and D. Jones, 1986, Microbial biomass, activity, and community structure in subsurface soils, *Ground Water* **24**:365–374.
Fierer, N., J. P. Schimel, and P. A. Holden, 2003, Variations in microbial community composition through two soil depth profiles, *Soil Biol. Biochem.* **35**:167–176.
Fliermans, C., and D. Balkwill, 1989, Microbial life in deep terrestrial subsurfaces, *BioScience* **39**:370–376.
Francis, A. J., J. M. Slater, and C. J. Dodge, 1989, Denitrification in deep subsurface sediments, *Geomicrobiology* **7**:103–116.
Franklin, R. B., D. R. Taylor, and A. L. Mills, 2000, The distribution of microbial communities in anaerobic and aerobic zones of a shallow coastal plain aquifer, *Microb. Ecol.* **38**:377–386.
Fredrickson, J. K., R. J. Hicks, S. W. Li, and F. J. Brockman, 1988, Plasmid incidence from deep subsurface sediments, *Appl. Environ. Microbiol.* **54**:2916–2923.
Fredrickson, J. K., T. R. Garland, R. J. Hicks, J. M. Thomas, S. W. Li, and K. M. McFadden, 1989, Lithotrophic and heterotrophic bacteria in deep subsurface sediments and their relation to sediment properties, *Geomicrobiology* **7**:53–66.
Fredrickson, J. K., D. L. Balkwill, J. M. Zachara, S. W. Li, F. J. Brockman, and M. A. Simmons, 1991, Physiological diversity and distributions of heterotrophic bacteria in deep Cretaceous sediments of the Atlantic coastal plain, *Appl. Environ. Microbiol.* **57**:402–411.
Fredrickson, J. K., F. J. Brockman, B. N. Bjornstad, P. E. Long, S. W. Li, J. P. McKinley, J. V. Wright, J. L. Conca, T. L. Kieft, and D. L. Balkwill, 1993, Microbiological characteristics of pristine and contaminated deep vadose sediments from an arid region, *Geomicrobiol. J.* **11**:95–107.
Fredrickson, J. K., J. P. McKinley, S. A. Nierzwicki-Bauer, D. C. White, D. B. Ringelberg, S. A. Rawson, S. Li, F. J. Brockman, and B. N. Bjornstad, 1995, Microbial community structure and biogeochemistry of Miocene subsurface sediments: implications for long-term microbial survival, *Molec. Ecol.* **4**:619–626.

Fredrickson, J. K., J. P. McKinley, B. N. Bjornstad, P. E. Long, D. B. Ringelberg, D. C. White, L. R. Krumholz, J. M. Suflita, F. S. Colwell, R. M. Lehman, T. J. Phelps, and T. C. Onstott, 1997, Pore-size constraints on the activity and survival of subsurface bacteria in a late Cretaceous shale–sandstone sequence, northwestern New Mexico., *Geomicrobiol. J.* **14**:183–202.

Fritze, H., J. Pietikainen, and T. Pennanen, 2000, Distribution of microbial biomass and phospholipid fatty acids in Podzol profiles under coniferous forest, *Eur. J. Soil Sci.* **51**:565–573.

Ghiorse, W. C., and J. T. Wilson, 1988, Microbial ecology of the terrestrial subsurface, *Adv. Appl. Microbiol.* **33**:107–173.

Godsy, E. M., D. F. Goerlitz, and D. Grbic Galic, 1992, Methanogenic biodegradation of creosote contaminants in natural and simulated groundwater ecosystems, *Ground Water* **30**:232–242.

Gold, T., 1992, The deep, hot biosphere, *Proc. Natl. Acad. Sci. USA* **89**:6045–6049.

Haldeman, D. L., and P. S. Amy, 1993, Bacterial heterogeneity in deep subsurface tunnels at Rainier Mesa, Nevada Test Site, *Microb. Ecol.* **25**:183–194.

Haldeman, D. L., P. S. Amy, D. B. Ringelberg, and D. C. White, 1993, Characterization of the microbiology within a 21 m^3 section of rock from the deep subsurface, *Microb. Ecol.* **26**:145–159.

Harvey, R. W., R. L. Smith, and L. George, 1984, Effect of organic contamination upon microbial distributions and heterotrophic uptake in a Cape Cod, Mass., aquifer, *Appl. Environ. Microbiol.* **48**:1197–1202.

Haveman, S. A., and K. Pedersen, 2002, Distribution of culturable microorganisms in Fennoscandian Shield groundwater, *FEMS Microbiol. Ecol.* **39**:129–137.

Haveman, S. A., K. Pedersen, and P. Ruotsalainen, 1999, Distribution and metabolic diversity of microorganisms in deep igneous rock aquifers of Finland, *Geomicrobiology* **16**:277–294.

Hersman, L. E., 1997, Subsurface microbiology: effects on the transport of radioactive waste in the vadose zone, in: *The Microbiology of the Terrestrial Deep Subsurface*, P. S. Amy and D. L. Haldeman, eds., CRC Press, Boca Raton, pp. 299–324.

Hicks, R. J., and J. K. Fredrickson, 1989, Aerobic metabolic potential of microbial populations indigenous to deep subsurface environments, *Geomicrobiology* **7**:67–77.

Hunt, C. B., 1986, *Surficial Deposits of the United States*. Van Nostrand Reinhold, New York.

Jimenez, L., 1990, Molecular analysis of deep-subsurface bacteria, *Appl. Environ. Microbiol.* **56**:2108–2113.

Jones, R. E., R. E. Beeman, and J. M. Suflita, 1989, Anaerobic metabolic processes in the deep terrestrial subsurface, *Geomicrobiology* **7**:117–130.

Keswick, B. H., 1984, Sources of groundwater pollution, in: *Groundwater Pollution Microbiology*, G. Bitton and C. P. Gerba, eds., Wiley-Interscience, New York, pp. 39–64.

Kieft, T. L., P. S. Amy, F. J. Brockman, J. K. Fredrickson, B. N. Bjornstad, and L. L. Rosacker, 1993, Microbial abundance and activities in relation to water potential in the vadose zones of arid and semiarid site, *Microb. Ecol.* **26**:59–78.

Kieft, T. L., J. K. Fredrickson, J. P. McKinley, B. N. Bjornstad, S. A. Rawson, T. J. Phelps, F. J. Brockman, and S. M. Pfiffner, 1995, Microbiological comparisons within and across contiguous lacustrine, paleosol, and fluvial subsurface sediments, *Appl. Environ. Microbiol.* **61**:749–757.

Kieft, T. L., E. M. Murphy, D. L. Haldeman, P. S. Amy, B. N. Bjornstad, E. V. McDonald, D. B. Ringelberg, D. C. White, J. Stair, R. P. Griffiths, T. S. Gsell, W. E. Holben, and D. R. Boone, 1998, Microbial transport, survival, and succession in a sequence of buried sediments, *Microb. Ecol.* **36**:336–348.

Kolbel-Boelke, J., E. -M. Anders, and A. Nehrkorn, 1988, Microbial communities in the saturated groundwater environment II: diversity of bacterial communities in a Pleistocene sand aquifer and their in vitro activities, *Microb. Ecol.* **16**:31–48.

Krumholz, L. R., J. P. McKinley, G. A. Ulrich, and J. M. Suflita, 1997, Confined subsurface microbial communities in Cretaceous rock, *Nature* **386**:64–66.

Krumholz, L. R., S. H. Harris, S. T. Tay, and J. M. Suflita, 1999, Characterization of two subsurface H_2-utilizing bacteria, *Desulfomicrobium hypogeium* sp. nov. and *Acetobacterium psammolithicum* sp. nov., and their ecological roles, *Appl. Environ. Microbiol.* **65**:2300-6.

Lawrence, J. R., Y. T. J. Kwong, and G. D. W. Swerhone, 1997, Colonization and weathering of natural sulfide mineral assemblages by *Thiobacillus ferroxidans*, *Can. J. Microbiol.* **43**:178–188.

Lawrence, J. R., M. J. Hendry, L. I. Wassenaar, J. J. Germida, G. M. Wolfaardt, N. Fortin, and C. W. Greer, 2000, Distribution and biogeochemical importance of bacterial populations in a thick clay-rich aquitard system, *Microb. Ecol.* **40**:273–291.

Lehman, R. M., F. S. Colwell, R. Smith, M. E. Delwiche, S. P. O'Connell, J. K. Fredrickson, F. J. Brockman, A. -L. Reysenbach, T. L. Kieft, T. J. Phelps, D. B. Ringelberg, and D. C. White, 1999, Longitudinal and vertical variations in the microbial ecology of a fractured basalt aquifer with respect to a contaminant plum. Presented at the International Symposium on Subsurface Microbiology – August 1999, Vail, CO.

Lehman, R. M., F. F. Roberto, D. Earley, D. F. Bruhn, S. E. Brink, S. P. O'Connell, M. E. Delwiche, and F. S. Colwell, 2001, Attached and unattached bacterial communities in a 120-meter corehole in an acidic, crystalline rock aquifer, *Appl. Environ. Microbiol.* **67**:2095–2106.

Lehman, R. M., S. P. O'Connell, A. Banta, J. K. Fredrickson, A. -L. Reysenbach, T. L. Kieft, and F. S. Colwell, 2004, Microbiological comparison of core and groundwater samples collected from a fractured basalt aquifer with that of dialysis chambers incubated *in situ*, *Geomicrobiol. J.* **21**:169–182.

Lovley, D. R., 1991, Dissimilatory Fe(III) and Mn(IV) reduction, *Microbiol. Rev.* **55**:259–287.

Lovley, D. R., and F. H. Chapelle, 1995, Deep subsurface microbial processes, *Rev. Geophys.* **33**:365–381.

Ludvigsen, L., H. Albrechtsen, D. B. Ringelberg, F. Ekelund, and T. H. Christensen, 1999, Distribution and composition of microbial populations in a landfill leachate contaminated aquifer (Grindsted, Denmark), *Microb. Ecol.* **37**:197–207.

McMahon, P. B., and F. H. Chapelle, 1991, Microbial production of organic acids in aquitard sediments and its role in aquifer geochemistry, *Nature* **349**:233–235.

McKinley, J. P., T. O. Stevens, J. K. Fredrickson, J. M. Zachara, F. S. Colwell, K. B. Wagnon, S. C. Smith, S. A. Rawson, and B. N. Bjornstad, 1997, Biogeochemistry of anaerobic lacustrine and paleosol sediments within an aerobic unconfined aquifer, *Geomicrobiol. J.* **14**:23–39.

Madigan, M. T., J. M. Martinko, and J. Parker, 1997, *Brock Biology of Microorganisms*, 8th Edition. Prentice-Hall, Upper Saddle River, NJ.

Martino, D. P., E. L. Grossman, G. A. Ulrich, K. C. Burger, J. L. Schlichenmeyer, J. M. Suflita, and J. W. Ammerman, 1998, Microbial abundance and activity in a low-conductivity aquifer system in East-Central Texas, *Microb. Ecol.* **35**:224–234.

Mouser, P. J., D. M. Rizzo, W. F. M. Roling, and B. M. van Breukelen, 2005, A multivariate statistical approach to spatial representation of groundwater contamination using hydrochemistry and microbial community profiles, *Environ. Sci. Technol.* **39:**7551–7559.

Murphy, E. M., J. A. Schramke, J. K. Fredrickson, H. W. Bledsoe, A. J. Francis, D. S. Sklarew, and J. C. Linehan, 1992, The influence of microbial activity and sedimentary organic carbon on the isotope geochemistry of the Middendorf aquifer, *Water Resour. Res.* **28:**723–740.

Musslewhite, C. L., M. J. McInerney, H. Dong, T. C. Onstott, M. Green-Blum, D. Swift, S. J. MacNaughton, D. C. White, C. J. Murray, and Y. -J. Chien, 2003, The factors controlling microbial distribution and activity in the shallow subsurface, *Geomicrobiol. J.* **20:**245–261.

Onstott, T. C., T. J. Phelps, F. S. Colwell, D. B. Ringelberg, D. C. White, D. R. Boone, J. P. McKinley, T. O. Stevens, P. E. Long, D. L. Balkwill, W. T. Griffin, and T. L. Kieft, 1998, Observations pertaining to the origin and ecology of microorganisms recovered from the deep subsurface of Taylorsville Basin, Virginia, *Geomicrobiology* **15:**353–385

Onstott, T. C., T. J. Phelps, T. Kieft, F. S. Colwell, D. L. Balkwill, J. K. Fredrickson, and F. J. Brockman, 1999, A global perspective on the microbial abundance and activity in the deep subsurface, in: *Enigmatic Microorganisms and Life in Extreme Environments*, J. Seckbach, ed., Kluwer Academic, Dordrecht, The Netherlands, pp. 489–500.

Palumbo, A. V., J. C. Schryver, M. W. Fields, C. E. Bagwell, J. -Z. Zhou, T. Yan, X. Liu, and C. C. Brandt, 2004, Coupling of functional gene diversity and geochemical data from environmental samples, *Appl. Environ. Microbiol.* **70:**6525–6534.

Pedersen, K., 1993, The deep subterranean biosphere, *Earth Sci. Rev.* **34:**243–260.

Pedersen, K., 1997, Microbial life in deep granitic rock, *FEMS Microbiol. Rev.* **20:**399–414.

Pedersen, K., 2000, Exploration of deep intraterrestrial microbial life: current perspectives, *FEMS Microbiol. Lett.* **185:**9–16.

Pedersen, K., and S. Ekendahl, 1990, Distribution and activity of bacteria in deep granitic groundwaters of southeastern Sweden, *Microb. Ecol.* **20:**37–52.

Pedersen, K., J. Arlinger, L. Hallbeck, and C. Pettersson, 1996, Diversity and distribution of subterranean bacteria in groundwater at Oklo in Gabon, Africa, as determined by 16S rRNA gene sequencing, *Molec. Ecol.* **5:**427–436.

Phelps, T. J., E. G. Raione, D. C. White, and C. B. Fliermans, 1989, Microbial activities in deep subsurface environments, *Geomicrobiology* **7:**79–91.

Phelps, T., E. Murphy, S. Pfiffner, and D. White, 1994, Comparison between geochemical and biological estimates of subsurface microbial activities, *Microb. Ecol.* **28:**335–349.

Phelps, T. J., S. M. Pfiffner, K. A. Sargent, and D. C. White, 1994, Factors influencing the abundance and metabolic capacities of microorganisms in eastern coastal plain sediments, *Microb. Ecol.* **28:**351–364.

Potts, M., 1994, Dessication tolerance of prokaryotes, *Microbiol. Rev.* **58:**755–805.

Richter, D. D., and D. Markewitz, 1995, How deep is soil? *Bioscience* **45:**600–614.

Roling, W. F. M., B. M. van Breukelen, M. Braster, B. Lin, and H. W. van Verseveld, 2001, Relationships between microbial community structure and hydrochemistry in a landfill leachate-polluted aquifer, *Appl. Environ. Microbiol.* **67:**4619–4629.

Russell, C. E., R. Jacobson, D. L. Haldeman, and P. S. Amy, 1994, Heterogeneity of deep subsurface microorganisms and correlations to hydrogeological and geochemical parameters, *Geomicrobiol. J.* **12:**37–51.

Shi, T., R. H. Reeves, D. A. Gilichinsky, and E. I. Friedmann, 1997, Characterization of viable bacteria from Siberian permafrost by 16S rDNA sequencing, *Microb. Ecol.* **33:**169–179.

Sinclair, J. L., and W. C. Ghiorse, 1989, Distribution of aerobic bacteria, protozoa, algae, and fungi in deep subsurface sediments, *Geomicrobiology* **7**:15–31.

Sinclair, J. L., S. J. Rantke, J. E. Denne, L. R. Hathaway, and W. C. Ghiorse, 1990, Survey of microbial populations in buried-valley aquifer sediments from northeastern Kansas, *Ground Water* **28**:369.

Smith, R. L., R. W. Harvey, and D. R. LeBlanc, 1991, Importance of closely spaced vertical sampling delineating chemical and microbiological gradients in ground water studies, *J. Contam. Hydrol.* **7**:285–300.

Soil Survey Staff, 1975, *Soil taxonomy.* U.S. Department of Agriculture, Washington, DC.

Stevens, T. O., and B. S. Holbert, 1995, Variability and density dependence of bacteria in terrestrial subsurface samples: implications for enumeration, *J. Microbiol. Methods* **21**:283–292.

Stevens, T. O., and J. P. McKinley, 1995, Lithotrophic microbial ecosystems in deep basalt aquifers, *Science* **270**:450–454.

Szewzky, U., R. Szewzky, and T. -A. Stenstrom, 1994, Thermophilic, anaerobic bacteria isolated from a deep borehole in granite in Sweden, *Proc. Natl. Acad. Sci. USA* **91**:1810–1813.

Takai, K., M. R. Mormile, J. P. McKinley, F. J. Brockman, W. E. Holben, W. P. Kovacik, and J. K. Fredrickson, 2003, Shifts in Archaeal communities associated with lithological and geochemical variations in subsurface Cretaceous rock, *Environ. Microbiol.* **5**:309–320.

Tate, R. L., 1979, Microbial activity in organic soils as affected by soil depth and crop, *Appl. Environ. Microbiol.* **37**:1085–1090.

Taunton, A. W., S. A. Welch, and J. F. Banfield, 2000, Microbial controls on phosphate weathering and lanthanide distributions during granite weathering and soil formation, *Chem. Geol.* **169**:371–382.

Taylor, J. P., B. Wilson, M. S. Mills, and R. G. Burns, 2002, Comparison of microbial numbers and enzymatic activities in surface soils and subsoils using various techniques, *Soil Biol. Biochem.* **34**:387–401.

Ulrich, G. A., D. Martino, K. Burger, J. Routh, E. L. Grossman, J. W. Ammerman, and J. M. Suflita, 1998, Sulfur cycling in the terrestrial subsurface: commensal interactions, spatial scales, and microbial heterogeneity, *Microb. Ecol.* **36**:141–151.

Vreeland, R. H., W. D. Rosenzweig, and D. W. Powers, 2000, Isolation of a 250 million-year-old halotolerant bacterium from a primary salt crystal, *Nature* **407**:897–900.

Walvoord, M. A., P. Pegram, F. M. Phillips, M. Person, T. L. Kieft, J. K. Fredrickson, J. P. McKinley, and J. B. Swenson, 1999, Groundwater flow and geochemistry in the southeastern San Juan Basin: implications for microbial transport and activity, *Water Resour. Res.* **35**:1409–1424.

Whitman, W. B., D. C. Coleman, and W. J. Wiebe, 1998, Prokaryotes: the unseen majority, *Proc. Natl. Acad. Sci. USA* **95**:6578–6583.

Wilson, J. T., J. F. McNabb, D. L. Balkwill, and W. C. Ghiorse, 1983, Enumeration and characterization of bacteria indigenous to a shallow water-table aquifer, *Ground Water* **21**:134–142.

Winograd, J. I., and F. N. Robertson, 1982, Deep oxygenated groundwater: anomaly or common occurence? *Science* **216**:1227–1230.

Wood, B. D., C. K. Keller, and D. L. Johnstone, 1993, In situ measurement of microbial activity and controls on microbial CO_2 production in the unsaturated zone, *Water Resour. Res.* **29**:647–659.

Zhang, C., R. M. Lehman, S. M. Pfiffner, S. P. Scarborough, A. V. Palumbo, T. J. Phelps, J. J. Beauchamp, and F. S. Colwell, 1997, Spatial and temporal variations of microbial properties at different scales in shallow subsurface sediments, *Appl. Biochem. Biotechnol.* **63–65:**797–808.

Zhou, J., B. Xia, H. Huang, A. V. Palumbo, and J. M. Tiedje, 2004, Microbial diversity and heterogeneity in sandy subsurface soils, *Appl. Environ. Microbiol.* **70:**1723–1734.

Chapter 7

SPATIAL ORGANISATION OF SOIL FUNGI
How fungi are organised at different spatial scales and the consequences of this for soil function

Karl Ritz
National Soil Resources Institute, Cranfield University, Silsoe, Bedfordshire MK45 4DT, UK

Abstract: Filamentous fungi are unique and significant "spatial integrators" of soil systems. By virtue of their indeterminate and mycelial form, they are able to occupy large volumes of the soil matrix and influence a panoply of soil-based services that underpin the functioning of terrestrial ecosystems. In this chapter, factors which affect the spatial organisation filamentous fungi are described and reviewed at the scale of the hypha, the mycelium, and the community. Environmental factors such as the architecture of the soil and the spatial distribution of nutrient resources play pivotal roles in determining the form and organisation of mycelia. Biotic interactions between fungi and other soil-dwelling organisms further pattern the fungal colonies and the resultant communities. Some of the consequences of such spatial organisation for soil function relate particularly to soil structural dynamics, biotic regulation of plant and microbial communities, and the transport of nutrient elements through the soil system.

Keywords: fungi, environmental spatial heterogeneity, biotic interactions, nutrient cycling, nutrient transport, soil

1. INTRODUCTION

Fungi are ubiquitous members of soil microbial communities, but comprise a varying proportion of the biomass in different systems. They tend to dominate in soils containing high proportions of organic matter and low pH, and generally constitute a smaller proportion in intensively managed mineral soils. They are involved in a plethora of functional roles encompassing a wide range of biological, chemical, and physical processes, and are consequently of great ecological significance.

Two fundamental growth-forms are manifest by fungi, a discrete spheroid cellular form known as holocarpic, and a filamentous form termed eucarpic. Holocarpic forms predominate in the yeasts, where growth occurs via the genesis and expansion of adjacent cells ("budding") and results in the formation of short fragmentary chains of largely disconnected cells. Eucarpic forms, which predominate within the kingdom, involve the formation of filamentous tubes or hyphae, which grow by apical extension and thus elongate into the environment in which the fungi grow. Enzymes present on the surface of the hyphae are released into their vicinity and degrade substrates into assimilable forms which are subsequently absorbed and used in growth and metabolism. Hyphae periodically branch, which leads to the development of an interconnected, network of filaments termed a mycelium (Fig. 7-1). This is an indeterminate structure which is very efficient in exploring and filling space, and means that mycelia are exceptionally well adapted to growing in spatially structured environments such as soils (Figs. 7-1; 7-2).

The aims in this chapter are to review how eucarpic fungi in soils are spatially organised, what governs such organisation and what the consequences are for ecosystem function. These issues will be addressed at three spatial scales, namely, the sub-colony, the mycelium, and the community.

2. SUB-MYCELIAL SCALE

This is the basic level at which mycelial organisation is controlled, and it is remarkable how the two simple processes of apical extension and branching, augmented by higher-order aggregation of hyphae in some species, can lead to such a diversity of form and variety of function that is manifest across the fungal kingdom. Fundamentally, these properties have a genetic basis and are therefore intrinsic to individual fungi. The molecular genetic basis of fungal morphogenesis, although the subject of much study, remains poorly understood. Hyphae show directional growth in response to external stimuli such as electrical fields (Gow, 1990), topographical cues (Read et al., 1997), nutrient gradients (Crawford et al., 1993), toxin gradients (Fomina et al., 2000), and chemotopic signals (Sbrana and Giovannetti, 2005). Hyphal extension rates appear to be governed by the rate of delivery of substrate, via vesicles, to the hyphal tip. A sub-cellular structure termed the Spitzenkörper, which is a highly dynamic phase-dark body found at the tip of elongating hyphae, apparently plays a major role in apical growth and hyphal orientation (LopezFranco and Bracker, 1996; Riquelme et al., 1998). The precise mechanisms of branching in eucarpic fungi are also not fully comprehended. Current thinking is that if the rate of vesicle supply (i.e., substrate for building new biomass) to the tip exceeds the capacity for incorporation, it apparently triggers branching, which leads to the formation of a new tip and therefore an increased sink size that can accommodate the substrate delivery rate (Katz et al., 1972; Trinci, 1984; Watters and Griffiths, 2001). However, whilst apical and sub-apical branching is common, branches do not necessarily arise in the vicinity of hyphal tips. In some species, branches are often associated with septae (cross-walls), but this is not a prerequisite, as coenocytic (aseptate) fungi also form branches in a consistent manner. Branch angles can vary, which leads to different spatial characteristics of mycelia even at the earliest stages of development (e.g., Fig. 7-1; see also Fig. 1 in Regalado et al., 1997), and is a prerequisite if morphologically complex higher-order structures are to be formed. A significant consequence of the relationship between vesicle supply rate and branching is that there are feedback mechanisms invoked whereby branch frequency is more likely to be increased where available substrate concentrations are high, and increased branching results in the development of a greater surface area for further substrate uptake. This provides a mechanistic basis for the foraging strategies that are apparent in many fungi and discussed in more detail below. As connected networks, mycelia are highly coordinated living systems, and their indeterminate nature makes them phenotypically very plastic. Within a single mycelium, a variety of physiologically distinct processes can be occurring simultaneously in different regions, as well as substantial morphological and functional differences being manifest (Cairney and Burke, 1996).

3. MYCELIAL SCALE

Whatever the intrinsic genetic ground rules, the extrinsic environment plays a significant governing role upon the development and form of fungal colonies. Since mycelia are spatially distributed systems, their organisation often reflects the spatial structure of the environment in which they are growing, particularly with respect to habitat space and resource distribution.

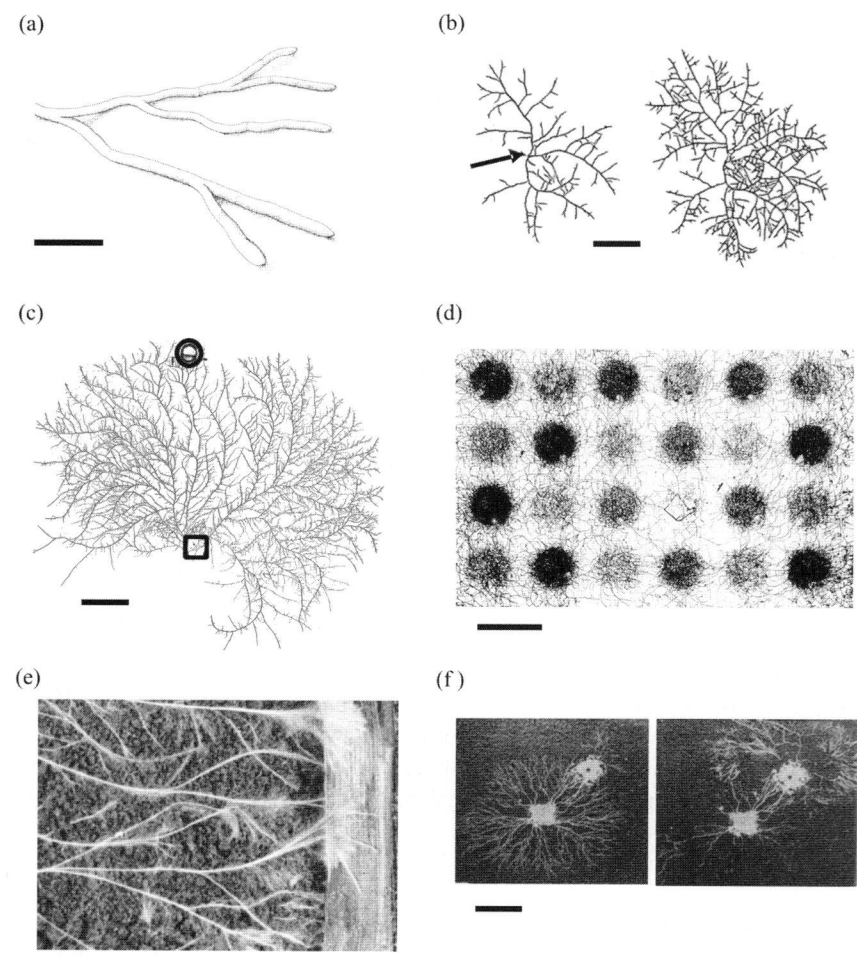

Figure 7-1. The fungal mycelium and its spatial organisation up to the centimetre scale. (a) The fundamental structural unit of eucarpic fungi – the branching hypha. (From Jennings and Lysek, 1999.) (b) Early developmental stages of a mycelium of *Trichoderma viride* 44 h (left) and 48 h (right) after germination of a spore (arrow). Scale bar = 100 μm. (From Ritz and Crawford, 1990.) (c) Foraging growth of *T. viride* in the presence of a discrete nutrient resource (circle), from the point of inoculation (square). Scale bar = 5 mm. (d) Spatial patterns in a single mycelium of *T. viride* reflecting local underlying concentrations of substrate. Three such concentrations are present: high and low in an alternating tessellation of circular domains, and very low in the interstitial regions. Note relatively greater density (branch frequency) of mycelium on both high- and low-nutrient domains that are distal from the point of inoculation (denoted by square). Scale bar = 5 mm. (From Ritz et al., 1996.) (e) Foraging behaviour of *Agrocybe gibberosa*, changing from explorative cords to exploitative colonising fans at tips of foraging cords of when piece of straw (to right of image) is encountered. Scale bar = 5 mm. (From Robinson et al., 1993.) (f) Foraging system of *Hypholoma fasciculare* in

3.1. Habitat space

Soil pore networks define the physical framework in and through which hyphae must grow and mycelia develop. The mycelial form is well suited to growth in porous structures since hyphae in contact with surfaces can absorb nutrients from the substratum, and the branched nature of the mycelium presents a large surface area/volume ratio. Apical extension of hyphae allows growth through air gaps and a concomitant bridging between surfaces (Fig. 7-2). The limits of such aerial growth are unknown, but distances of 10 mm for undifferentiated hyphae are clearly attained. When considered as a single entity, the aerial hypha appears to be an efficient means of exploring space, but there appears to be some form of metabolic cost associated with it at the mycelial level. Otten et al. (2004) studied how the width of experimentally prescribed gaps in sand and a sandy loam affected the spatial extent of mycelia of *Rhizoctonia solani* and the ability of mycelia that had bridged such gaps to colonise substrate bait comprising sterile *Papaver* seeds. The average radial expansion rate of the colonies was approximately 67% faster along the direction of a gap compared to other directions. However, colonisation efficiency was reduced, but not totally curtailed, by increasing gap width up to 6.5 mm (Fig. 7-3a). The orientation of the gaps also significantly affected colonisation efficiency, with the gap either perpendicular or parallel (relative to the growing front of the mycelium) to the direction of the target bait. The colonisation efficiency was either reduced by up to 50% or enhanced by up to 140% (Fig. 7-3b). Gaps can therefore act as barriers or preferential pathways depending on the spatial geometry of both fungus and environment.

The spatial dimension in which mycelia develop also appears to be an important factor in determining the degree to which they fill space. This has rarely been considered in any detail, possibly because experimental mycologists have been wedded to the Petri dish for too long. When growing in a predominantly two-dimensional mode, i.e., across a plane surface, the amount of resource needed to fill space is considerably less than if three-dimensional volumes are to be filled, with a power-law relationship relative to radius. In experimental systems, mycelia have been demonstrated to show preferential growth over sand and soil surfaces compared to growth into the

---

presence of wood bait. Inoculum was placed on lower wood block, and a remote bait has been located by foraging cords (left); over time, the mycelium prevails in the newly colonised resource and regresses on the older domain. Scale bar = 2 cm. (From Dowson et al., 1988.)

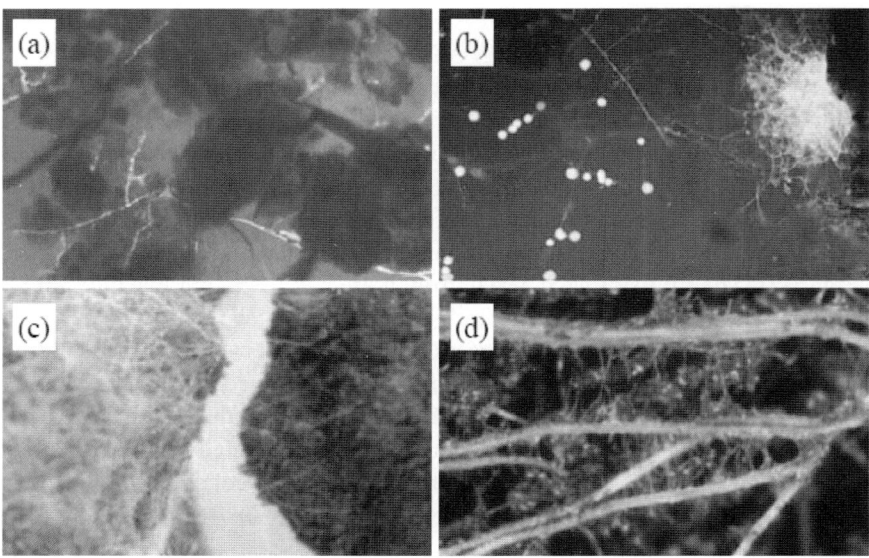

Figure 7-2. Fungal mycelia in the soil environment. (a) Mycelium of *Rhizoctonia solani* growing in sterilised arable soil, visualised in a thin section stained with SCRI Renaissance 2200. UV epifluorescent illumination. Image width = 150 μm. (b) Unidentified mycelium growing in soil pore, visualised in a thin section of undisturbed pasture soil, stained with Fluorescent Brightener 28. Note proliferation of hyphae on pore wall in left of image. Bright spherical objects are sporangia. UV epifluorescent illumination. Image width = 150 μm. (c) Hyphae of *Fusarium oxysporum* f. sp. *raphani* colonising a pair of adjacent soil aggregates. Aggregate on left is sterile, hence extensive mycelial development. Aggregate on right is non-sterile; reduced mycelial growth is due to competitive effects of indigenous microflora and reduced nutrient levels therein. Image width = 1 cm. (d) Unidentified hyphae bridging roots of *Plantago lanceolata* growing in non-sterile field soil. Note abundance of mucilage films. Image width = 2 cm.

soil matrix (Otten et al., 2004), and this has large implications for the epidemiological development of soil-borne fungal diseases (Otten and Gilligan, 2006). Even in natural environments, the foraging cord systems of wood-decomposing basidiomycetes and mycelial fans of ectomycorrhizal (ECM) species appear to be virtually two-dimensional, since the litter layers which represent the primary matrix through which they grow are generally relatively shallow. Many experimental systems utilise two-dimensional systems largely for convenience, but with the advent of non-destructive tomographic imaging techniques, this should change. Spatially explicit modelling of mycelial growth is also increasing in sophistication and many of the contemporary approaches (e.g., Meskauskas et al., 2004; Falconer et. al., 2005; Boswell et al., 2007) can, in principle, be extrapolated to three dimensions.

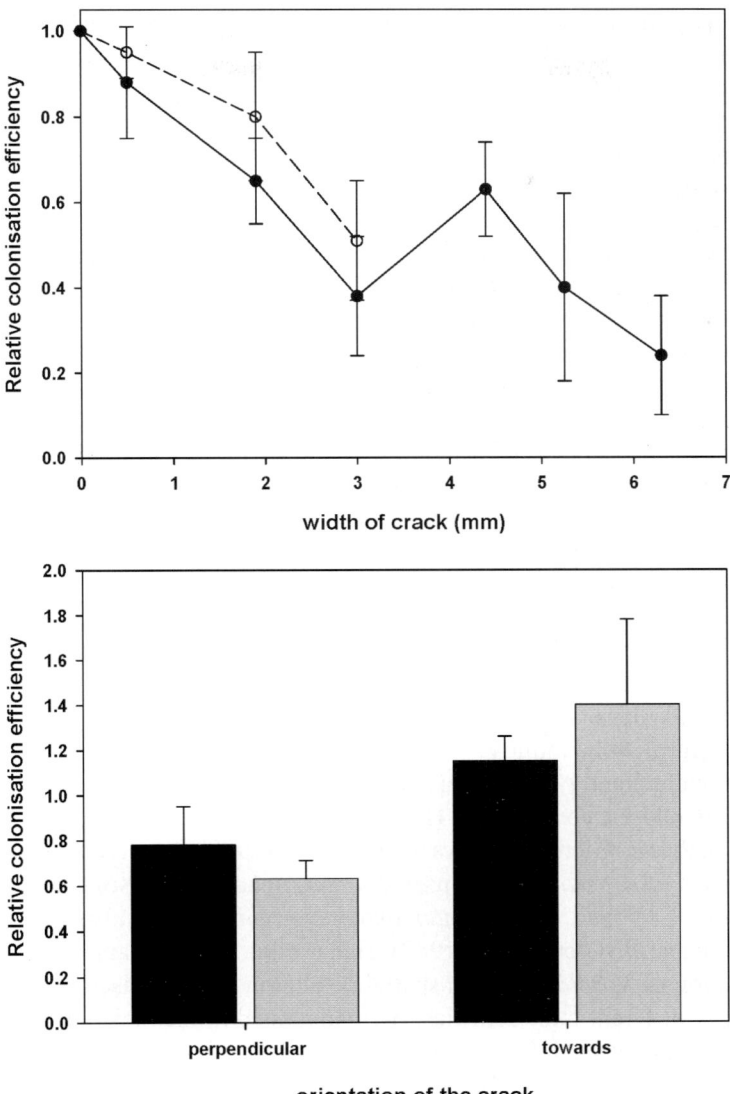

Figure 7-3. Colonisation efficiency, measured as the proportion of replicates in which *Rhizoctonia solani* spreading over soil (black symbols) or sand (open/grey symbols) surfaces colonises a target seed at 30 mm distance from the point of inoculation 13 days after such inoculation, (a) in relation to the width of a gap midway between inoculum and bait and (b) in relation to the orientation of a 2 mm wide gap perpendicular or aligned to the predominant growth direction of the mycelium. (Adapted from Otten et al., 2004.)

In organic soils and litter horizons, purely physical constraints are unlikely to occur due to the inherent friability of the matrix and the fact that many fungi inhabiting such substrates possess enzymes capable of degrading constituent organic material. However, in mineral soils, physical constraints are of more significance, both directly by presenting barriers to passage, and indirectly via control of water distribution and associated transport of gases. In structured mineral soils, the pore network is a highly complex three-dimensional labyrinth with varying degrees of connectivity and tortuosity and ultimately defines the available paths which hyphae can take. In principle, hyphae should be unable to penetrate pores narrower than their diameters, although they may be able to physically separate soil particles to create paths available for growth. This mechanism is known to occur with roots, where radial expansion can create a cracking front that eases forward extension (Hettiara, 1973), and with earthworms, which can similarly exert soil-distorting pressures (McKenzie and Dexter, 1988). It is unclear whether these mechanisms occur with fungi. Individual hyphae can exert pressures of 50 mN m^{-2} (Money, 2004), and, where hyphal aggregation occurs, pressures up to 1.3 kN m^{-2} can be generated and may be spectacularly manifest, for example, as the emergence of basidiocarps through pavements (Niksic et al., 2004). Schack-Kirchner et al. (2000) did not detect evidence for such behaviour when studying thin sections of ECM hyphal distribution in spruce forest soil, and argue that such hyphae are inefficient at creating pore space by direct physical means.

Soil bulk density influences the extent to which soils are colonised by mycelia. It might be intuitively expected that the higher the bulk density, and hence concomitantly smaller relative volume of pore space, the more fungal growth would be constrained. However, there is little evidence to support this hypothesis. Glenn and Sivasithamparam (1990) measured higher concentrations of *R. solani* in compacted versus uncompacted soils, and greater densities of hyphae of *Gauemannomyces graminis* were also apparent in compacted soils (Glenn et al., 1987). Harris et al. (2003) mapped the hyphal distribution of *R. solani* at high spatial resolution in a sterilised mineral soil, packed at different bulk densities, but maintained at identical air-filled pore volumes. They found fungal extent increased concomitant with soil bulk densities from 1.2 to 1.4 Mg m^{-3} but stabilised thereafter up to a density of 1.6 mg m^{-3} (Fig. 7-4).

However, such effects of bulk density may be contingent on the nature of the community present in the soil. In pure-culture experiments, Toyota et al. (1996b) detected a reduction in the number of propagules produced by

Spatial Organisation of Soil Fungi 187

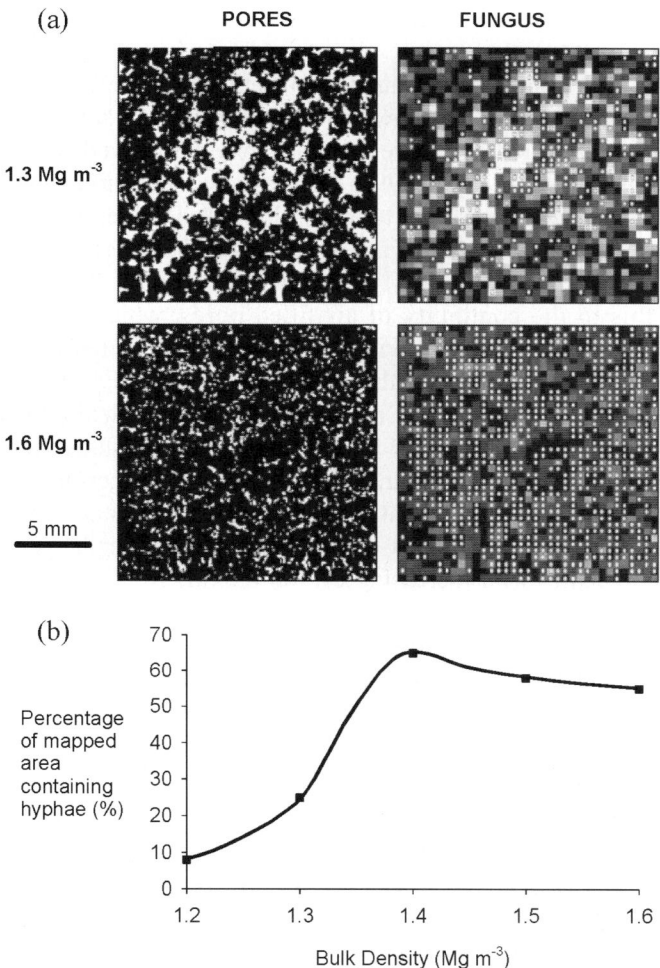

Figure 7-4. Effect of bulk density upon spatial organisation of *Rhizoctonia solani* grown in sterilised soil. (a) Maps of pore networks (left column) and associated distribution of hyphae (right column) as visualised in thin sections of soil. In fungal maps, spots denote presence of hyphae in underlying region of the section; grey scale relates to mean porosity in associated region (black = >90% porosity, white = <10% porosity). (b) Relationship between bulk density and fungal extent. (Adapted from Harris et al., 2002.)

Fusarium oxysporum as soil bulk density was increased from 1.3. to 1.5 Mg m^{-3}, but no such effect when a competing species of fungus (same genus) or a bacterial species was co-inoculated. However, there was an allied reduction in propagule numbers of the competing fungus at the higher bulk density, but not the bacteria. This demonstrates how soil structure can affect microbial interactions.

There is also a significant interaction between the soil water content, the pore network and fungal growth. Filamentous fungi are predominantly aerobes and water-filled pores are generally oxygen-starved, which curtails hyphal extension. Otten et al. (1999) demonstrated that the extent to which mycelia of *R. solani* were able to grow through a mineral soil was very sensitive to air-filled porosity. They determined that there were threshold values of matric potential below which growth was rapidly curtailed, and attributed this to the continuity of air-filled pores, which is governed by the three-dimensional distribution of water in the pore network. Schack-Kirchner et al. (2000) mapped the spatial location of ECM hyphae in a spruce forest soil as a function of matric potential. They determined that the mean hyphal length density was independent of water status and that the majority of hyphae were located in pores significantly larger than the hyphal diameter. They also recorded that the percentage of hyphae "embedded in the soil matrix" (i.e., in volumes of soil containing fewer large pores) decreased significantly with increasing carbon dioxide (CO_2) concentration in the soil air. They attributed this to the high requirement for oxygen (O_2) by the mycelia, resulting in a preferential growth in larger pores where gas diffusion rates were less constrained. These observations suggest that hyphae penetrate water-filled pores only to a very limited extent.

3.2. Resource distribution

The distribution of nutrient resources has a strong effect upon the spatial organisation of fungal mycelia. Historically, this was rarely appreciated by experimental mycologists, again perhaps because of a long-held preoccupation with the Petri dish and uniform culture media. However, in, natural circumstances, most nutrient resources are heterogeneously distributed. On woodland and forest floors, fallen twigs and branches are discrete and dispersed, and dung patches and cadavers represent isolated nutrient deposits. Within the fabric of soils, the pore network enhances spatial isolation of organic matter within its tortuous paths. Ritz (1995) demonstrated the impact of spatial heterogeneity in nutrients upon mycelial development by essentially discretising the agar plate concept. He devised a system whereby spatial patterns of nutrient resources could be prescribed using tessellations of agar tiles of differing composition. Diffusion of nutrients into adjacent domains was precluded by an air gap between the tiles, which offered little barrier to the growth of hyphae. Tessellations involving a variety of spatial patterns of high- and low-nutrient status domains were established, and inoculated with a single colony of a number of moulds. Colony development was qualitatively mapped and demonstrated a number of characteristic responses by different species. In general, hyphal density reflected the underlying nutrient status, being high and low in nutrient-rich and nutrient-poor domains, respectively, although there was evidence for growth in low-nutrient tiles being greater when high-nutrient tiles were present in the tessellation. Different colony extension rates within different regions of the same mycelium were often observed, being greater on low-nutrient tiles than on high domains. Patterns of formation of reproductive structures were strongly influenced by the underlying heterogeneity in the tessellation. Such structures tended to be formed only in low-nutrient tiles by *T. viride* and *R. solani*, and only in high-nutrient tiles by *Alternaria alternata*. Sclerotia were only produced by *R. solani* in tessellations where there was some heterogeneity, being absent in the nutritionally uniform arrangements. Furthermore, sclerotial production was strongly asymmetric in nutritionally symmetric, but heterogeneous, tessellations. Many subsequent studies have since demonstrated the effects of spatial heterogeneity upon the organisation of fungal colonies (see reviews by Boddy, 1999; Ritz and Crawford, 1999). Some general principles can be drawn. Hyphae tend to proliferate and mycelia are generally dense in the presence of nutrients, and are sparse in nutrient-impoverished zones. This relates to contrasting explorative (searching for substrate) and exploitative (utilising a located resource) growth phases, and demonstrates that fungi have distinct foraging strategies. These are particularly well developed amongst cord-forming basidiomycetes. Such strategies appear to be commensurate with the natural distribution of substrates utilised by different fungi. Foraging strategies expressed by different species can be classified as close-range, short-range, and long-range, and are each suited to the exploitation of different types of spatial distribution in resource units (Boddy, 1993; Rayner, 1994). Close-range foraging is apposite where resource units are small and spatially frequent, and is exemplified by species which produce fairy rings in litters or grasslands. Short-range foraging involves the outgrowth of dense mycelia from resource bases; outgrowth is then curtailed following contact with fresh resources. Degeneration of the older mycelium then occurs as the new material is colonised. This strategy is effective where resources are relatively small and frequent, but spatially separated. Long-range foraging is appropriate where resources are large and spatially distant. Here, outgrowth of distinctly sparse, rapidly extending mycelial cords or rhizomorphs occurs followed by localised proliferation of assimilative hyphae when a resource is encountered. The explorative mycelium can show some degree of persistence and has the consequence of allowing some ability to forage in time as well as space, since resources, which may subsequently be deposited in the vicinity of persistent cords, will become colonised by this inoculum.

There have been many studies into such foraging strategies by both saprotrophic and ectomycorrhizal basidiomycetes (e.g., Boddy, 1999; Donnelly and Boddy, 2001; Zakaria and Boddy, 2002; Donnelly et al., 2004; Harris and Boddy, 2005).

Since the fungal mycelium is essentially an interconnected network of tubes, there is the potential to transport materials between different regions within it (Jennings et al., 1974; Cairney, 1992; Lindahl and Olsson, 2004). This ability to translocate materials within the mycelium has profound consequences both for the fungal organism and ecosystem function (Falconer et al., 2005; Ritz, 2006). This is how hyphae are able to grow through nutrient-impoverished regions – if resources are absent locally, they can be imported from elsewhere in the mycelium to the growing tips and be used to drive explorative growth. A wide variety of elements can be transported within mycelia, predominant amongst them are the major nutrients C, N, P, and S. Mycelial connections between patchy resources set up source–sink relationships that affect transport within mycelia and are reviewed in detail by Ritz (2006). Since such transport occurs within hyphae, it is independent of the soil matrix and not subject to diffusive constraints invoked by adsorption to soil components or tortuous paths, or assimilation by other organisms. Fungi therefore affect the transport and distribution of elements in soil systems in ways rarely considered by soil scientists and certainly ignored in the majority of nutrient cycling and soil transport models. Intriguingly, non-nutrient elements such as cesium and uranium are also mobile within some mycelia, and may be concentrated in fruiting bodies, with implications for transmission of these materials into higher trophic levels if they are consumed by fauna.

Faunal grazing of mycelia affects their patterns of morphogenesis. Hedlund et al. (1991) showed that when *Mortierella isabellina* mycelia were subject to grazing by the collembolan *Onychiurus armatus*, the mycelial form switched from appressed growth and sporulating hyphae to fast-growing fan-shaped sectorial growth, production of extensive aerial mycelium and a cessation of sporulation. Collembolan grazing of *Hypholoma fasciculare* growing from woodblocks was shown to reduce the extent and growth rate of mycelia, but also induced the development of point growth of cords with distinct fanned margins (Harold et al., 2005). Increased grazing pressure tends to increase such effects, and different species of grazer also may have different effects on such mycelial extension rates and coverage, but apparently not on fractal mass or, fractal surface dimension (Kampichler et al., 2004). To date, all such grazing effects have been studied in experimental microcosms.

3.3. Community scale

A biological community is defined as an assemblage of populations of organisms, with populations being further defined as a group of similar organisms, nominally of the same species. Originally devised in the context of large, determinate organisms and generally well-defined species, the soil fungal community is less readily defined and certainly less readily characterised. Since fungal mycelia are indeterminate systems, the definition of what constitutes an individual is not straightforward (Rayner, 1991). Detecting precisely which fungi are present in a soil has exercised mycologists for over a century, and the debate as to the definitive approach is far from over. Again, this is challenged by the indeterminate nature of the organism, and issues relating to what the functional unit is – the hypha, the mycelium, the propagule or combinations thereof?

Hundreds of studies have characterised the constitution of fungal assemblages found in soils and litters from a taxonomic perspective, with species being assessed using a wide variety of survey techniques ranging from presence of charismatic basidiocarps, though culture-dependent isolation on various media, to sophisticated analyses of nucleic acids directly extracted from soil samples. These studies generally show that there are characteristic assemblages of fungi found under particular ecological circumstances and contexts, and more or less consistent ecological successions of fungi (see Frankland, 1992; Dighton, 2005), akin to those that occur for other classes of organisms at larger ecosystem scales. But these views must always be taken with the caveat that the description of the community is contingent on the methods used to assess it. The spatial organisation of fungal communities in soils has received considerably less attention than their taxonomic composition.

What then defines the spatial organisation of a soil fungal community? At any moment in time, the spatial arrangement of the community will be defined by a combination of a range of historical and prevailing factors. Fundamentally, it must relate to three main factors:

1. *The presence or arrival of inoculum within the domain under consideration.*

Notwithstanding that, Beijerink's principle that "everything is everywhere; the environment selects" is both a spatially and a temporally scale-dependent concept; this is clearly not the case for fungi at the field scale. For example, Boerner et al. (1996), in studying patterns of arbuscular mycorrhizal (AM) fungal infectivity along a successional chronosequence in Ohio, detected isolated areas within disturbed sites that were devoid of AM infectiveness, and many similar bait studies show this phenomenon is common. Fungal inoculum arrival at a particular point in space could originate from a variety of sources including spore rain from aerial sources, propagule transport in soil water flows, import via carrier organisms such as fauna, ingrowth from mycelia on the boundaries of the domain, residue import from above-ground litter by fauna or via residue and soil management (e.g., tillage) in farmed systems. Kaldorf et al. (2004) analysed ECM colonisation of aspen hybrids (*Populus* spp.) planted in a non-uniform pattern by morphotyping and nucleic acid analysis and detected contrasting patterns of spatial homogeneity for some ECM types, but for others, a correlation was observed with distance from contrasting tree species bordering the experimental field. They interpreted this data in terms of "vicinal" invasion, where incursion of mycorrhizal inoculum is derived from proximal trees versus "random" invasion where inoculum arises from dispersed sources. Whether such inoculum germinates and becomes established will then be contingent on further factors, including those discussed below.

2. The patterning of the niches in which the constituent species are adapted to growing in the context thereof.

The niche is an all-encompassing concept that is defined by a variety of abiotic and biotic factors. The spatial distribution of niches seen by soil fungi can be characterised by spatially explicit sampling and associated statistical analyses of fungal parameters. Mottonen et al. (1999) characterised ergosterol concentrations and a variety of other soil properties in a 1 ha *Pinus sylvestris* stand in a notably thorough study based on sampling across a regularly spaced grid and two scales. They detected strong spatial autocorrelation (90% of variance) with a 4 m range apparent in the semi-variogram (Fig. 7-5). This spatial pattern was most closely related to pH, organic matter thickness and total carbon content. The spatial location of individual trees was also mapped in this study, and there was no apparent spatial association between ergosterol and trees. Fungal biomass is therefore not necessarily affected by proximity to the primary autotrophs in the system at the stand scale, but detailed studies of community structure suggest that at higher taxonomic resolution, such structures may be apparent. Lilleskov et al. (2004) synthesised eight studies involving detailed surveys of spatial structure in ECM communities at the stand scale (about 50 m) within US forests. They utilised a variety of statistical methods for assessing spatial structure and determined that in most of the sites, community similarity decreased with distance. There was a general agreement between the different tests used, but they were not congruent. They concluded that most of the dominant fungal taxa had variogram ranges of approximately <3 m, but these went up to 17 m. Berglund et al. (2005) studied the temporal dynamics of the composition of wood-decaying fungal communities at the scale of individual downed logs and a plot scale of 0.1 ha over 8 years.

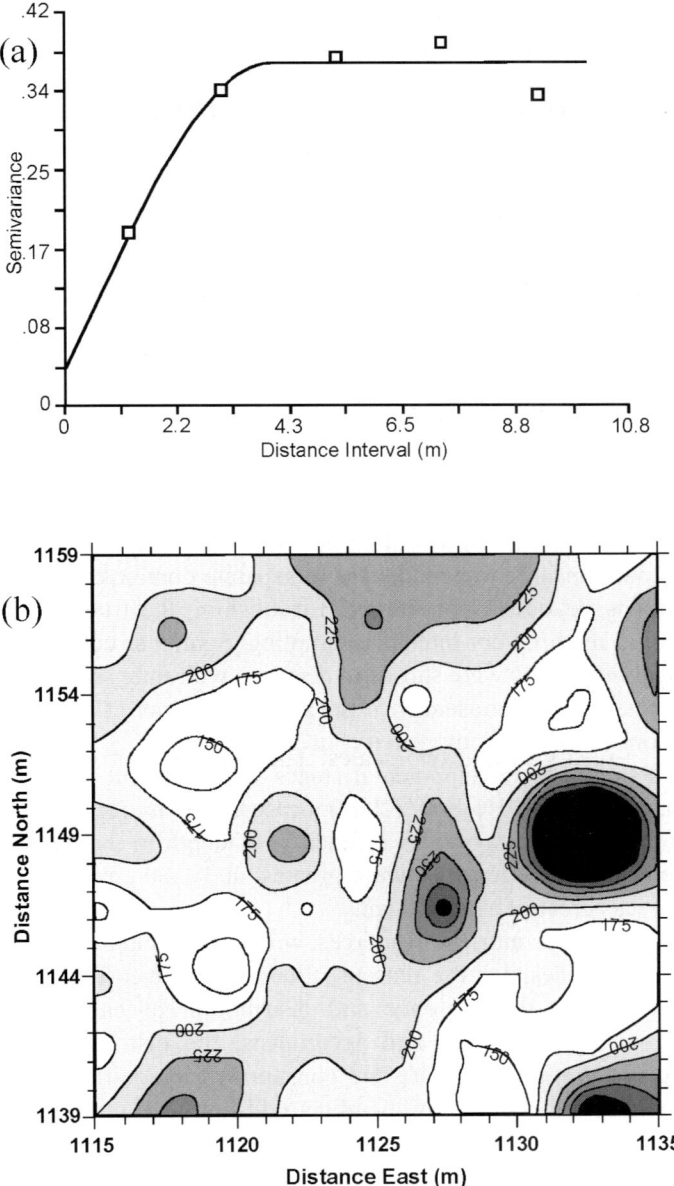

Figure 7-5. Spatial structure in fungal biomass in a *Pinus sylvestris* forest in Finland. (a) Semi-variogram of ergosterol concentration (milligram per kiligram) within a 1 ha plot. (b) Kriged map of ergosterol concentration in a sub-plot within the main plot. (Adapted from Mottonen et al., 1999.)

By monitoring 70 logs and seven plots, they detected marked changes in species richness and community composition within the logs for both short-lived corticoid species and perennial polypore species. At the stand scale there was no change in the species richness of polypore or corticoid communities, but the latter showed changes in community composition between years, reflecting their short-lived fruit bodies. Whilst not based on soil-inhabiting fungi, this survey is informative in terms of an ecological demonstration; however, this survey was based on frequency of occurrence of fruiting bodies and therefore did not consider mycelial phases. However, not all ECM communities show such strong dynamics. For example, strong spatial stability in the genetic structure of a *Phialocephala fortinii* community was measured in a 9 m^2 plot over 3 years by Queloz et al. (2005). Brundrett and Abbott (1995) measured AM infectiveness of soil cores sampled from a sclerophyllous forest in Western Australia. A high degree of small-scale spatial variability was found in inoculum levels, with variation between adjacent cores as great as that from cores taken 15 m apart. By assaying both AM and ECM inoculum potential, they detected evidence that these two types appeared to be localised in separate domains, and there were further domains devoid of either type. Establishing the true scale of such domains would require very intensive sampling. Spatial structure in patterns of AM fungal infectivity were shown to decrease with time since disturbance along a successional chronosequence in Ohio by Boerner et al. (1996), with a progression to more homogeneity in infectiveness as the succession developed. Patterns of developmental stages within fungal types may also show spatial differences. He et al. (2002) measured the frequency of vesicles, arbuscules, hyphae, and spores as a function of depth in the Negev desert, and detected that spore densities were greatest at 10–20 cm depth, but root colonisation was greatest at 20–30 cm.

It is arguable that nutrient resources will be particularly important in defining fungal niches, for the reasons alluded to above. Thus, for saprophytes, this will be the presence and distribution of energy-containing substrates, and for mutualists and antagonists, the distribution of hosts. However, given the indeterminate and consummate space-filling properties of the mycelium, the community members will not necessarily be confined to substrate-rich zones. Communities of ECM fungi typically show patterns of association related to distance from their host trees (Last et al., 1984; Matsuda and Hijii, 1998; Fig. 7-6), and this apparently applies at a plant-community level as well. For example, Dickie and Reich (2005) showed marked changes in fungal community structure as a function of distance from forest edges. The abundance of mycorrhizae was greatest near the forest

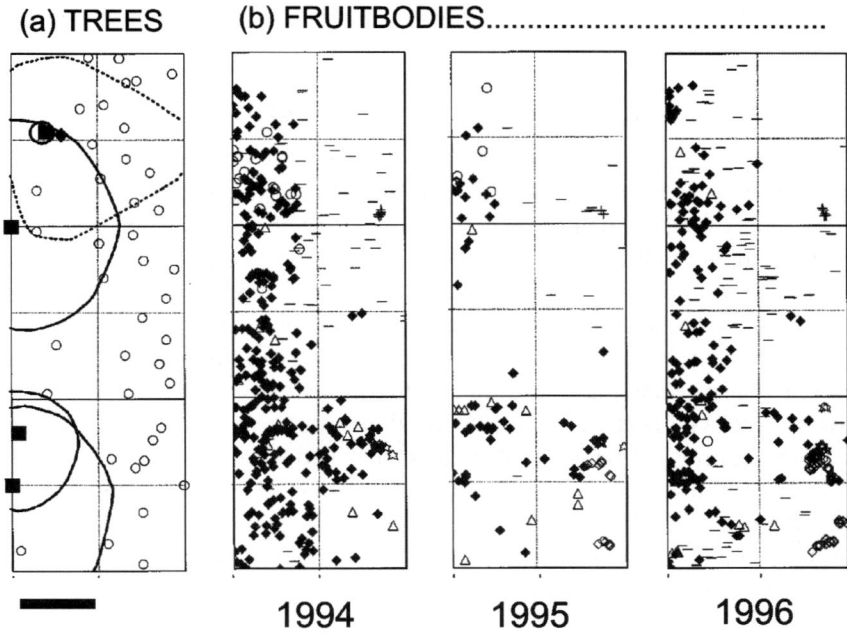

Figure 7-6. Maps of basidiocarp distribution in vicinity of trees in a forest in Japan. (a) Distribution of standing trees and crown projection areas of the trees *Abies firma* and *Carpinus laxiflora*. (b) Location of individual fruit bodies within this area over 3 successive years. Each symbol represents a different species, consistently within the figure. Scale bar = 5 m. (From Matsuda and Hijii, 1998.)

edge, and declined markedly at around 15 m into abandoned agricultural fields. They found high infectivity and diversity near the forest edge, high infection but a lower diversity at intermediate distances and both low infection and fungal diversity distant from trees. This appeared to impact the degree to which pioneer tree seedlings were infected and therefore may influence the rate of successional development.

Whilst there is often a strong spatial relation between the location of mutualistic hosts and their fungal partners, such relationships can be affected by other community-level properties, such as the proximity of other fungi and plants. Mummey et al. (2005) showed the components of the AM fungal community colonising *Dactylis glomerata* was strongly controlled by the presence of *Centaurea maculosa* in the vicinity. Agerer et al. (2002) made detailed maps of the locations of individual fungal species within communities in the vicinity of *Picea abies* trees, and demonstrated a wide range of spatial association, and apparent disassociation, between the variety of species present. The "biological" niches seen by fungi will therefore also be defined by the spatial organisation of the broader soil community. The mechanistic bases of this are discussed in the next section.

3. *The outcome of interactions with other organisms present in the domain.*

When mycelia grow through soil, they inevitably encounter other organisms. The interactions that ensue will determine whether passage or proliferation is realised or not, and this will affect the spatial distribution of mycelial units within the soil. Toyota et al. (1996a) studied the ability of *F. oxysporum* to colonise aggregates of soil that had been sterilised and colonised by single strains of fungi and bacteria compared to sterile controls. There was a wide range in the degree of inhibition to colonisation, with other genera of *Fusarium* and known antibiotic-producing bacteria presenting the greatest resistance. It was notable that *Phanerochaete magnoliae*, a fungus from an entirely different ecological niche, had very little impact on the colonising ability of the soil-inhabiting *Fusarium*. In a further experiment, they demonstrated that greater levels of general microbial diversity in the target aggregates inhibited the colonisation ability of the *Fusarium*, but this could be mitigated by increasing the nutrients available exclusively to the colonising fungus (Toyota et al., 1996b).

Interactions between fungal mycelia have been extensively studied both in vitro and in soil-based microcosms (e.g., Stahl and Christensen, 1992; Boddy, 2000). When two mycelia encounter one another, the outcome of the interaction can follow four outcomes: (i) if the mycelia are somatically compatible, the hyphae may fuse together (anastomose) and effectively a larger-scale individual is formed, (ii) intermingling, where mycelia ramify together with no apparent outcome, (iii) deadlock, where the mycelia lay down a resistant barrier at the interaction front and there is no further passage of either fungus, (iv) replacement, where one fungus kills the other, assimilating the dead mycelium. The outcomes of such interactions are affected by a variety of genetic and environmental factors, including the relative nutritional status and biomass of the individuals involved, and the area of the interaction zone. Most studies have involved binary (i.e., single pair) interactions, which can hardly be construed as a community, but they serve to demonstrate that the extant spatial configuration of a soil fungal community will be contingent on many factors over and above the genetic identity and somatic compatibility of the confronting mycelia. Using microcosm systems based on the tessellated agar tile concept, White et al. (1998) studied to what extent the temporal dynamics of a three-species model fungal community was contingent upon the initial spatial arrangement of the constituents, and whether the outcome of interactions between three fungal species could be predicted on the basis of knowledge of outcomes from binary interactions. They demonstrated that there was a high degree of sensitivity in starting arrangement and stochasticity in the spatial dynamics of the model communities. Identical starting configurations sometimes led to similar outcomes amongst replicates, and sometimes widely divergent outcomes (Fig. 7-7). It was not possible to predict the interaction outcome of three-species systems on the basis of the binary component interactions. In further work, it was demonstrated that the development of the model fungal communities was related both to the spatial scale of the individual mycelia relative to the scale of the entire microcosm system, and the spatial contexts of the mycelia at the outset (Sturrock et al., 2002). A cellular automaton model based on these experimental systems and calibrated using laboratory-derived results further demonstrated the scale-dependency of the fungal community dynamics, and suggested that system-level properties were emergent from the context of the whole system (Bown et al., 1999).

Encounters with fungal-grazing fauna will not inevitably result in consumption – there are feeding preferences and differing degrees of palatability amongst fungal grazers (Bonkowski et al., 2000; Jorgensen et al., 2003; Sabatini et al., 2004; Kaneda and Kaneko, 2004; Saleh-Lakha et al., 2005), and thus the community structure of fungal grazers will affect both the compositional and spatial structure of the soil fungal community.

3.4. Conclusions

Filamentous fungi are unique and significant "spatial integrators" of soil systems. By virtue of their indeterminate and mycelial form, they are able to occupy large volumes of the soil matrix and influence a panoply of soil-based services that underpin the functioning of terrestrial ecosystems. There are many interactions between fungi and soil structure, both in terms of how fungi drive structural dynamics, and the impacts of such structure upon fungi (Ritz and Young, 2004). It is noteworthy that the architecture of the soil pore networks will have a significant modulating influence on many of the factors governing the spatial aspects of fungal community structure. The topology of the pore network will regulate the transmission of inocula through the soil matrix. Soil structure will define the patterning of both abiotic and biotic niches in many ways. And it will regulate organismal interactions directly, for example by controlling the spatial extent of interactions between mycelial fronts, or via constraining accessibility of mycelia and hyphae to grazers via size-exclusion mechanisms. The location of inoculum, whether mycelial- or propagule-based, and mutualist or antagonist, will affect where plants can establish in successions and where they prevail within communities. Source–sink relationships between fungi will affect nutrient transport through the soil fabric. Even in mycology, relatively few studies to date have explicitly dealt with these issues, and terrestrial ecologists generally

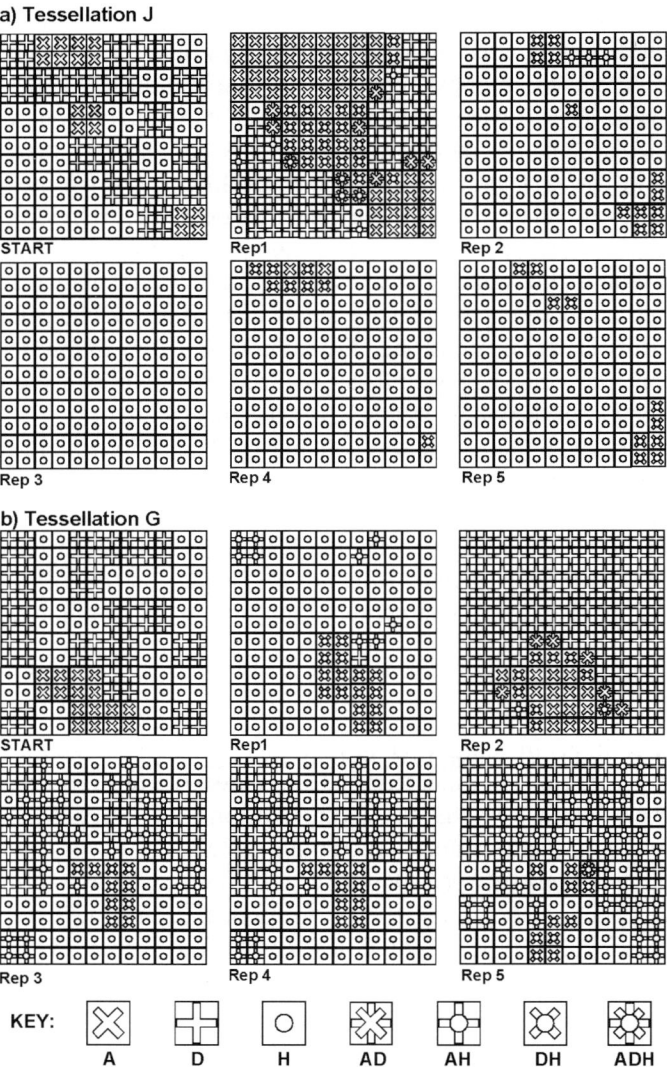

Figure 7-7. Maps showing spatial patterns in fungal communities in model systems based on tessellated arrays of agar tiles. The system involved inoculating sterile tiles with prescribed patterns of three fungal species (A – *Poria placenta*; D – *Coniophora marmorata*; H – *Paecilomyces variotii*) as shown in the "START" map. Each array comprised a matrix of 6×6 tiles each separated by an air gap of 3 mm; the symbols denote which species were present within each quadrant of each domain. The associated five maps are examples of the emergent spatial configuration in precise replicates of the communities 7 weeks after incubation. Note how the tessellation in (a) resulted in a much more consistent spatial pattern in the emergent community that the tessellation in (b). (From White et al., 1998.)

must take more cognisance of the underground networks that fungi establish and run.

REFERENCES

Agerer, R., R. Grote, and S. Raidl, 2002, The new method 'micromapping', a means to study species-specific associations and exclusions of ectomycorrhizae, *Mycol. Prog.* **1**:155–166.
Berglund, H., M. Edman, and L. Ericson, 2005, Temporal variation of wood-fungi diversity in boreal old-growth forests: Implications for monitoring, *Ecol. Appl.* **15**:970–982.
Boddy, L., 1993, Saprotrophic cord-forming fungi: warfare strategies and other ecological aspects, *Mycol. Res.* **97**:641
Boddy, L., 1999, Saprotrophic cord-forming fungi: meeting the challenge of heterogeneous environments, *Mycologia* **91**:13–32.
Boddy, L., 2000, Interspecific combative interactions between wood-decaying basidiomycetes, *FEMS Microbiol. Ecol.* **31**:185–194.
Boerner, R. E. J., B. G. Demars, and P. N. Leicht, 1996, Spatial patterns of mycorrhizal infectiveness of soils long a successional chronosequence, *Mycorrhiza* **6**:79–90.
Bonkowski, M., B. S. Griffiths, and K. Ritz, 2000, Food preferences of earthworms for soil fungi, *Pedobiologia* **44**:666–676.
Boswell, G. P., H. Jacobs, K. Ritz, G. M. Gadd, and F. A. Davidson, 2007, The development of fungal networks in complex environments, *Bull. Math. Biol.* **69**:605–634.
Bown, J. L., C. J. Sturrock, W. B. Samson, H. J. Staines, J. W. Palfreyman, N. A. White, K. Ritz, and J. W. Crawford, 1999, Evidence for emergent behaviour in the community-scale dynamics of a fungal microcosm, *Phil. Trans. Roy. Soc. Lond. B* **266**:1947–1952.
Brundrett, M. C., and L. K. Abbott, 1995, Mycorrhizal fungus propagules in the jarrah forest .2. Spatial variability in inoculum levels, *New Phytol.* **131**:461–469.
Cairney, J. W. G., 1992, Translocation of solutes in ectomycorrhizal and saprotrophic rhizomorphs, *Mycol. Res.* **96**:135–141.
Cairney, J. W. G., and R. M. Burke, 1996, Physiological heterogeneity within fungal mycelia: an important concept for a functional understanding of the ectomycorrhizal symbiosis, *New Phytol.* **134**:685–695.
Crawford, J. W., K. Ritz, and I. M. Young, 1993, Quantification of fungal morphology, gaseous transport and microbial dynamics in soil: an integrated framework utilising fractal geometry, *Geoderma* **56**:157–172.
Dickie, I. A., and P. B. Reich, 2005, Ectomycorrhizal fungal communities at forest edges, *J. Ecol.* **93**:244–255.
Dighton, J., 2005, *The Fungal Community: Its Organisation and Role in the Ecosystem*. Marcel Dekker, New York.
Donnelly, D. P., and L. Boddy, 2001, Mycelial dynamics during interactions between *Stropharia caerulea* and other cord-forming, saprotrophic basidiomycetes, *New Phytol.* **151**:691–704.
Donnelly, D. P., L. Boddy, and J. R. Leake, 2004, Development, persistence and regeneration of foraging ectomycorrhizal mycelial systems in soil microcosms, *Mycorrhiza* **14**:37–45.
Dowson, C. G., and A. D. M. Rayner, 1988, The form and outcome of mycelial interactions involving cord-forming decomposer basidiomycetes in homogeneous and heterogeneous environments, *New Phytol.* **109**:423–432.

Falconer, R. E., J. L. Bown, N. A. White, and J. W. Crawford, 2005, Biomass recycling and the origin of phenotype in fungal mycelia, *Proc. Royal Soc. Lond., B, Biol. Sci.* **272**:1727–1734.

Fomina, M., K. Ritz, and G. M. Gadd, 2000, Negative fungal chemotropism to toxic metals, *FEMS Microbiol. Lett.* **193**:207–211.

Frankland, J. C., 1992, Mechanisms of fungal succession, in: *The Fungal Community: Its Organisation and Role in the Ecosystem*, G. C. Carroll, and D. T. Wicklow, eds., Marcel Dekker, New York, pp. 383–402.

Glenn, O. F., J. M. Hainsworth, C. A. Parker, and K. Sivasithamparam, 1987, Influence of matric potential and soil compaction on the growth of the take-all fungus through soil, *Trans. Br. Mycol. Soc.* **88**:83–89.

Glenn, O. F., and K. Sivasithamparam, 1990, The effect of soil compaction on the saprophytic growth of *Rhizoctonia solani*, *Plant Soil* **121**:282–286.

Gow, N. A. R., 1990, Growth and guidance of the fungal hypha, *Microbiology* **140**:3193–3205.

Harris, K., I. M. Young, C. A. Gilligan, W. Otten, and K. Ritz, 2003, Effect of bulk density on the spatial organisation of the fungus *Rhizoctonia solani* in soil, *FEMS Microbiol. Ecol.* **44**:45–56.

Harris, M. J., and L. Boddy, 2005, Nutrient movement and mycelial reorganization in established systems of *Phanerochaete velutina*, following arrival of colonized wood resources, *Microb. Ecol.* **50**:141–151.

Harold, S., G. M. Tordoff, T. H. Jones, and L. Boddy, 2005, Mycelial responses of *Hypholoma fasciculare* to collembola grazing: effect of inoculum age, nutrient status and resource quality, *Mycol. Res.* **109**:927–935.

He, X. L., S. Mouratov, and Y. Steinberger, 2002, Spatial distribution and colonization of arbuscular mycorrhizal fungi under the canopies of desert halophytes, *Arid Land Res. Manag.* **16**:149–160.

Hedlund, K., L. Boddy, and C. M. Preston, 1991, Mycelial responses of the soil fungus, *Mortierella isabellina*, to grazing by *Onychiurus armatus* (Collembola), *Soil Biol. Biochem.* **23**:361–366.

Hettiara, D. R., 1973, Stress deformation behaviour of soil in root growth mechanics, *J. Agric. Sci.* **18**:309–320.

Jennings, D. H., and G. Lysek, 1999, *Fungal Biology*. Bios Scientific Publishers, Oxford, UK.

Jennings, D. H., J. D. Thornton, M. F. J. Galpin, and C. R. Coggins, 1974, Translocation in fungi. In *Transport at the Cellular Level*, M. A. Sleigh, and D. H. Jennings, eds., Cambridge University Press, Cambridge, UK, pp. 139–156.

Jorgensen, H. B., S. Elmholt, and H. Petersen, 2003, Collembolan dietary specialisation on soil grown fungi, *Biol. Fertil. Soils* **39**:9–15.

Kaldorf, M., C. Renker, M. Fladung, and F. Buscot, 2004, Characterization and spatial distribution of ectomycorrhizas colonizing aspen clones released in an experimental field, *Mycorrhiza* **14**:295–306.

Kampichler, C., J. Rolschewski, D. P. Donnelly, and L. Boddy, 2004, Collembolan grazing affects the growth strategy of the cord-forming fungus *Hypholoma fasciculare*, *Soil Biol. Biochem.* **36**:591–599.

Kaneda, S., and N. Kaneko, 2004, The feeding preference of a collembolan (*Folsomia candida* Willem) on ectomycorrhiza (*Pisolithus tinctorius* (Pers.)) varies with mycelial growth condition and vitality, *Appl. Soil Ecol.* **27**:1–5.

Katz, D., D. Goldstein, and R. F. Rosenberger, 1972, Model for branch initiation in *Aspergillus nidulans* based on measurements of growth parameters, *J. Bacteriol.* **109**:1097–1100.

Last, F. T., P. A. Mason, K. Ingleby, and L. V. Fleming, 1984, Succession of fruitbodies of sheathing mycorrhizal fungi associated with *Betula-Pendula*, *For. Ecol. Manage.* **9**:229–234.

Lilleskov, E. A., T. D. Bruns, T. R. Horton, D. L. Taylor, and P. Grogan, 2004, Detection of forest stand-level spatial structure in ectomycorrhizal fungal communities, *FEMS Microbiol. Ecol.* **49**:319–332.

Lindahl, B. D., and S. Olsson, 2004, Fungal translocation – creating and responding to environmental heterogeneity, *Mycologist* **18**:79–88.

LopezFranco, R., and C. E. Bracker, 1996, Diversity and dynamics of the Spitzenkorper in growing hyphal tips of higher fungi, *Protoplasma* **195**:90–111.

Matsuda, Y., and N. Hijii, 1998, Spatiotemporal distribution of fruitbodies of ectomycorrhizal fungi in an *Abies firma* forest, *Mycorrhiza* **8**:131–138.

McKenzie, B., and A. R. Dexter, 1988, Radial pressures generated by the earthworm *Aporrectodea rosea*, *Biol. Fertil. Soils* **5**:323–327.

Meskauskas, A., M. D. Fricker, and D. Moore, 2004, Simulating colonial growth of fungi with the Neighbour-Sensing model of hyphal growth, *Mycol. Res.* **108**:1241–1256.

Money, N. P., 2004, The fungal dining habit – a biomechanical perspective, *Mycologist* **18**:71–76.

Mottonen, M., E. Jarvinen, T. J. Hokkanen, T. Kuuluvainen, and R. Ohtonen, 1999, Spatial distribution of soil ergosterol in the organic layer of a mature Scots pine (*Pinus sylvestris* L.) forest, *Soil Biol. Biochem.* **31**:503–516.

Mummey, D. L., M. C. Rillig, and W. E. Holben, 2005, Neighboring plant influences on arbuscular mycorrhizal fungal community composition as assessed by T-RFLP analysis, *Plant Soil* **271**:83–90.

Niksic, M., I. Hadzic, and M. Glisic, 2004, Is *Phallus impudicus* a mycological giant? *Mycologist* **18**:21–22.

Otten, W., and C. A. Gilligan, 2006, Soil structure and soil-borne diseases: using epidemiological concepts to scale from fungal spread to plant epidemics, *Eur. J. Soil Sci.* **57**:26–37.

Otten, W., C. A. Gilligan, C. W. Watts, A. R. Dexter, and D. Hall, 1999, Continuity of air-filled pores and invasion thresholds for soil-borne fungal plant pathogen, *Rhizoctonia solani*, *Soil Biol. Biochem.* **31**:1803–1810.

Otten, W., K. Harris, I. M. Young, K. Ritz, and C. A. Gilligan, 2004, Preferential spread of the pathogenic fungus *Rhizoctonia solani* through structured soil, *Soil Biol. Biochem.* **36**:203–210.

Queloz, V., C. R. Grunig, T. N. Sieber, and O. Holdenrieder, 2005, Monitoring the spatial and temporal dynamics of a community of the tree-root endophyte *Phialocephala fortinii sl.*, *New Phytol.* **168**:651–660.

Rayner, A. D. M., 1991, The challenge of the individualistic mycelium, *Mycologia* **83**:48–71.

Rayner, A. D. M., 1994, Pattern-generating processes in fungal communities, in: *Beyond the Biomass: Compositional and Functional Analysis of Soil Microbial Communities*, K. Ritz, J. Dighton, and K. E. Giller, eds., Wiley, Chichester, UK, pp. 247–258.

Read, N. D., L. J. Kellock, T. J. Collins, and A. M. Gundlach, 1997, Role of topographic sensing for infection-structure differentiation in cereal rust fungi, *Planta* **202**:163–170.

Regalado, C. M., B. D. Sleeman, and K. Ritz., 1997, Aggregation and collapse of fungal wall vesicles in hyphal tips: a model for the origin of the Spitzenkörper, *Proc. Royal Soc. Lond., B.* **352**:1963–1974.

Riquelme, M., C. G. Reynaga-Pena, and G. Gierz, 1998, What determines growth direction in hyphae? *Fungal Genet. Biol.* **24**:101–109.

Ritz, K., 1995, Growth responses of some soil fungi to spatially heterogeneous nutrients, *FEMS Microbiol. Ecol.* **16**:269–280.

Ritz, K., 2006, Fungal roles in transport processes in soils, in: *Fungi in Biogeochemical Cycles*, G. M. Gadd, ed., Cambridge University Press, Cambridge, UK, pp. 51–73.

Ritz, K., and J. W. Crawford, 1990, Quantification of the fractal nature of colonies of *Trichoderma viride*, *Mycol. Res.* **94**:1138–1142.

Ritz, K., and J. W. Crawford, 1999, Colony development in nutritionally heterogeneous environments, in: *The Fungal Colony*, N. A. Gow, G. Robson, and G. M. Gadd, eds., Cambridge University Press, Cambridge, UK, pp. 49–74.

Ritz, K., and I. M., Young, 2004, Interactions between soil structure and fungi, *Mycologist* **18**:52–59.

Ritz, K., S. M. Millar, and J. W. Crawford, 1996, Detailed visualisation of hyphal distribution in fungal mycelia growing in heterogeneous nutritional environments. *J. Microbiol. Meth.* **25**:23–28.

Robinson, C.H., J. Dighton, and J.C. Frankland, 1993, Resource capture by interacting fungal colonies on straw, *Mycol. Res.* **97**:547–558.

Sabatini, T., M. Ventura, and G. Innocenti, 2004, Do Collembola affect the competitive relationships among soil-borne plant pathogenic fungi? *Pedobiologia* **48**:603–608.

Saleh-Lakha, S., M. Miller, R. G. Campbell, K. Schneider, P. Elahimanesh, M. M. Hart, and J. T. Trevors, 2005, Microbial gene expression in soil: methods, applications and challenges, *J. Microbiol. Meth.* **63**:1–19.

Sbrana, C., and M. Giovannetti, 2005, Chemotropism in the arbuscular mycorrhizal fungus *Glomus mosseae*, *Mycorrhiza* **15**:539–545.

Schack-Kirchner, H., K. V. Wilpert, and E. E. Hildebrand, 2000, The spatial distribution of soil hyphae in structured spruce-forest soils, *Plant Soil* **224**:195–205.

Stahl, P. D., and M. Christensen, 1992, *In vitro* mycelial interactions among members of a soil micro-fungal community, *Soil Biol. Biochem.* **24**:309–316.

Sturrock, C. J., K. Ritz, W. B. Samson, J. L. Bown, H. J. Staines, J. W. Palfreyman, J. W. Crawford, and N. A. White, 2002, The effects of fungal inoculum arrangement (scale and context) on emergent community development in an agar model system, *FEMS Microbiol. Ecol.* **39**: 9–16.

Toyota, K., K. Ritz, and I. M. Young, 1996a, Microbiological factors affecting the colonisation of soil aggregates by *Fusarium oxysporum* F. sp. *Raphani*, *Soil Biol. Biochem.* **28**:1513–1521.

Toyota, K., K. Ritz, and I. M. Young, 1996b, Survival of bacterial and fungal populations following chloroform-fumigation: effects of soil matric potential and bulk density, *Soil Biol. Biochem.* **28**:1545–1547.

Trinci, A. P. J., 1984, Regulation of hyphal branching and hyphal orientation, in: *The Ecology and Physiology of the Fungal Mycelium*, H. H. Jennings, and A. D. M. Rayner, eds., Cambridge University Press, Cambridge, UK, pp. 24–52.

Watters, M. K., and A. J. F. Griffiths, 2001, Tests of a cellular model for constant branch distribution in the filamentous fungus *Neurospora crassa, Appl. Environ. Microbiol.* **67**:1788–1792.

White, N. A., C. J. Sturrock, K. Ritz, W. B. Samson, J. L. Bown, H. J. Staines, J. W. Palfreyman, and J. W. Crawford, 1998, Interspecific fungal interactions in spatially heterogeneous systems, *FEMS Microbiol. Ecol.* **27**:21–32.

Zakaria, A. J., and L. Boddy, 2002, Mycelial foraging by *Resinicium bicolor*: interactive effects of resource quantity, quality and soil composition, *FEMS Microbiol. Ecol.* **40**:135–142.

Chapter 8

SPATIAL HETEROGENEITY OF PLANKTONIC MICROORGANISMS IN AQUATIC SYSTEMS
Multiscale patterns and processes

Bernadette Pinel-Alloul[1] and Anas Ghadouani[2]
[1]*Groupe de Recherche Interuniversitaire en Limnologie (GRIL), Université de Montréal, Montréal, Québec, Canada;* [2]*Aquatic Ecology and Ecosystem Studies, School of Environmental Systems Engineering, The University of Western Australia, Perth, Western Australia, Australia*

Abstract: Patchiness of planktonic microorganisms may have important implications in microbial communities not only at small scale within habitats but also at large scales within lake basins and districts in landscapes, and within oceanic regions and biogeographical provinces. However, studies are generally limited to one specific planktonic entity (bacterio-, phyto-, or zooplankton) or one spatial scale and extent (across oceans or freshwater systems, or within systems), and there is still no functional perspective on multiscale patchiness patterns of microbial communities and their generative processes. This review presents some of the key aspects of plankton spatial heterogeneity including concepts, patterns, and processes in the context of a multiscale perspective. The ecological significance of spatial heterogeneity for planktonic microorganisms is presented with a functional perspective relating distribution patterns to environmental processes. The importance of abiotic and biotic forces and that of the biophysical coupling in structuring microbial community in aquatic systems at scales relevant to ecological states or processes of organisms, populations, and ecosystems is discussed. The importance of the application of new and advanced technology, as well as statistical approches is presented and their spatial relevance discussed.

Keywords: Spatial heterogeneity, microorganisms, plankton communities, marine and freshwater ecosystems

1. INTRODUCTION

Environmental heterogeneity is fundamental to the structure and dynamics of ecosystems (Levin, 1992). Natural ecosystems are heterogeneous at scales ranging from microhabitats to landscapes (Sparrow, 1999). In aquatic systems, environmental heterogeneity arises as the result of vertical and horizontal structuring of marine and freshwater habitats, and is reflected in the distribution of planktonic microorganisms across multiple spatial scales (Giller et al., 1994; Raffaelli et al., 1994; Pinel-Alloul, 1995). Unlike terrestrial systems where the spatial distribution of microorganisms is driven by variations in soil properties and plant distribution in a solid medium (Franklin and Mills, 2003), in aquatic systems, spatial variation in planktonic microbial communities is influenced by the physical, chemical, and biological properties of a fluid medium (Neill, 1994). Environmental drivers of spatial heterogeneity in planktonic microorganisms vary hierarchically according to spatial scales and microbial entities. At small and fine scales, spatial heterogeneity of bacteria, protists and algal cells is due, firstly, to physical processes such as local turbulent mixing, thermal stratification, light transparence, and nutrient layering of the water column (Reynolds, 1994; Harvey et al., 1997; Seymour et al., 2004), and secondly to biological processes such as depth-stratified grazing by zooplankton (Kettle et al., 1987; Williamson et al., 1996). For instance, fine-scale vertical physical stratification of aquatic systems create deep, thin layers of microbial plankton in small, wind-sheltered lakes (Pick et al., 1984; Pedrós-Alió et al., 1987; Lindholm, 1992; Gervais, 1998), and in continental shelves and coastal fjords of oceans (Cowles, 2003). Larger and more motile microorganisms such as zooplankton are also affected by physical processes but they mainly respond to biological processes related to spatial variations in biological drivers such as food resources and/or predators. These larger organisms usually experience medium- to large-scale spatial heterogeneity and aggregate along oceanic fronts or in lake-water strata that provide high concentrations of food and low vulnerability to predators (Pinel-Alloul et al., 1999; Harvey et al., 2001; Clark et al., 2001; Masson et al., 2001; Thackeray et al., 2004; Masson et al., 2004). In landscapes, large-scale spatial heterogeneity of microbial planktonic communities in freshwater systems reflects regional biogeographical and historical processes related to geology, topography, and climate, and local changes in the morphometry, water chemistry, trophic status, and disturbance of lakes and rivers (Pinel-Alloul et al., 1990a, b, Pinel-Alloul et al., 1995; Stemberger and Lazorchak, 1994; Stemberger et al., 1996; Pinto-Coelho et al., 2005). Studying large-scale spatial heterogeneity of lake plankton in landscapes helped develop concepts of metacommunities (Liebold et al., 2004) and meta-ecosystems (Loreau et al., 2003). Freshwater plankton microbial community is now perceived as a "complex adaptive system" (CAS theory; Leibold and Norberg, 2004; Norberg, 2004) in which emergent spatial patterns result not only from local processes (within and among ecosystems) but also from regional processes related to plankton dispersal and hydrological connectivity within lake and river networks (Kratz et al., 1997; Forbes and Chase, 2002; Shurin et al., 2000; Cottenie et al., 2003; Cottenie and De Meester, 2003; Cohen and Shurin, 2003). In oceans, large-scale patchiness of microbial plankton is related to biogeographical provinces (Haury et al., 1978; Longhurst, 1998; Platt and Sathyendranath, 1999) and major advective forces such as gyres and frontal zones (Clark et al., 2001; Garçon et al., 2001; Platt et al., 2005), upwelling currents and eddies (Crawford et al., 2005), and coastal freshwater runoffs (Seppälä and Balode, 1999).

Patchiness is a common feature in plankton distribution in aquatic systems and is now viewed as the rule rather then the exception (Pinel-Alloul, 1995). Plankton spatial heterogeneity occurs on a hierarchical continuum of scales in aquatic systems ranging in size from the smallest lakes to the largest oceanic provinces (Mackas et al., 1985; Giller et al., 1994; Pinel-Alloul, 1995; Long and Azam, 2001) and this patchiness is fundamental to population dynamics, community organization, and stability (Mehner et al., 2005). Functional heterogeneity is the concept that links spatial patchiness to environmental processes operating hierarchically over different scales of the environment and different levels of biological organization (Kolasa and Pickett, 1991). At the organismal level, the focus is on the morphological, physiological, and behavioural responses to small-scale heterogeneity in resource availability and threats from competitors, predators, and parasites. At the population and community levels, the focus is on the responses to large-scale environmental heterogeneity and spacing of suitable natural habitat patches in ecosystems and landscapes. Therefore, a multiscale perspective is essential to identify and characterize the pertinent scales of spatial dependency of aquatic microbial entities, to appreciate the nature and magnitude of sources of variation, and to determine the underlying abiotic and biotic processes governing the spatial distribution of aquatic microorganisms, and their relative forces at multiple scales.

In this chapter, we have limited the scope of the review to planktonic organisms starting from bacteria up to zooplankton. The aim of the review is to attract attention to some key aspects of plankton spatial heterogeneity including concepts, patterns, and processes in the context of a multiscale perspective. Firstly, we briefly examine the concepts and the ecological significance of spatial heterogeneity for planktonic microorganisms with a functional perspective relating distribution patterns to environmental processes. Secondly, we present a multiscale perspective of patchiness patterns of planktonic microorganisms and their generative processes. Then, we discuss the importance of abiotic and biotic forces and that of the biophysical coupling in structuring microbial community in aquatic systems at scales relevant to ecological states or processes of organisms, populations, and communities. We then examine how biophysical coupling scales up through planktonic food webs. We, especially, compare the relative influence of abiotic and biotic processes on spatial distribution of microbial planktonic communities in oceans, lakes, and rivers across a continuum of scales. Finally, we present and discuss the applicability of recent advanced sampling technologies to assess plankton spatial heterogeneity along a continuum of scales, as well as the relevance of geostatistic, multivariate, and multifractal analyses for describing multiscale spatial patterns of aquatic microorganisms and modelling their environmental control.

2. CONCEPTS AND ECOLOGICAL SIGNIFICANCE OF SPATIAL HETEROGENEITY IN PLANKTON

The concepts of scale and spatial heterogeneity is very promising as the integrative basis for modern ecology, since spatio-temporal heterogeneity, organizational hierarchies and body size are the main scaling factors for ecological patterns and processes (Azovsky, 2000). As ecological systems are always hierarchically organized (O'Neill et al., 1986), spatial patchiness is now recognized as an essential property of nature (Kolasa and Pickett, 1991).

It is well established that spatial distribution of planktonic organisms is strongly heterogeneous (Giller et al., 1994; Pinel-Alloul, 1995). This phenomenon is commonly called plankton patchiness. Patchiness is a fundamental property inherent to all plankton communities, and elucidation of its origin is of significant importance for both research and practical applications. Traditionally, studies on spatial heterogeneity of aquatic organisms focused on the estimation and comparison of patchiness and aggregation patterns among different communities or environments without referring to their generative processes. This concept of "measured heterogeneity" based on the variance mean ratio (the variance function: see Downing et al., 1987; Pace et al., 1991) and other indices of spatial heterogeneity (Downing, 1991) provided powerful tools for comparing patchiness patterns at different spatial scales and across ecological entities and ecosystems. However, it lacks biological relevance since it reveals little on how organisms are organized in relation to environmental processes (Pinel-Alloul, 1995). The alternative concept of "functional heterogeneity" arose from a new perspective in which ecological interactions between organisms and their environment are studied at spatial scales at which individuals, populations, and communities perceive habitat variation and operate with environmental factors (Kolasa and Rollo, 1991). In aquatic systems, microbial planktonic entities can be bacterial and algal cells, zooplankton animals, or populations and communities. The degree of functional heterogeneity increases with habitat complexity and scale extent, and with the level of biological organization, from individuals to communities. Recent developments in advanced sampling technologies and spatial modelling allowing simultaneous analysis of spatial patterns of microorganisms and of physical, chemical, and biological properties of the environment, greatly helped to investigate functional heterogeneity in microbial planktonic communities in marine and freshwater ecosystems (Sprules et al., 1992; Postel et al., 2000; Seuront and Strutton, 2003).

The patchy distribution is fundamental for all microbial communities from bacteria to phytoplankton and zooplankton. Aquatic microorganism's patchiness influences population dynamics, nutrient uptake and recycling, trophic interactions, feeding, mating, and predation, and overall, plankton microbial community structure and function. Heterotrophic bacteria constitute an important share of total plankton biomass similar or higher than that of primary producers (Cho and Azam, 1990). Bacterioplankton plays a major role in biogeochemical processes and carbon cycling in marine and freshwater environments, regulating the entry of algal exudates from primary producers into the pool of DOC and decomposing the vertical flux of particles and organic matter (Riemann and Søndergaard, 1986; Cho and Azam, 1988; Cole et al., 1988). Thus, spatial distribution patterns of heterotrophic bacteria play a critical role in the functioning of aquatic systems including oceans, lakes, and rivers (Kirchman, 2000; Wetzel, 2001).

Phytoplankton patchiness is a well-documented phenomenon in oceanography and limnology (Charlson et al., 1987; Harris, 1994). Plankton community patchiness (or variability) is of current interest because of its impact on water quality, aquatic productivity, eutrophication, and global climate changes. Chlorophyll a (Chl-a) biomass is the most comprehensive descriptor of phytoplankton patchiness across multiple scales in oceans (Platt and Sathyendranath, 1999; Li and Harrison, 2001; Lovejoy et al., 2001; Seymour et al., 2005) and freshwaters (Harris, 1994; Reynolds, 1994; Kling et al., 2000; Jones and Knowlton, 2005). Phytoplankton patchiness is coupled with bacterial production and zooplankton grazing in both marine and freshwaters, manifesting the dependence of bacteria on resources supplied to them by phytoplankton, either directly through exudation of labile organic photosynthates or cell lysis, or indirectly through sloppy feeding of zooplankton grazers (White et al., 1991; Cowles et al., 1998; Lie et al., 2004).

Zooplankton heterogeneity across multiple spatial scales is also an important focus of aquatic ecological research because zooplankton serves a key role in pelagic food webs as they transfer energy from primary producers to higher trophic levels (Pinel-Alloul, 1995). The ecological importance of zooplankton spatial heterogeneity for the structure and functioning of aquatic systems has been highlighted with respect to various processes: species reproduction, population dynamics, and prey–predator interactions in marine water (Legendre and Michaud, 1998; Clark et al., 2001) and freshwater (Sprules and Munawar, 1986; Sprules et al., 1991; Pinel-Alloul, 1995).

Patchiness of planktonic microorganisms may have important implication in microbial communities not only at small scale within habitats but also at large scales within lake basins and districts in landscapes, and within oceanic regions, and biogeographical provinces. However, studies are generally limited to one specific planktonic entity (bacterio-, phyto-, or zooplankton) or one spatial scale and extent (across oceans or freshwater systems, or within systems), and there is still no functional perspective on multiscale patchiness patterns of microbial communities and their generative processes.

3. MULTISCALE PERSPECTIVE OF MICROBIAL SPATIAL PATTERNS AND PROCESSES

Plankton patchiness occurs along a hierarchical continuum of spatial scales in aquatic systems (Fig. 8-1). Traditionally, scales of plankton patchiness have been defined independently in oceans and lakes. In marine ecosystems, Haury et al. (1978) proposed six spatial scales of plankton patchiness ranging from large to small extents: mega-scale and macroscale (10^3–10^4 km), mesoscale (10^2–10^3 km), coarse-scale (100 m–10^2 km), fine-scale (1 m–1 km), and microscale (1 cm–10 m). In lakes, Malone and McQueen (1983) and Pinel-Alloul (1995) proposed four types of spatial scales of plankton patchiness: large-scale (>1 km), coarse-scale (10 m–1 km), fine-scale or small-scale (1–10 m), and microscale (<1 m). Recently, the range of spatial scales of interest for plankton patchiness has expanded. On one hand, landscape and latitudinal perspectives applied to lake studies (Kratz et al., 1997; Pinto-Coelho et al., 2005), allowed assessing spatial heterogeneity of freshwater plankton communities at regional (10^2–10^3 km) and global (10^3–10^4 km) scales of similar extent than the largest scales (meso- and mega-scale) for marine plankton. On the other hand, new technologies such as *in situ* fluorometry (Beutler et al., 2002; Ghadouani and Smith, 2005), video-recording (Tiselius, 1998; Davis et al., 2004), high-resolution digital camera recording (Benfield et al., 2003) and new fine-scale sampling techniques such as rapid freezing (Krembs et al., 1998a, b) or pneumatically operated sampling (Seymour et al., 2000; Lunven et al., 2005) enable the study of microbial patchiness at microscale (1 cm–1 m) and nanoscales (1 µm–1 cm). All these scales for plankton patchiness in marine and fresh waters are nested along a continuum of spatial and time scales (Fig. 8-1). This scale-continuum represents an integrative framework for assessing multiscale patterns and processes of plankton patchiness in aquatic systems.

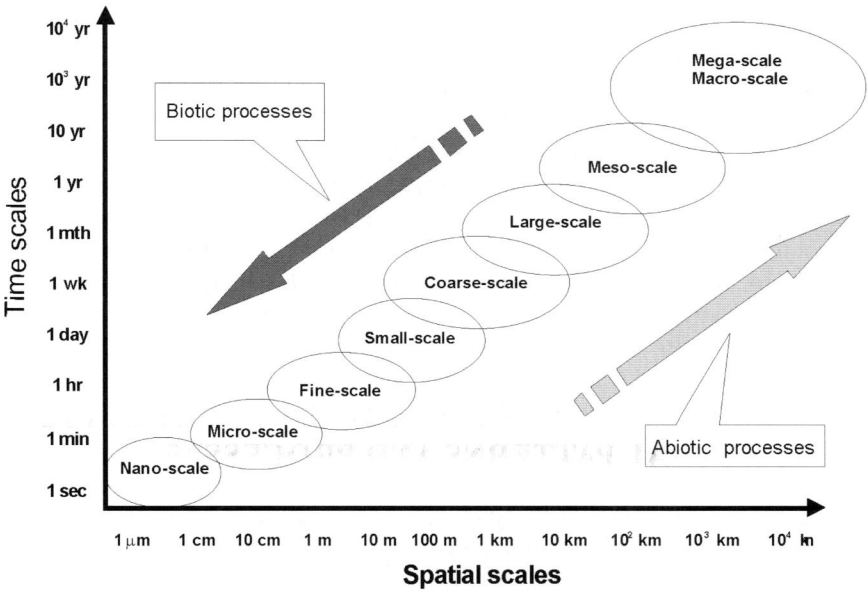

Figure 8-1. Categories of spatial scales in marine and freshwater systems.

Currently, there is no integrative review on the multiscale characteristics of spatial heterogeneity of all planktonic communities in marine and freshwater systems. The most recent review (Pinel-Alloul, 1995) focused on zooplankton patchiness and its generative processes in oceans and lakes, and proposed a multiscale perspective based on the "multiple driving forces" hypothesis. The multiscale perspective implies that the hierarchical levels of spatial scales are nested so that large-scale plankton patches consist of aggregations of small-scale patches. The physical and biological mechanisms responsible for structuring zooplankton in space refer to the multiple driving forces hypothesis (Pinel-Alloul, 1995). The multiple forces hypothesis states that neither abiotic processes nor biotic processes alone, can explain the complexity of plankton spatial structure in aquatic systems. It rather indicates that plankton patchiness patterns are driven by many abiotic processes interacting with many biotic processes, and that the relative influence of abiotic and biotic processes varies along the scale continuum (Fig. 8-1). The relative importance of these processes corresponds to a gradation in effects over scales, the physical effects predominating at broad spatial scales, while biological effects predominate at finer scales (Pinel-Alloul, 1995).

Aquatic microbial communities are organized at a variety of spatial scales, which likely reflect the scales of heterogeneity in microbial distribution and physical, chemical, and biological properties of the fluid environment (Neill, 1994). However, as the importance of microorganisms to ecosystems is in terms of their mediation of ecological processes, spatial scales investigated in studies generally reflect the scales thought to be appropriate for a specific community (bacterioplankton, phytoplankton, or zooplankton) and a specific generative process (physical or biological forces). Most of studies conducted on spatial patterns of microbial organisms in aquatic environments have considered coarse community univariate attributes (e.g., total abundance, biomass, activity, and production) (Li and Harrison, 2001; Kling et al., 2000; Bode et al., 1996; Seymour et al., 2000; Gazol and del Giorgio, 2000), although some studies have examined finer multivariate attributes such as species assemblages (Pinel-Alloul et al., 1990a, b; Harvey et al., 1997), functional guilds (Pinel-Alloul et al., 1996; Reynolds and Petersen, 2000), biomass size classes (Masson et al., 2000, 2004), algal spectral groups (Feitz et al., 2005), and bacterial morphotypes (cocci, rods, vibrio) or genotypes based on molecular techniques (5′nuclease assays, 16 SrRNA genes, ARISA profiles) (Fisher and Triplett, 1999; Jochem, 2001; Suzuki et al., 2001). Although it has been suggested that, because of the hierarchical nature of spatial variability, multiscale analysis is crucial to fully represent the complexity of natural systems, many of the studies on spatial variability of microbial plankton focused on a single scale. Given that environmental factors do not necessarily operate independently, or at distinct spatial scales, but on a continuum of scales, studying microbial systems using a single analytical scale cannot provide a complete understanding of community dynamics (Franklin and Mills, Chapter 1, this volume). Multiscale comparisons, in which patterns are analysed at different spatial scales, may be more useful when trying to identify the factors that control microbial community distribution. Given that the mechanisms responsible for developing and maintaining heterogeneity among aquatic microbial communities vary according to scale, the extent and nature of biological variability will also differ with scale. Small- and microscale variability in microbial abundance can often be as great as, or greater than the variability at the largest scales. For instance, abundances of bacteria and phytoplankton can be less variable across large-scale features in oceanic provinces or regional lake districts, than across small-scale features of deep layers of microbial organisms in oceans and lakes. The nature and pattern of variability in microbiological community will also differ with scale. Over large scales, spatial variability is characterized by distribution patterns that remain relatively stable and are often predictable over time. In contrast, microscale processes are generally more intermittent, and are likely to generate spatial distributions characterized by the ephemeral existence of discrete spots of high patchiness and low patchiness (Seymour et al., 2004).

The idea of considering the spatial variability of planktonic communities as a multiscale process was introduced into oceanography by Denman and Platt (1976) and Haury et al. (1978) and later into limnology by Pinel-Alloul (1995). However, aquatic scientists have only recently begun to focus on multiscale comparisons, and started to find evidence for nested scales of spatial structure in microbial communities. Multiscale studies on microbial patchiness in marine and freshwater systems are still rare but constitute exiting new directions for the study of patchiness processes. For bacteria, spatial variation in community composition in Wisconsin lakes was analysed at small (within-lake) and large (among-lake) spatial scales (Yannarell and Triplet, 2004). The magnitude of variation in bacterial community was found small within lake compared to among-lake variation. Spatial heterogeneity at large scale was driven by regional gradients in lake productivity and concentration of bacterial substrates, while small-scale variation was due to spatial isolation of lake subbasins. For phytoplankton, the best documented multiscale study has been conducted in marine systems over scales of millimetres to thousand of kilometres using satellite remote sensing, and fluorescence and ocean colour as proxy measurements of chlorophyll biomass (Lovejoy et al., 2001). This multiscale study gave birth to the multi-fractal cascade picture now accepted as a good approximation to link spatial patterns of oceanic turbulence and phytoplankton distribution across scales. Under this perspective, mesoscale patchiness associated with algal bloom events occur in response to large-scale oceanographical features; embedded within these large-scale processes are small-scale and microscale patchiness processes occurring across distances of micrometres to centimetres due to small-scale horizontal and vertical variations in dissolved organic and inorganic substrates, sinking particulate matter, turbulence, and water stability (Seymour et al., 2005). Studies assessing multiscale variations in phytoplankton are still scarce, and the multifractal cascade picture has not been fully explored yet. Multiscale analysis of spatial variation in the biomass of autotrophic picoplankton and nanoplankton showed that large-scale variation among lakes driven by gradients in nutrients, conductivity and pH, was higher than small-scale variation within lakes among stratified water layers (Pinel-Alloul et al., 1996). Multiscale studies of zooplankton distribution were conducted in marine systems across scales ranging from tens, hundreds, and thousands of kilometres using continuous plankton recording (Piontkovski and Williams, 1995). In coastal areas, multiscale distribution of marine calanoids differed from this of purely passive chemical variables (i.e., temperature and salinity) and of phytoplankton cells, indicating that biological forces such as food and predators are strong drivers of calanoid distribution patterns (Seuront and Lagadeuc, 2001). In coastal lagoon, zooplankton community spatial distribution was studied from microhabitat to ecosystem scales (Avois-Jacquet, 2002). Local hydrodynamics, phytoplankton distribution, and zooplankton behaviour were found the main processes generating and maintaining the spatial patchiness of zooplankton across scales. In freshwater systems, multiscale studies of spatial heterogeneity of zooplankton within lakes indicated higher patchiness on the vertical than on the horizontal dimensions in well stratified lakes (Pinel-Alloul and Pont, 1991; Pinel-Alloul et al., 2004). Other factors, zooplankton body size, sampling scale, and depth, also accounted for significant variation in zooplankton patchiness across scales within lakes (Pinel-Alloul et al., 1988). Multiscale studies conducted across scales on zooplankton spatial heterogeneity in lake districts showed the greatest variability among lakes in relation to productivity gradients, although small-scale variability among water layers in thermally stratified lakes can also be of importance (Masson et al., 2004).

4. MULTISCALE PATTERNS AND PROCESSES OF MICROBIAL PLANKTON PATCHINESS

To adequately review patterns and processes of spatial distribution of planktonic microbial organisms across multiple spatial scales, first we should describe common patchiness patterns of microbial populations and communities, and then determine the driving forces that control their distribution patterns at each spatial scale in marine (Table 8-1) and freshwater (Table 8-2) systems.

Table 8-1. Multiscale perspective of the main abiotic and biotic processes driving microbial patchiness in marine systems. Categories of spatial scales follow the continuum of scales defined by Haury et al. (1978) for marine systems.

Spatial scales	Marine systems	Dominant patterns	Microbial attributes	Ecological processes	
				Abiotic	Biotic
Mega scale Macroscale 10^3–10^4 km	Open oceans[1-6]	Marine biomes Biogeographical provinces Abiotic control	Bacterioplankton Phytoplankton	Physical forcing Water circulation, sea temperature, nutrient fluxes	Bacterial–algal coupling
	Open oceans[7]	Latitudinal gradients Abiotic control	Zooplankton	Physical Forcing Latitude, sea temperature, energy availability	Primary production Primary–secondary producers coupling
Mesoscale 10^2–10^3 km	Coastal zones, Gulfs, Bays, Straits, Estuaries[8-14]	Frontal zones, hydrological features, topographic features, shelf to deep-ocean patterns, basin-wide horizontal patterns Abiotic control	Bacterioplankton Phytoplankton	Physical forcing Mesoscale eddies, upwelling fronts, shelf-break fronts, cyclonic gyres, tidal currents, river outflows, salinity gradients, nutrients inputs, organic carbon inputs	Bacterial–algal coupling Surface chlorophyll distribution
	Coastal zones, gulfs, bays, straits, estuaries[15-29]	Inshore–offshore gradients Abiotic control	Zooplankton	Physical forcing Basin geomorphology, river runoff, upwelling currents, estuarine circulation, temperature circulation, temperature and salinity gradients	Plankton diel migration, Plankton food web interactions
Coarse scale 1–100 km	Coastal bays and straits[30-32]	Inshore–offshore patterns, horizontal patterns Abiotic control	Sea-ice algae Phytoplankton	Physical forcing River plumes, salinity gradient, nutrient gradient	*In situ* algal growth Zooplankton grazing

continued

Spatial scales	Marine systems	Dominant patterns	Microbial attributes	Ecological processes	
				Abiotic	Biotic
	Estuaries, fjords[33-35]	Inshore–offshore patterns Horizontal patterns Abiotic and biotic control	Zooplankton Copepods	Wind regime, water discharge and residence time, marine inflows, tidal currents, freshwater runoff, exchange of water masses, physical vertical structure, salinity gradients	Food distribution Zooplankton vertical distribution
Fine scale 1–100 m	Coastal bays[36-38]	Horizontal patterns Abiotic control	Phytoplankton *Alexandrium*, *Gymnodinium*	Physical forcing Coastal currents, retentive gyres	Algal blooms Reseeding of resting cysts
	Open oceans[39-50]	Vertical patchiness Deep chlorophyll maximum: DCM Abiotic and biotic control	Phytoplankton Picocyanobacteria *Synechococcus* Mixotrophic flagellates *Chrysochromulina* Dinoflagellates *Gymnodinium* *Gyrodinium*	Strong density gradients Frontal features, historical wind stress upwelling currents, euphotic depth, light gradients, nutrient gradients, nutricline depth, pycnocline depth, nutrient limitation in surface waters, nutrient diffusion across pycnocline	Phytoplankton adaptation to low light Phytoplankton sinking Algal mixotrophy Zooplankton grazing
	Oceans Coral reefs[51-64]	DVM patterns Abiotic and biotic control	Zooplankton Copepods Euphausids	Light stimuli, temperature gradient, oxygen gradient, depth of pycnocline, depth of oxycline, oxygen minimum zone, salinity stratification	Predation risk Body size Color and morphology, Chlorophyll layers, Negative light phototaxis
Microscale 1 mm–10 cm		Patch distribution Aggregation patterns Abiotic control	Bacteria subpopulations	Microscale features: water viscocity, micro turbulence, organic substrate patches, nutrient patches, polymer dynamics	Phytoplankton cell lysis Phytoplankton particles coagulation processes

continued

Spatial scales	Marine systems	Dominant patterns	Microbial attributes	Ecological processes	
				Abiotic	Biotic
	Oceans[65-68]	Abiotic control	Phytoplankton	Microscale turbulence	Swarm formation
	Oceans[69-73]	Microscale patterns Abiotic and biotic control	Flow cytometry *Synechococcus*	Passive transport Density gradient	School formation Swimming behaviour
	Oceans[74-78]	Microscale patterns Biotic control[74-77]	Zooplankton		Food searching Mate-searching Sexual encounter Predator avoidance

[1]Longhurst (1998); [2]Li and Harrison (2001); [3]Li et al. (2004); [4]Platt and Sathyendranath (1999); [5]Platt et al. (2005); [6]White et al. (1995); [7]Woodd-Walker et al. (2002); [8]Sherr et al. (2001); [9]Hansell and Ducklow (2003); [10]Nagata et al. (2000); [11]Jochem (2001); [12]Martin (2003); [13]Crawford et al. (2005); [14]Batten and Crawford (2005); [15]Harvey et al. (1997); [16]Bode et al. (1996); [17]Rantajärvi et al. (1998); [18]Clark et al. (2001); [19]Hays et al. (2001); [20, 21]Beaugrand (2004a, b); [22]Ashjian et al. (2005); [23]Coyle (2005); [24]Hunt and Hosie (2005); [25]Simard and Mackay (1989); [26]Fernandez et al. (1993); [27]Lo et al. (2004); [28]Roman et al. (2005); [29]Descroix et al. (2005); [30]Monti et al. (1996); [31]Harvey et al. (2001); [32]Lovejoy et al. (2001); [33]Lindahl and Henroth (1988); [34]Falkenhaug et al. (1997); [35]Froneman (2004); [36]Tada et al. (2001); [37]Martin et al. (2005); [38]Townsend et al. (2005); [39]Cullen (1982); [40]Gould (1987); [41]Cowles et al. (1998); [42]Cowles (2003); [43]Ediger and Yilmaz (1996); [44]Holm-Hansen and Hewes (2004); [45]Holm-Hansen et al. (2005); [46]Basterretxea et al. (2002); [47]Figueroa et al. (1998); [48]Lunven et al. (2005); [49]Fennel and Boss (2003); [50]Karlson et al. (1996); [51]Ringelberg (1995); [52]Atkinson et al. (1996); [53]Steele and Henderson (1998); [54]Tarling et al. (1999); [55]Pearre (2003); [56]Bonnet et al. (2005); [57]Smith and Madhupratap (2005); [58]Yahel et al. (2005); [59]Besiktepe et al. (1998); [60]Hays et al. (1994); [61]Irgoien et al. (2004); [62]Hansen et al. (2004); [63]Ashjian et al. (2005); [64]Hidalgo et al. (2005); [65]Mitchell and Fuhrman (1989); [66]Duarte and Vaqué (1992); [67]Seymour et al. (2000); [68]Long and Azam (2001); [69]Waters and Mitchell (2002); [70]Waters et al. (2003); [71]Seuront et al. (2001); [72]Seymour et al. (2005); [73]Cowles (2003); [74]O'Brien (1988); [75]Price (1989); [76]Tiselius (1992); [77, 78]Seuront et al. (2004a, b).

Table 8-2. Multiscale perspective of the main abiotic and biotic processes driving microbial patchiness in freshwater systems. Categories of spatial scales follow the continuum of scales defined by Pinel-Alloul (1995) for freshwater systems.

Spatial scales	Freshwater systems	Dominant patterns	Microbial attributes	Ecological processes	
				Abiotic	Biotic
Large scale >1 km	Lake districts Among-lake patterns[1–11]	Metacommunity patterns Landscape patterns Abiotic control	Bacterioplankton	Abiotic forcing Regional factors, geographic location, landscape factors, lake area, lake topographic position Trophic gradient Total phosphorus (TP) Hydrological retention time, pH, water clarity, dissolved organic carbon (DOC) Landscape factors: lake	Chl*a*, bacteria–phytoplankton coupling
	Lake districts Among-lake patterns[12–20]	Landscape patterns Abiotic control	Phytoplankton	topography, upland and lowland lakes, geological substrate, downslope accumulation of solutes and nutrients, water residence time, catchment/lake area ratio, water chemistry Total phosphorus	Zooplankton grazing
	Lake districts Among-lake patterns[21–36]	Landscape patterns Metacommunity patterns Abiotic and biotic control	Zooplankton	Landscape factors Lake topography, lake morphometry, lake size and depth, lake trophic gradient, Total phosphorus, watershed land use, water transparency,	Species dispersal and colonisation, local biotic factors, competition, predation, riparian and aquatic vegetation

continued

Spatial Heterogeneity of Planktonic Microbes

Spatial scales	Freshwater systems	Dominant patterns	Microbial attributes	Ecological processes	
				Abiotic	Biotic
Large scale >1 km	Large lakes Within-lake patterns[37-41]	Lakewide patterns Abiotic control	Bacteria Phytoplankton	Water chemistry, pH, conductivity, calcium, salinity gradients Physical forcing Wind-induced circulation, water temperature, water stability, non-algal seston, inflow of rivers, onshore–offshore gradient	Bacteria–phytoplankton coupling
	Large lakes Within-lake patterns[37,42-52]	Lakewide patterns Inshore–offshore transects Abiotic and biotic control	Zooplankton	Water stratification Water temperature River water input Water chemistry Conductivity, pH, calcium	Fish predation Invertebrate predation
Large scale >1 km	River districts Among-river patterns[53,54]	Landscape patterns Abiotic control	Bacterioplankton Phytoplankton	Trophic gradients, Total Phosphorus, Allochthonous sources of carbon	Autochthonous sources of carbon
	Large rivers Within-river patterns[55-65]	Longitudinal pattern Transversal patterns Abiotic control	Bacterioplankton Phytoplankton	River discontinuities, lake edge structure, water retention time, mixing conditions, nutrient supply, light limitation, turbidity	Macrophyte beds Resuspension of periphyton
	Large rivers Within-river patterns[57,58,61,66,67]	Longitudinal pattern Transversal patterns Abiotic and biotic control	Zooplankton	Water retention time	Macrophyte beds Chl-*a* Fish predation

continued

Spatial scales	Freshwater systems	Dominant patterns	Microbial attributes	Ecological processes	
				Abiotic	Biotic
Coarse scale 10 m–1km	Large lakes[68]	Horizontal patterns Abiotic control	Phytoplankton	River plumes Drainage basin inflow Major nutrient inputs	
	Deep lakes[69–87]	Vertical patterns Diel vertical migration: DVM Biotic and abiotic control	Zooplankton	Water stratification Vertical trade-offs Temperature profile Oxygen profile Transparency to UV Oxygen in deep waters	Light-negative phototaxis Fish predation Food availability Invertebrate predation Surface avoidance behaviour Thermal preference
	Shallow lakes[88–90]	Horizontal patterns Diel horizontal migration: DHM Biotic control	Zooplankton		Macrophyte beds Fish predation Invertebrate predation
Coarse scale 10m-1km	Small rivers[91]	Horizontal patterns Abiotic control	Phytoplankton	Riverine circulation zone, river flow, local geometry of river channel, dead zones and groynes, water retention time	
Fine scale 1–10 m	Lakes[92–94]	Vertical patchiness Surface algal blooms Abiotic control	Phytoplankton *Microcystis* *Ceratium*	Flow field, wind direction, wind sheltering, current gyres, light gradient	Algal blooms Cyanobacteria blooms Littoral vegetation
	Stratified lakes and reservoirs, meromictic lakes, dimictic lakes[95–98]	Vertical patterns Bacterial deep layers Abiotic control	Green and purple photosynthetic bacteria *Chlorochromatium*, *Chlorobium*, *Lamprobacter*	Water stratification Vertical gradients Thermocline depth Chemocline depth Light gradients Sulphur–oxygen interface	Zooplankton grazing

continued

Spatial scales	Freshwater systems	Dominant patterns	Microbial attributes	Ecological processes	
				Abiotic	Biotic
	Stratified lakes, oligotrophic lakes, meso-eutrophic lakes[99–121]	Phytoplankton deep layers DCM Abiotic and biotic control	Chrysomonads Dynobrion, Ochromonas Cryptophytes Cryptomonas Diatoms Stephanodiscus, Cyclotella, Asterionella, Fragilaria Picocyanobacteria Chroococcus Cyanobacteria Microcystis, Planktothrix, Limnothrix, Pseudanabaena	Water stability Vertical gradients Trophic status Thermocline depth Light penetration Euphotic depth Nutrient availability Nutrient depth gradients Water turbulence Oxygen in deep waters Stream rich-nutrient inflow in deep waters	Algal *in situ* growth Algal low-light adaptation Algal buoyancy Algal motility Algal mixotrophy Algal bacterivory Algal migratory behaviour Zooplankton grazing Zooplankton nutrient cycling
Vertical patterns					
	Stratified, oligo-phic, and mesoeutro-phic lakes[122–125]	Biotic and abiotic control	Zooplankton Rotifers	Vertical gradients Water stratification Oxic–anoxic boundary	Food-rich interface
	Stratified, oligotro-phic, and mesoeutro-ophic lakes[126–129]	Vertical patterns Biotic control	*Daphnia*		DCM layers Food searching
	Deep lakes[130]	Vertical patterns Abiotic control	Copepods	Temperature and oxygen gradients	Deep habitat preference
					continued

Spatial scales	Freshwater systems	Dominant patterns	Microbial attributes	Ecological processes	
				Abiotic	Biotic
	Shallow lakes[131-135]	Horizontal patterns Abiotic and biotic control	Cladocerans	Littoral zone habitat partition and refuge Oxygen levels in shallow and deep waters	Habitat preference Macrophytes beds Sediments and rocks Open water Migration from plant to plant Spatial structure of plants Predation avoidance
Microscale 1–10 cm	Lakes[136-140]	Littoral zone	Zooplankton Rotifers *Brachionus* Cladocerans *Polyphemus* *Daphnia* Copepods *Arctodiaptomus*	Water illumination Water body type Meteorology	Swarm and schools formation Grazing groups Food searching Mating groups Mate-searching behaviour Sexual encounter Predator avoidance Clone or species variation
Nanoscale 1–100 μm	Lakes[141-142]		Bacteria		

[1]Cole et al. (1993); [2]Weisse and MacIsaac (2000); [3]Yannarell and Triplet (2004); [4]Yannarell and Triplet (2005); [5]Reche et al. (2005); [6]Méthé and Zher (1999); [7]Lindström (2000); [8]Lindström (2001); [9]Lindström (2006); [10]Horner-Devine et al. (2003); [11]Bird and Kalff (1984); [12]Moss et al. (1989); [13]Masson et al. (2000); [14]Smith (2003); [15, 25]Pinel-Alloul et al. (1990a, b); [16]Pinel-Alloul et al. (1996); [17]Beisner et al. (2003); [18]Krast and Frost (2000); [19]Kling et al. (2000); [20]Jones and Knowlton (2005); [21]McCauley and Kalff (1981); [22]Yan (1986); [23]Pace (1986); [24]Keller and Colon (1994); [26]Pinel-Alloul et al. (1995); [27]Stemberger and Lacorchak (1994); [28]Masson et al. (2004); [29]Pinto-Coelho et al. (2005); [30]Gyllström et al. (2005); [31]Hessen et al. (1995); [32]Waervagen et al. (2002); [33]Attayde and Bozelli (1988); [34]Shurin et al. (2000); [35]Cottenie et al. (2003); [36]Shurin and Havel (2003); [37]Pinel-Alloul et al. (1999); [38]Bertoni et al. (2004); [39]Fietz et al. (2005); [40]Hicks et al. (2004); [41]Hall et al. (2003); [42]Patalas and Salki (1992); [43]Jones et al. (1995); [44]Lacroix and Lescher-Moutoué (1995); [45]Pinel-Alloul et al. (2004); [46]Thackeray et al. (2004); [47]Pothoven et al. (2004); [48]Masson and Pinel-Alloul (1998); [49]Masson et al. (2001); [50]Van de Meutter et al. (2004); [51]Johannson et al. (1991); [52]Dejen et al. (2004); [53]Findlay et al. (1991); [54, 60]Basu and Pick (1997a); [55]Schultz

et al. (2003); [56]Maranger et al. (2005); [57, 58]Basu et al. (2000a, b); [59]Engelhardt et al. (2004); [61]Walks and Cyr (2004); [62]Hudon (2000); [63]Hudon et al. (1996); [64]Köhler (1994); [65]Wissel et al. (2005); [66]Basu and Pick (1986); [67]Jack and Thorp (2002); [68]Bertoni et al. (2004); [69]Gliwicz (1986); [70]Lampert (1989); [71]Lampert (1993); [72]Ringelberg et al. (1991); [73]Dodson (1990); [74]Stirling et al. (1990); [75]Brancelj and Blejec (1994); [76]Angeli et al. (1995); [77, 78]Ghan et al. (1998a, b); [79]Ringelberg (1999); [80]Voss and Mumm (1999); [81]Ringelberg and Van Gool (2003); [82]Easton and Gophen, 2003; [83]Chang and Hanazato (2004); [84]Winder et al. (2004); [85]Leech et al. (2005); [86]Gerhardt et al. (2006); [87]Havel and Lampert (2006); [88]Burks et al. (2002); [89]Wojtal et al. (2003); [90]Van de Meutter et al. (2004); [91]Engelhardt et al. (2004); [92]Schernewski et al. (2005); [93]Pearl (1996); [94]Jacoby et al. (2000); [95]Camacho et al. (2002); [96]Chapin et al. (2004); [97]Selig et al. (2004); [98]Goddard et al. (2005); [99]Fee (1976); [100]Moll and Stoermer (1982); [101]Fahnenstiel and Glime (1983); [102]Priscu and Goldman (1983); [103]Pick et al. (1984); [104]Abbott et al. (1984); [105]Bird and Kalff (1989); [106]Lindholm (1992); [107]Barberio and McCair (1996); [108]Gervais (1997); [109]Gervais (1998); [110]Gross et al. (1997); [111]Perez et al. (2002); [112, 113]Knapp et al. (2003a, b); [114]Camacho et al. (2003b); [115]Barberio and Tuchman (2004); [116]Wojciechowska et al. (2004); [117]Adler et al. (2000); [118]Miracle et al. (1993); [119]Wurtsbaugh et al. (2001); [120]Teubner et al. (2003); [121]Kürmayer and Jüttner (1999); [122]Miracle and Alfonso (1993); [123]Armengol et al. (1998); [124]Weithoff (2004); [125]Ignoffo et al. (2005); [126]Kesslert and Lampert (2004); [127]VanGool and Ringelberg (1996); [128]Jensen et al. (2001); [129]Lauren-Määta et al. (1997); [130]Kasprzak et al. (2005); [131]DiFonzo and Campbell (1988); [132]Sakuma and Hanazato (2002); [133]Chang and Hanazato (2005); [134]Kuczynska-Kippen and Nagengast (2006); [135]Olson et al. (2004); [136]Butorina (1986); [137]Charoy (1995); [138]Weber and Van Noorwijk (2002); [139]Schabetsberger and Jersabek (2004); [140]Nihongi et al. (2004); [141, 142]Krembs et al. (1998a, b).

4.1 Mega-scale patterns and processes of microbial patchiness in oceans

4.1.1 Patterns

Mega- or macroscale patterns of plankton patchiness occur over latitudinal gradients among the biogeographical provinces of oceans, showing high variability in bacterial and phytoplankton abundance and production (Ducklow, 2000; Li et al., 2004), Mega-scale distribution patterns of surface chlorophyll based on satellite remote sensing of ocean colour partition the oceans in four biomes and 57 biogeographical provinces (Longhurst, 1998) (Fig. 8-2). These marine ecological provinces represent large patches of similar level of plankton biomass with consistent phytoplankton community structure (Platt and Sathyendranath, 1999; Platt et al., 2005). This geographical perspective is an essential step in describing mega-scale functional heterogeneity of planktonic microorganisms in oceans which has a great potential for the understanding of the long-term effects of global climate change on plankton distribution in oceans. Recently, this geographical scheme was reinforced by the inclusion of bacteria and picophytoplankton to give a macroecological view of microbial spatial heterogeneity among seven ecological provinces of the North Atlantic Ocean (Li and Harrison, 2001). Spatial pattern of marine biomes and ecological provinces shown to be robust for bacterial and algal communities also appears to influence the diversity and abundance of higher trophic levels such as mesozooplankton copepods (Woodd-Walker et al., 2002).

4.1.2 Processes

Physical forces of water circulation in open oceans are the ultimate causes of the spatial structuring of marine ecological provinces and drive mega-scale distribution patterns of plankton seen in satellite images (Martin, 2003). Marine provinces with common physical forcing have phytoplankton communities that respond in a similar and predictable fashion to changes in local environmental forcing (Platt et al., 2005). This physical forcing constitutes the first step of a hierarchy of abiotic (sea temperature, nutrient fluxes) and biotic (microbial biological coupling) processes which are the proximal forces driving plankton patchiness in open oceans. Especially, biotic interactions within microbial food webs have a strong influence on spatial structuring of marine plankton. Mega-scale studies showed strong coupling between

Spatial Heterogeneity of Planktonic Microbes 223

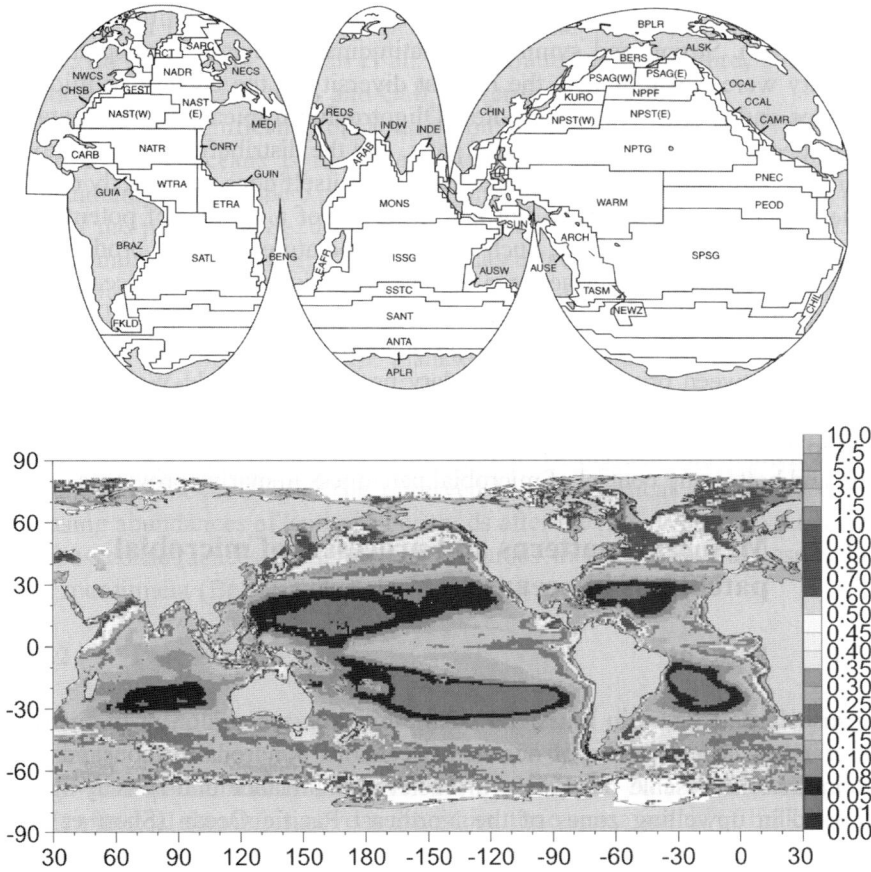

Figure 8-2. Mega-scale patterns of plankton patchiness in oceanic provinces: (top) Longhurst's oceanic provinces (Longhurst, 1995); (bottom) remote sensing spatial pattern of chlorophyll distribution in oceanic provinces. Data are six-year annual values compiled for 2003. (Modifed from http://earthobservatory.nasa.gov/Newsroom/NasaNews/2005/2005030218443.html.)

spatial patterns of heterotrophic bacteria and chlorophyll biomass in various ecological provinces (Li et al., 2004), and a general coherence in-depth profiles of chlorophyll and bacteria (Li and Harrison, 2001). The link between sinking particulate organic matter, bacteria, protists, and algae in microbial food webs has been proposed as the main scheme of the carbon cycle and carbon dioxide (CO_2) biological pump in deep oceans (Nagata et al., 2000; Hansell and Ducklow, 2003; Yamaguchi et al., 2004). Mega-scale

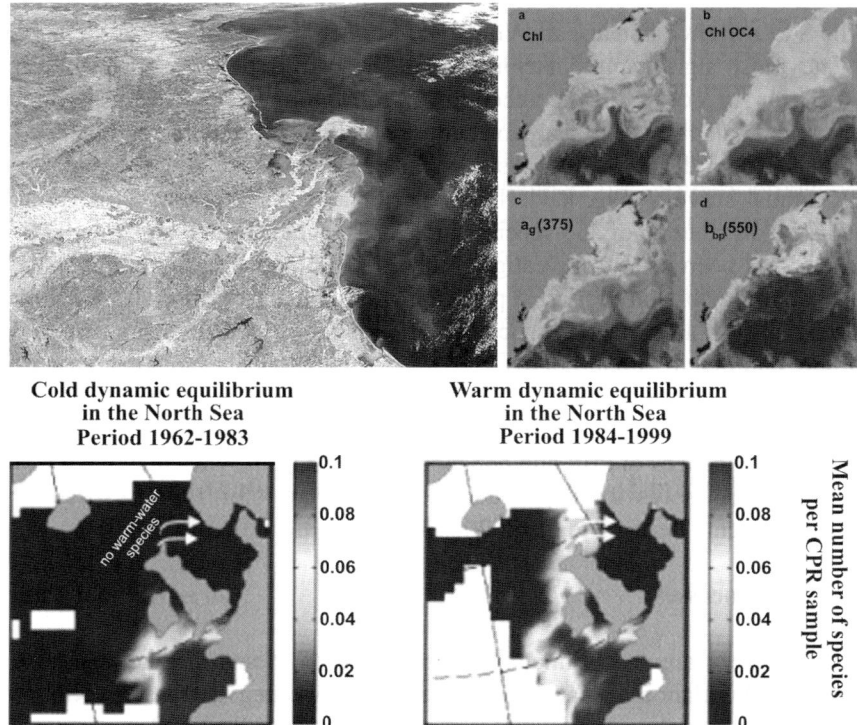

Figure 8-3. Long-term changes in mesoscale patterns of copepod patchiness in the North Atlantic Ocean. (Beaugrand, 2004c.)

abundance and production in near-bottom layers (Nagata et al., 2000). Biotic forces interact with abiotic forces in governing microbial patchiness at mesoscales because the hydrodynamic forcing is transmitted in the microbial food web through a strong coupling between bacterial and algal communities (Lie et al., 1992; Jiao and Ni, 1997; Robarts et al., 1996; Pedrós-Alió et al., 1999; Gasol and Duarte, 2000; Sherr et al., 2001; Jochem, 2001). These common spatial patterns can generate subsurface patches of bacteria and phytoplankton decoupled from zooplankton grazing, allowing local phytoplankton biomass to increase, and toxic algae to bloom. For instance, high surface abundance of bacteria in marine waters has been related to dense blooms of the filamentous cyanobacterium *Trichodesmium* sp. (Biddanda and Benner, 1997).

Mesoscale spatial patterns in zooplankton distribution are also closely related to major hydrographic and topographic features, river runoffs, upwelling currents, and advection of marine waters (Lindahl and Hernroth, 1988; Clark et al., 2001; Harvey et al., 2001; Lo et al., 2004; Roman et al., 2005; Pedersen et al., 2005; Smith and Madhupratap, 2005). Oceanic fronts separated hydrographical regions based on temperature/salinity properties, and may function ecologically either as a barrier which separates distinct zooplankton communities or as a semi-diffuse front across which crossfrontal exchange of the distinct communities may occur (Ashjian et al., 2005; Coyle, 2005). Six distinct zooplankton assemblages strongly correlated with frontal/oceanographic zones were detected along a 2,150 km CPR transect in the Southern Ocean (Hunt and Hosie, 2005). Mesoscale distribution of euphausiids and amphipods in the northeast Atlantic Ocean are also associated with frontal features (Hays et al., 2001). In the St. Lawrence Estuary and Gulf, distribution of krill and copepods is mostly controlled by the advective forcing of the estuarine circulation, the strong vertical currents, and the cyclonic gyre in the Gulf (Descroix et al., 2005).

Overall, recent studies showed that hydrographic and topographic features in open and coastal oceans play a significant role in shaping spatial distribution of all plankton species, and suggested a good coupling of mesoscale distribution patterns of ocean temperature, chlorophyll, and calanoid copepods (Batten and Crawford, 2005). Mesoscale patchiness of marine zooplankton may have profound impact on fisheries. Oceanic zones associated with mesoscale anticyclonic currents, hydrographic or bathymetric fronts, and freshwater nutrient-rich outflows represent hot spots for secondary production, because higher zooplankton biomass can lead to more optimal foraging by fish and greater trophic transfer efficiency (Pinca and Dallot, 1995; Roman et al., 2005). In coastal zones of the Northern Pacific, complex interactions of mesoscale physical features and diel vertical migration behaviour shape the spatial patchiness of euphausiids, a major food item for important large marine fish and mammals (Simard and Mackas, 1989; Ressler et al., 2005). In the Bay of Biscay in northern Atlantic, the slope current and its associated shelf-break frontal structure are crucial in controlling phytoplankton production, zooplankton abundance and grazing, distribution of larvae of fishes and benthic invertebrates, and ultimately in governing the structure and function of the whole pelagic food web (Fernandez et al., 1993).

4.3 Large-scale patterns and processes of microbial patchiness in freshwaters

4.3.1 Patterns

Large-scale patterns of microbial community patchiness occur among lakes and rivers in the landscape at a scale extent (10^2–10^3 km) similar to mega-scale patterns in marine systems. Spatial heterogeneity in bacterial community was reported within lake districts in temperate zones of North America (Letarte and Pinel-Alloul, 1991; Letarte et al., 1992; Cole et al., 1993; Weisse and MacIsaac, 2000; Smith and Prairie, 2004; Yannarell and Triplett, 2004, 2005; Thorpe and Jones, 2005), Europe (Lindström, 2000, 2001, 2006; Reche et al., 2005) and New Zealand (Freidrich et al., 1999). Overall, mean abundance of bacterioplankton in freshwater systems vary by three orders of magnitude (10^5–10^8 cells·ml^{-1}) (Servais et al., 1995; Kalff, 2002). Bacterial communities displayed coherent spatial patterns in biomass, production and community composition in relation to lake trophic gradients at regional scale (Cole et al., 1993; Weisse and MacIsaac, 2000; Yannarell and Triplett, 2004, 2005; Reche et al., 2005). Bacterial communities also exhibit important spatial variation in abundance (1–13 10^6 cells·ml^{-1}) among rivers (Findlay et al., 1991; Basu and Pick, 1997a; Schultz et al., 2003). Large-scale patterns of heterogeneity in chlorophyll biomass, and phytoplankton communities were also detected in landscapes among lakes and rivers (Pinel-Alloul et al., 1990b, 1996; Basu and Pick, 1996; Kratz et al., 1997; Kratz and Frost, 2000; Kling et al., 2000; Jones and Knowlton, 2005; Chételat et al., 2006). Across large trophic gradients, Chl-a biomass ranges from 0.01 to 500 µg L^{-1} among lakes (TP: 1–500 µg L^{-1}; Wetzel, 2001) and from 2 to 257 µg L^{-1} among rivers (TP: 5–280 µg L^{-1}; Chételat et al., 2006). Even on smaller trophic gradient in lakes (TP: 3–34 µg L^{-1}) and rivers (TP: 7–212 µg L^{-1}), Chl-a biomass varies by more than 10-folds in lakes (0.3–7.6 µg L^{-1}; Masson et al., 2000), as well as in rivers (1.8–27.6 µg L^{-1}; Basu and Pick, 1996). Total phosphorus (TP) and Chl-a biomass in lakes and rivers are well correlated and considered the best indicators of cultural eutrophication and change in catchment geology, topography, and land use (Håkanson and Peters, 1995; Basu and Pick, 1996;, Smith et al., 2006; Chételat et al., 2006). Spatial changes in algal biomass along lake trophic gradients are associated to changes in phytoplankton community composition. Pico- and nanophytoplancton such as small diatoms and chrysophytes are dominant in oligo-mesotrophic lakes while large algae, especially harmful and toxic cyanobacteria, predominate in eutrophic and hypereutrophic lakes (Seip and

Reynolds, 1995; Reynolds and Petersen, 2000; Wetzel, 2001; Kalff, 2002). Large-scale latitudinal studies on spatial variation in zooplankton community recently conducted in the USA (Pinto-Coelho et al., 2005) and Europe (Gyllström et al., 2005) also revealed a strong relation with trophic status. TP concentration was found the most important predictor of large-scale variation in zooplankton biomass and community structure, before Chl-a biomass (Pinto-Coelho et al., 2005). However, the strength of the relation with TP may vary across climate zones (as expressed by the ice cover duration), as cold and warm lakes showed different zooplankton responses to trophic gradient (Gyllström et al., 2005). Large-scale (10^2–10^3 km) studies of spatial patchiness of zooplankton were conducted at regional scales in lake districts in Canada (Yan, 1986; Pinel-Alloul et al., 1990a, 1995; Keller and Colon, 1994; McNaught et al., 2000; Swadling et al., 2000; Rusak et al., 2002; Masson et al., 2004), USA (Arnolt and Vanni, 1994; Stemberger and Lacorzak, 1994; Dodson et al., 2005), South America (Attayde and Bozelli, 1998) and Northern Europe (Hessen et al., 1995; Waervagen et al., 2002). In small undisturbed north-temperate lakes, the majority of large-scale variation in zooplankton abundance is spatial in origin; zooplankton patchiness is related primarily to spatial variation in trophic status among lakes and regions rather than to temporal variation among years (Rusak et al., 2002).

Large-scale microbial patchiness is also a common phenomenon reported at scale extents of several kilometres within large lakes and rivers. Lakewide patterns of bacterial and/or algal patchiness were studied in large lakes such as Lake Baikal (Fietz et al., 2005), Lake Superior (Hicks et al., 2004), Lake Maggiore (Bertoni et al., 2004), Lake Ontario (Hall et al., 2003), and Lake Geneva (Pinel-Alloul et al., 1999) (Fig. 8-4a). Zooplankton patchiness at whole-lake scale or along inshore–offshore transects were documented in the North American Great Lakes (Johannson et al., 1991, Pothoven et al., 2004), in the alpine lakes in France (Pinel-Alloul et al. 1999; (Fig. 8-4b); Masson et al., 2001), in glacial lakes in Scotland (Jones et al., 1995; Thackeray et al., 2004) and Canada (Patalas and Salki, 1992, 1993), and in large tropical lakes (Pinel-Alloul et al., 2004; Dejen et al., 2004). Longitudinal and transversal distribution patterns of microbial communities were studied in large rivers in North America (Maranger et al., 2005; Schultz et al., 2003; Basu et al., 2000a, b; Basu and Pick, 1997b; Hudon et al., 1996), and Germany (Köhler, 1994), and in small lake–stream systems (Walks and Cyr, 2004).

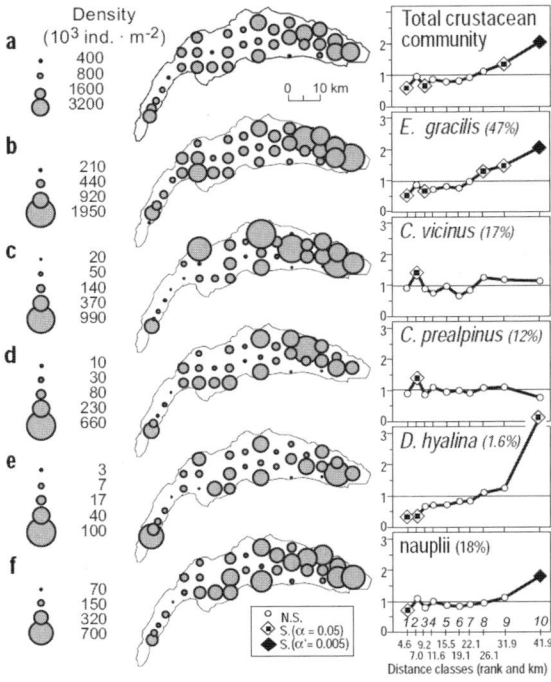

Figure 8-4a. Large-scale patterns of community patchiness within a large lake (Lake Geneva). (Reproduced from Pinel-Alloul et al., 1999.)

4.3.2 Processes

Studies of large-scale patterns of distribution in planktonic communities among lakes gave birth to two different models of driving forces: The metacommunity and the landscape models.

The metacommunity model that was developed in ecology and evolution implies that lakes are isolated or interconnected patches (Wiens, 1997); it suggests that both local environmental processes within lakes and regional processes related to microorganisms' dispersal among lakes govern plankton spatial patchiness at the scale of landscapes (Shurin et al., 2000; Mitchels et al., 2001a, b; Shurin and Havel, 2002; Cohen and Shurin, 2003; Cottenie et al., 2003; Havel and Shurin, 2004; Cottenie, 2005). A recent study on biological invasions of freshwater systems by cladoceran zooplankton suggests that effective dispersal at global scale among continents is rare but greatly increased in the past century with international commerce; however, since an exotic species has invaded, its dispersal at regional and continental

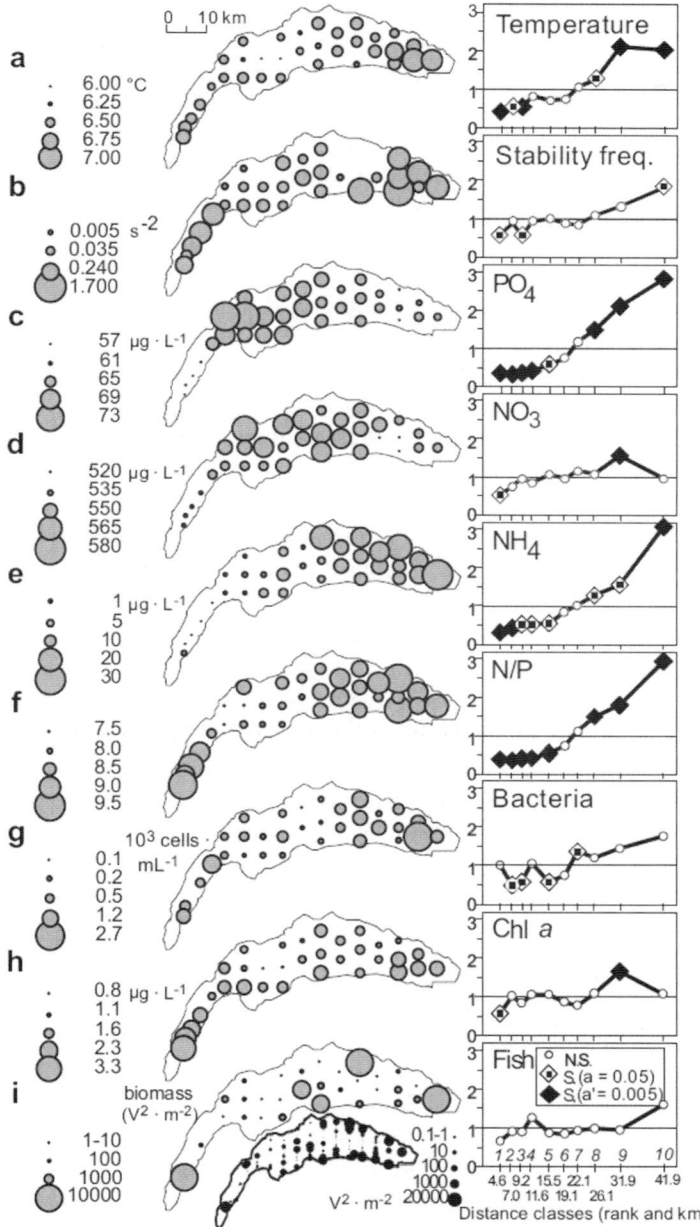

Figure 8-4b. Large-scale patterns of community patchiness within a large lake (Lake Geneva). (Reproduced from Pinel-Alloul et al., 1999.)

scale can be rapid if not limited by local environmental factors (Havel and Medley, 2006). The dispersal pathways among freshwater habitat patches and the probability that dispersing microorganisms will reach suitable patches are not only affected by the spatial distance between the different habitats but also by their hydrological connectivity within the surrounding landscape. Microbial communities in interconnected habitats close to each other should be more similar than in distant and isolated habitats. However, dispersal and colonization in local habitats may be limited by environmental conditions due to physical (drought of ponds, low temperature, anoxic conditions, disturbances, etc.) and biological (competition and predation by local species) constraints.

The landscape model that was developed in limnology considers that topographic position of lakes in landscapes has profound impact on various features of lakes, from trophic status to microbial and fish communities (Kratz et al., 1997; Kratz and Frost, 2000). The main drivers of large-scale landscape variation in lake features are the water residence time and the catchment/lake area ratio (drainage ratio) which increase along the altitudinal and longitudinal gradients. Seepage lakes (also called upland lakes) with higher topographic position in the landscape tend to be more isolated, have higher water retention time, lower drainage ratio, and are less productive. In contrast, drainage lakes (also called lowland lakes) with lower landscape position are hydrologically well connected, have shorter water residence time, and higher drainage ratio, and they tend to be more productive. As topographic position of lakes in the landscape controls their trophic status, seepage and drainage lakes would support different microbial communities.

The application of metacommunity and landscape models to studies of microbial patchiness is still in its infancy. However, spatial patterns of microbial communities described in lake districts support the predictions of both the metacommunity and landscape models. For instance, bacterial phylogenetic diversity is related to lake area and spatial distribution of lakes in a high mountain landscape; higher diversity was recorded in large lakes which offer more complex habitats and favour higher colonization rate, and lakes closer in the landscape had similar bacterial fingerprints (Reche et al., 2005). In temperate lakes in Northern America and Europe, spatial patterns in bacterial genotypic diversity are related to regional and landscape factors such as the geographic location (northern versus southern lakes), topographic position (upland and lowland lakes), and hydrological retention time of lakes (Yannarell and Triplett, 2004, 2005; Lindström, 2006). Spatial patchiness in phytoplankton community is also related to the topographic position of lakes which controls nutrient input, water chemistry, and algal biomass (Krast and Frost, 2000; Beisner et al., 2003). The consistent and directional processing of materials along the hydrological networks of lakes and streams produces large-scale spatial patterns in many limnological variables (alkalinity, conductivity, nutrients, Chl-*a*) which may govern distribution patterns of microbial communities (Kling et al., 2000). In general, zooplankton spatial patterns support the landscape model and reflect regional gradients in lake topography, morphometry and trophic status, as well as in watershed land use via its effects on nutrient inputs, water chemistry, and development of riparian and aquatic vegetation (Pinel-Alloul et al., 1990a, 1995; Stemberger and Lazorchak, 1994). Support for the metacommunity model comes from studies of zooplankton patchiness and dispersal in isolated or interconnected ponds (Cohen and Shurin, 2003; Louette and De Meester, 2005). These studies have shown that differences in pond environments at the local scale lead to divergence in zooplankton communities and higher patchiness, while the dispersal of individuals among ponds at the regional scale leads to convergence in communities and lower patchiness (Michels et al. 2001a, b; Cottenie et al., 2003; Cottenie and De Meester, 2003, 2004).

Overall, abiotic forces such as geographic and trophic gradients are the main driving forces structuring large-scale patchiness in microbial communities in lake districts. Regional heterogeneity in landscape position (seepage and drainage lakes) and resulting gradients in lake productivity, TP, pH, water clarity, and dissolved organic carbon (DOC) interact to determine bacterial assemblages and diversity (Méthé and Zehr, 1999; Lindström, 2000, 2001; Horner-Devine et al., 2003; Yannarell and Triplet, 2004, 2005). In north temperate lakes and rivers, TP input, a good indicator of lake trophic status and landscape position, appears to be the most significant driver of spatial variation in Chl-*a* biomass (Moss et al., 1989; Basu and Pick, 1997b; Masson et al., 2000; Smith, 2003) and autotrophic pico- and nanophytoplankton biomass (Pinel-Alloul et al., 1996) despite the fact that other mechanisms (including DOC input and herbivore grazing) drive the phytoplankton dynamics (Beisner et al., 2003). Even in reservoirs and shallow lakes subjected to high input or resuspension of non-algal seston, TP accounts for 79% of Chl-*a* variation across systems (Jones and Knowlton, 2005). Large-scale variation in zooplankton communities is also driven by abiotic factors, especially those associated to morphometry of lakes, water chemistry, and nutrients (Pinel-Alloul et al., 1990a, 1995; Keller and Conlon, 1994; Masson et al., 2004). TP concentration was found the best predictor of the biomass of crustacean zooplankton across latitudinal and regional scales (McCauley and Kalff, 1981; Pace, 1986; Yan, 1986; Masson et al., 2004; Pinto-Coelho et al., 2005; Gyllström et al., 2005). However, water chemistry (pH, conductivity and calcium content) can also have profound effects on crustacean species distribution (Pinel-Alloul et al.,

1990a, 1995; Hessen et al., 1995; Wærvagen et al., 2002). In particular, ambient Ca concentrations appeared to influence the distribution of *Daphnia* species: *Daphnia longispina* often occurred in Ca-rich lakes with low fish predation pressure while the smaller *D. cristata* often occurred in opposite conditions. In coastal tropical lakes, zooplankton communities are less sensitive to trophic gradients than to salinity gradients; rotifer species assemblages were good indicators of mesohaline and eutrophic conditions (Attayde and Bozelli, 1998).

Although physical forcing is the primary driver of large-scale microbial patchiness in freshwaters, biotic forcing produced by the ecological interactions between microbial communities sustains microbial patchiness. Bacterial abundance and production covary with phytoplankton abundance and production across broad scales in lakes (Bird and Kalff, 1984; Cole et al., 1988, 1993; White et al., 1991; Letarte and Pinel-Alloul, 1991; Weisse and MacIsaac, 2000; Smith and Prairie, 2004) and rivers (Basu and Pick, 1997a). However, the apparent coupling between bacterial and algal communities can be caused by common regulating factors such as temperature and phosphorus that are thought to act on both communities independently (Friedrich et al., 1999). Indeed, bacterial–algal coupling can be weak within large rivers and estuaries where allochthonous sources of carbon are more important than autochthonous algal sources to sustain bacterial biomass (Basu et al., 2000a; Findlay et al., 1991; Maranger et al., 2005).

Abiotic factors are also the main driving forces of microbial patchiness within lakes and rivers. Wind-induced circulation, water temperature and stability, water masses structured by lake morphology and river inflows explain lakewide spatial distribution of bacteria, phytoplankton, zooplankton, and even of large invertebrates such as mysids (Webster, 1990; Patalas and Salki, 1992; Jones et al., 1995; Lacroix and Lescher-Moutoué, 1995; Pinel-Alloul et al., 1999; Thackeray et al., 2004; Pothoven et al., 2004; Fietz et al., 2005). In tropical systems, inflowing rivers, which carried loads of suspended solids into lakes and reservoirs, also influence zooplankton patchiness (Dejen et al., 2004). Biotic factors affect onshore–offshore horizontal distribution of zooplankton prey, which are often inversely related to the distribution of fish and invertebrate planktivores (Masson and Pinel-Alloul, 1998; Masson et al., 2001; Wotjal et al., 2003; Pinel-Alloul et al., 2004; Van de Meutter et al., 2004). Within large rivers and estuaries, the higher complexity of habitats due to tributaries confluence, fluvial lakes, and littoral macrophyte development makes microbial spatial patterns more complex. Stable microbial populations develop in impoundment, shallow fluvial lakes and littoral dead zones, basins, pools along the main course of the river (Köhler, 1994). Lake edge structure, transition from lacustrine to riverine conditions and presence of aquatic vegetation in the littoral zone or at the outlet of fluvial lakes greatly govern microbial distribution patterns (Basu et al., 2000a, b; Walks and Cyr, 2004). Phytoplankton distribution patterns are driven by hydrodynamic processes, change in water residence time (Engelhardt et al., 2004), light limitation and turbidity, and resuspension of periphyton algae from macrophytes in littoral zones (Hudon et al., 1996; Basu et al., 2000b). Physical forcing by faster flushing and shorter residence time in the river course increase primary production, but decrease algal biomass (Wissel et al., 2005), whereas discharge reduction, longer residence time, and low water levels in weedy littoral zones coincide with blooms of algal taxa that generate noxious smells and odours (Hudon, 2000). In large temperate rivers, spatial variation in zooplankton biomass is related to water residence time, Chl-a concentration (Basu and Pick, 1996), development of littoral macrophyte beds (Basu et al., 2000b), and fish predation (Jack and Thorp, 2002).

4.4 Coarse-scale patterns and processes of microbial patchiness in aquatic systems

4.4.1 Patterns

Coarse-scale patterns of microbial patchiness correspond to large-scale extent (1–100 km) in marine waters (Haury et al., 1978) and small-scale extent (10–1,000 m) in lakes (Pinel-Alloul, 1995). In marine systems, studies reported coarse-scale horizontal patchiness in the distribution of sea-ice microalgae along coastal transects of tens of kilometres in the Hudson Bay (Monti et al., 1996) and of the phytoplankton in the St. Lawrence Estuary (Lovejoy et al., 2001). Coarse-scale horizontal patchiness of marine zooplankton is well documented in coastal fjords in northern seas characterized by a high exchange of water masses with the ocean outside and important freshwater runoff from the drainage basin in spring (Lindahl and Hernoth, 1988; Falkenhaug et al., 1997). Spatial distribution of zooplankton follows clear gradients in abundance and community composition along the length of fjords. Oceanic and deep-water copepod species aggregate at the outer marine sites whereas neritic species concentrate in the innermost freshwater sites (Falkenhaug et al., 1997). As fjord systems are generally highly stratified with sharp pycnocline, physical vertical structuring also affects zooplankton distribution. For instance, during autumn, calanoid copepods are transported from the ocean to the fjord by inflows of surface oceanic waters; once inside the fjord, copepods descend in deep waters and are more or less completely washed out of the fjord by the internal renewal of deep water during winter and spring (Lindahl and Hernoth, 1988). In small estuaries temporarily open or closed to the sea by the development of a sandbar, riverine conditions during the periods of high rainfall and freshwater runoff, or lacustrine conditions during the period of freshwater retention create very different coarse-scale spatial patterns in zooplankton abundance and community structure (Froneman, 2004). In coastal bays and straits, coarse-scale spatial patterns in zooplankton distribution and composition are associated to environmental gradients in salinity and algal resources; inshore regions influenced by freshwater runoff from major rivers are characterized by aggregation of euryhaline copepod species, and offshore regions with high biomass of phytoplankton by aggregation of herbivorous marine calanoids (Harvey et al., 2001).

In lakes, coarse-scale patchiness of small extent (10 m–1 km) was reported for the phytoplankton in small rivers subjected to hydraulic management as river groynes and dead zones (Engelhardt et al., 2004) and in lakes receiving major river inflows (Bertoni et al., 2004). However, the most documented coarse-scale patterns are the diel vertical (DVM; Ringelberg et al., 1991) and horizontal (DHM; Burks et al., 2002) migrations of zooplankton in lakes. In stratified waters, zooplankton face a trade-off situation where a choice is made between conditions of warm temperature and high food, but high predation pressure in surface waters, and cold temperature and low food, but low predation pressure in deep waters. When exposed to these vertical trade-offs, zooplankton perform DVM. Many zooplankton species, especially *Daphnia*, a prey of choice for fish larvae, avoid the surface layers during daytime, migrating in swarms downward to deep layers and returning to surface waters at night (Gliwicz, 1986; Lampert, 1989; Angeli et al., 1995; Ringelberg and Van Gool, 2003). The amplitude of DVM varies from <1 m in shallow eutrophic lakes to >50 m in clear oligotrophic lakes. DVM patterns and amplitude are stronger in clear waters presumably because zooplankton must seek deeper water during the day to find light levels low enough to avoid visual fish predation (Dodson, 1990). The DVM behaviour in large zooplankton is a phenotypic plastic trait and it has been shown to be heritable and genetically variable (Ringelberg et al., 1991). DVM patterns of *Daphnia* in large oligotrophic lakes have been the most studied (Stirling et al., 1990; Angeli et al., 1995; Ringelberg, 1999; Wissel and Ramacharan, 2003; Alonso et al., 2004), although copepods and pelagic invertebrate predators also adopt DVM behaviour (Ghan et al. 1998a, b; Voss and Mumm, 1999; Chang and Hanazato, 2004). In Lake Geneva, *Daphnia* hybrids (*D. hyalina x galeata*) and diaptomids (*Eudiaptomus gracilis*) have distinct DVM and diel feeding patterns which lead to temporal and spatial segregation (Angeli et al., 1995) (Fig. 8-5). A strong day/night contrast in depth distribution and feeding activity was observed for the large daphnids while the small daphnids and the diaptomids had lower amplitudes of DVM and weaker diel changes in feeding activity. These distinct behaviours can be viewed as specific adaptive strategies developed by diaptomids and daphnids to limit interspecific competition and to compromise between avoidance of starvation in deep waters and avoidance of visual predators in surface layers.

In shallow lakes and reservoirs, lack of deep refuges does not enable zooplankton species to use DVM to avoid fish predation risk. DHM behaviour is used to find a spatial refuge from fish predation during daytime in littoral submerged macrophytes (Lauridsen et al., 1999; Burks et al., 2002). Large-bodied zooplankton (daphnids and copepods) generally swim towards the open water at dusk and toward submerged macrophytes in littoral zones at dawn (Wojtal et al., 2003). *Daphnia* can migrate in swarms on distances ranging up to 30 m. DHM patterns are strong when macrophytes density is high and associated piscivore fishes are abundant in the littoral zone to control planktivore fishes and restrict their habitat to open water, then pushing zooplankton prey to hide in macrophyte beds during daytime. The impact of fishes and pelagic invertebrate predators may be additive and increases the likelihood that daphnids use DHM to seek refuge in the littoral zone (Burks et al., 2002; Van de Meutter et al., 2004). Macrophyte beds in the littoral zone of lakes may also serve as a refuge against predation by large insect larvae such as damselfly larvae; however, DHM of *Daphnia* driven by littoral invertebrate predators is in the opposite direction with less prey in littoral vegetation at daytime instead of nighttime as seen for DHM patterns driven by fish predation (Van de Meutter et al., 2004).

4.4.2 Processes

Coarse-scale horizontal patchiness of phytoplankton in marine embayments and estuaries are related to salinity and nutrient gradients associated to river plumes (Monti et al., 1996). A multifractal model was proposed to explain coarse-scale distribution of phytoplankton in the St. Lawrence Estuary, involving both *in situ* growth and turbulence at large scale and grazing by zooplankton at smaller scale (Lovejoy et al., 2001). In small rivers, the local geometry of river channel, the non-uniformity of river flow, and the variation in water residence time can create cross-stream heterogeneity and zones of prolonged retention of water with enhanced phytoplankton abundance (Engelhardt et al., 2004). In lakes, major nutrient inputs from the drainage basin and rivers are the most important drivers of phytoplankton coarse-scale horizontal heterogeneity (Bertoni et al., 2004). Tidal currents, freshwater runoff and local wind regime govern zooplankton spatial patterns in fjords,

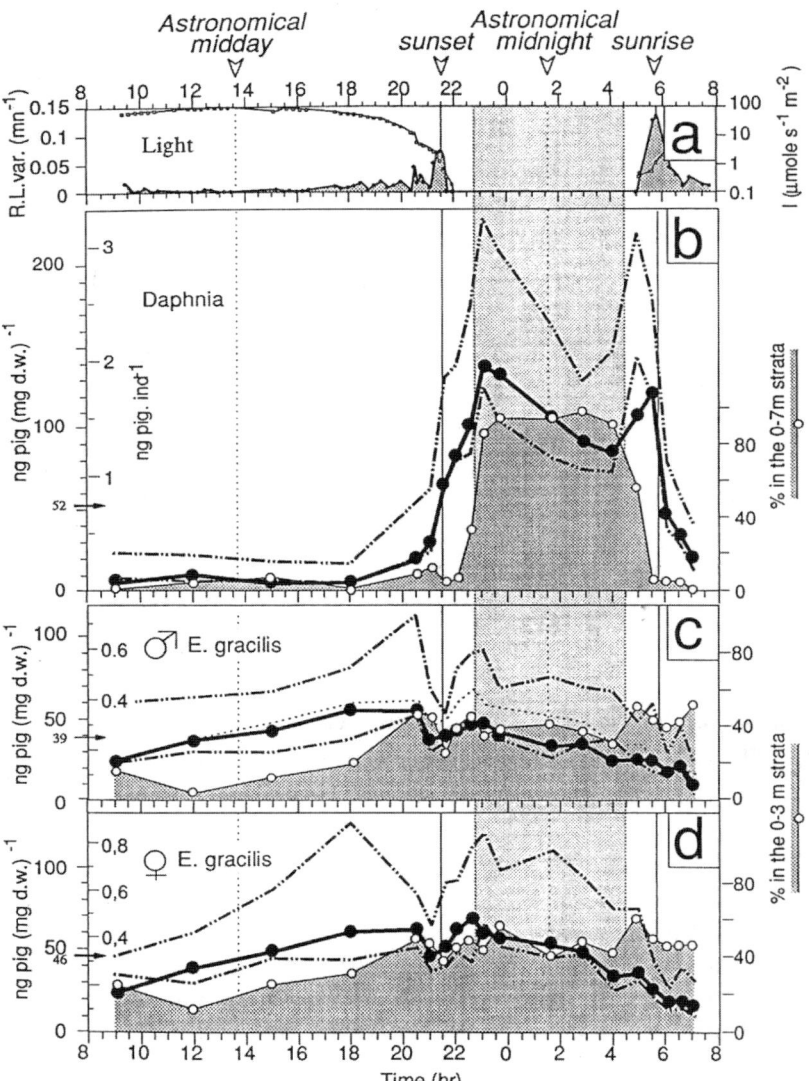

Figure 8-5. Coarse-scale patterns of patchiness associated to diel vertical migration (DVM) of freshwater zooplankton in Lake Geneva. (Adapted from Angeli et al., 1995.)

although exchange of zooplankton between fjords and outside oceans is strongly influenced by the vertical distribution of zooplankton in stratified water masses (Falkenhaug et al., 1995). In estuarine systems, coarse-scale distribution in phytoplankton food resources is not coupled with spatial patterns in calanoid abundance because copepod reproduction is negatively affect by water discharge and short residence time (Lawrence et al., 2004). In intermittently opening estuaries, zooplankton spatial distribution is determined by the interactive effects of freshwater inflow, water temperature and salinity, and estuary mouth status (open or closed) (Froneman, 2004). The spatial structure of plankton in coastal bays and straits is influenced by local hydrodynamic features, which through their action on surface water temperature, salinity, and stratification, lead to spatial differentiation of the phytoplankton and zooplankton communities (Harvey et al., 2001).

Many biotic and abiotic factors are recognized to induce DHM and DVM patterns in freshwater zooplankton and these factors may vary among lakes. The factors that control DHM behaviour are not yet as well understood as those controlling DVM behaviour. Biotic factors related to predation seem to be the most important driving forces for DVM and DHM in macrozooplankton (Gliwicz, 1986; Dodson, 1990; Lampert, 1993; Brancelj and Blejec, 1994; Angeli et al., 1995; Burke et al., 2002). It has been shown that DVM and DHM are induced by chemical substances (i.e., kairomones) exuded by fish or invertebrate predators (Lampert, 1989; De Meester et al., 1993; Van de Meutter et al., 2004). DVM patterns of *Daphnia galeata mendotae* were shown to be a demographic response to changes in planktivore abundance (Stirling et al., 1990). Moreover, *Daphnia* DVM frequency and amplitude are adjustable depending on kairomone concentrations, food availability, and vertical temperature profile (Winder et al., 2004). DHM patterns of *Daphnia* are stronger in lakes with littoral piscivores such as pikes which force young planktivores fish to forage in pelagic zones and *Daphnia* to seek refuge in littoral macrophyte beds during daytime (Burks et al., 2002). Spatial changes in DVM behaviour of copepods among lakes appears to be the result of a trade-off between predator avoidance in deep waters and food resource acquisition in surface waters (Ghan et al., 1998a). DVM patterns of large pelagic invertebrate predators (*Chaoborus, Leptodora*) are also induced by the presence of fish planktivores (Voss and Mumm, 1999; Chang and Hanazato, 2004). However, many abiotic factors such as diel light variation, water transparency to ultraviolet radiation (UVR), temperature, and oxygen levels in deep waters also control zooplankton DVM patterns. The reaction to relative changes in light intensity (negative phototaxis) is the primary physiological mechanism underlying endogenous rhythm of *Daphnia* DVM behaviour (Ringelberg, 1999; Angeli et al., 1995; Gerhardt et al., 2006). Water transparency and colour influences the strength of DVM in zooplankton which increases in clear waters (Dodson, 1990). In small lakes of different water transparency and colour, the mean daytime depth position of zooplankton varies from <1 m to >10 m, and is inversely related to water Secchi depth transparency (Wissel and Ramacharan, 2003). UVR may also influence zooplankton DVM in high-UV lakes (Alonso et al., 2004) inducing surface avoidance behaviour, while biotic factors, such as predation and food limitation, may be more important in low-UV lakes (Leech et al., 2005). Vertical gradients of both temperature and oxygen also influence DVM patterns of zooplankton in deep and shallow lakes (Easton and Gophen, 2003; Wissel and Ramacharan, 2003). Levels of oxygen and temperature found in deep waters may at times be so low as to set a lower limit for DVM amplitude. *Daphnia* species have very different DVM patterns and habitat partitioning depending of their thermal tolerances (Havel and Lampert, 2006). *Daphnia lumholtzi*, an exotic invader species from tropical lakes, has typical DVM and is able to use warm surface waters at night whereas *Daphnia mendotae*, a native species of temperate lakes, shows little tendency to migrate and remains in deep and cold waters.

4.5 Small-scale and fine-scale patterns and processes of microbial patchiness

4.5.1 Patterns

Fine-scale or small-scale patterns of microbial patchiness correspond to scale extent of one to several hundreds of metres in marine waters (Haury et al., 1978) and one to tens of metres in freshwaters (Pinel-Alloul et al., 1995). There is a considerable amount of variation in the abundance and activity of bacteria over depth and horizontal scales less than tens of metres in marine systems. A clear pattern for most of the sites is that of lower bacterial abundances and activity at depth below the euphotic zone. However isolated "hot spots" of bacterial abundance and activity can produce fine-scale surface or subsurface maxima at specific times and depths. Fine-scale vertical patchiness of phytoplankton due to algal blooms also occurs in surface waters of marine and freshwater systems. In coastal zones and marine bays, they are produced by species of toxic dinoflagellates (*Alexandrium fundyense, Gymnodinium splendens*) which can reach very high abundances (up to 2,400 cells·L^{-1}) and biomass (Chl-*a*: 1.7 µg·L^{-1}) (Tada et al., 2001; Martin et al., 2005; Townsend et al., 2005). In lakes, they are formed by intensive blooms of dinoflagellates (*Ceratium*) in oligotrophic lakes (Schernewski et al., 2005) or of filamentous and colonial cyanobacteria (*Anabaena planctonica,*

Microcystis aeruginosa, Aphanizomenon flosaquae) in eutrophic lakes (Paerl, 1996; Jacoby et al., 2000). However, the most ubiquitous feature of fine-scale vertical patchiness is the formation of deep microbial thin layers (Fennel and Boss, 2003; Cowles, 2003). Deep thin layers of photosynthetic bacteria are common in dimictic or meromictic stratified lakes and reservoirs (Camacho et al., 2002; Chapin et al., 2004; Selig et al., 2004; Goddard et al., 2005) (Fig. 8-6A). They are located below the thermocline or chemocline zones and associated to sharp vertical gradients in oxygen and sulphides. They are generally composed of a layer of green sulphur bacteria (*Chlorochromatium aggregatum* or *Chlorobium phaeobacteroides*) above a layer of purple sulphur bacteria (*Lamprobacter modestohalophilus*) (Chapin et al., 2004; Camacho et al., 2002, 2003a). Deep layers of photosynthetic bacteria are responsible for 20–90% of total photosynthetic production and can serve as a carbon source for oxic heterotrophic bacteria and copepods (Overmann et al., 1996, 1999).

Deep chlorophyll maxima (DCM) of phytoplankton are also very common in oceans (Cullen, 1982; Gould, 1987; Cowles et al., 1998; Cowles, 2003) and temperate lakes (Fee, 1976; Moll and Stoermer, 1982; Fahnenstiel and Glime, 1983; Priscu and Goldman, 1983; Pick et al., 1984; Barberio and McCair, 1996; Barberio and Tuchman, 2004) (Fig. 8-6B, C). In oceans, DCM established at the base of the euphotic zone (i.e., below 1% of surface irradiance) at depths varying between 40 and 100 m close to the pycnocline where concentrations of limiting nutrients increases (Ediger and Yilmaz, 1996; Holm-Hansen and Hewes, 2004; Holm-Hansen et al., 2005). The thickness of the DCM pigment band ranges from <1–80 m, but 20 or 30 m is common in oceans. Chlorophyll biomass in DCM is generally 3–10 times higher than at the surface, and primary production at DCM may account for 20–30% of total productivity in the water column. The species composition of DCM in oceans is associated to different groups, mainly autotrophic picoplankton (*Synechococcus*), mixotrophic flagellates (*Chrysochromulina*), and dinoflagellates (*Gymnodinium, Gyrodinium*) (Karlson et al., 1996; Lunven et al., 2005). In freshwaters, DCM of up to 30 $\mu g \cdot L^{-1}$ of Chl-*a* occur at the base of the euphotic zone (0.1–1.3% of surface irradiance) in stratified lakes (Fahnenstiel and Glime, 1983; Pick et al., 1984; Gervais, 1997; Pérez et al., 2002; Hedger et al., 2004; Barberio and Tuchman, 2004). The deepest DCM was reported in the clear ultra-oligotrophic Lake Tahoe (California) at 100 m deep, owing partly to sinking of diatoms and partly to *in situ* algal growth (Abbott et al., 1984). Chlorophyll concentration at DCM can reach levels as high as ten times that of surface waters. DCM size, vertical position and species composition vary among lakes depending of the thermocline depth, light penetration, lake fertility, nutrient clines, and oxygen status in deep waters (Lindholm, 1992). In oligotrophic lakes, DCM layers of phytoplankton can be formed by large blooms of phycoerythrin-rich picocyanobacteria (*Chroococcus*) (Perez et al., 2002; Camacho et al., 2003b; Pinel-Alloul et al., 2003), small diatoms (*Stephanodiscus, Cyclotella, Aulacoseira granulate, Asterionella formosa, Fragilaria crotonensis*) (Fahnenstiel and Grime, 1983; Ryan et al., 2005) and phytoflagellates (*Ochromonas, Cryptomonas, Synura, Mallomonas*) (Pick et al., 1984; Gervais, 1997, 1998; Knapp et al., 2003a, b) which represent an optimal food for zooplankton. In mesotrophic lakes, DCM are rather composed of inedible algae such large dinoflagellates (*Dinobryon, Ceratium hirundinella*) (Bird and Kalff, 1989; Gross et al., 1997) or phycocyanin-rich filamentous or colonial cyanobacteria (*Planktothrix rubescens, Limnothrix rosea, Pseudo-anabaena*) (Feuillade, 1994; Walsby et al., 1998; Kurmayer and Jüttner, 1999; Adler et al., 2000; Ernst et al., 2001; Humbert et al., 2001; Teubner et al., 2003; Knapp et al., 2003a; Jann-Para et al., 2004; Wojciechowska et al., 2004; Hedger et al., 2004). Phytoplankton DCM is generally associated with high numbers of anoxygenic photosynthetic bacteria and protozoan ciliates which may provide a large food supply for zooplankton (Kettle et al., 1987; Amblard et al., 1995; Adrian and Schipolowski, 2003).

Mesozooplankton usually show strong vertical fine-scale patchiness. In oceans, mesozooplankton biomass is considerably reduced at depths below the lower thermocline and the oxygen minimum zone (OMZ), which is a sharp subsurface gradient between oxic surface waters and suboxic deep waters. Some mesopelagic copepods show a subsurface abundance peak in the lower OMZ, a location of relatively high concentrations of organic particles associated to deep microbial layers promoting high feeding rates (Smith and Madhupratap, 2005). In lakes with steep vertical gradients, deep-hypolimnetic rotifer abundance maxima have been detected at the oxic–anoxic boundary, a food-rich interface retaining sinking dead organic material where layers of heterotrophic bacteria and microalgae (*Cryptomonas* and small flagellates) offer high food availability (Miracle and Alfonso, 1993; Armengol et al., 1998; Weithoff, 2004; Ignoffo et al., 2005). *Daphnia* populations change their migratory behaviour and fine-scale vertical distribution in response to the presence and algal composition of DCM (Kessler and Lampert, 2004). Daphnids can evaluate the food patches formed by DCM both quantitatively and qualitatively, possibly through perception of increased ingestion rate and algal associated odours (Van Gool and Ringelberg, 1996; Jensen et al., 2001). *Daphnia* may also perceive signals from toxic cyanobacteria as information associated with danger that may result in immediately inhibited feeding behaviour, reduced food intake, lower biomass and size, and altered swimming behaviour and reproduction (Haney et al., 1995; Hietala et al., 1997; Laurén-Määtä et al., 1997; DeMott,

Spatial Heterogeneity of Planktonic Microbes

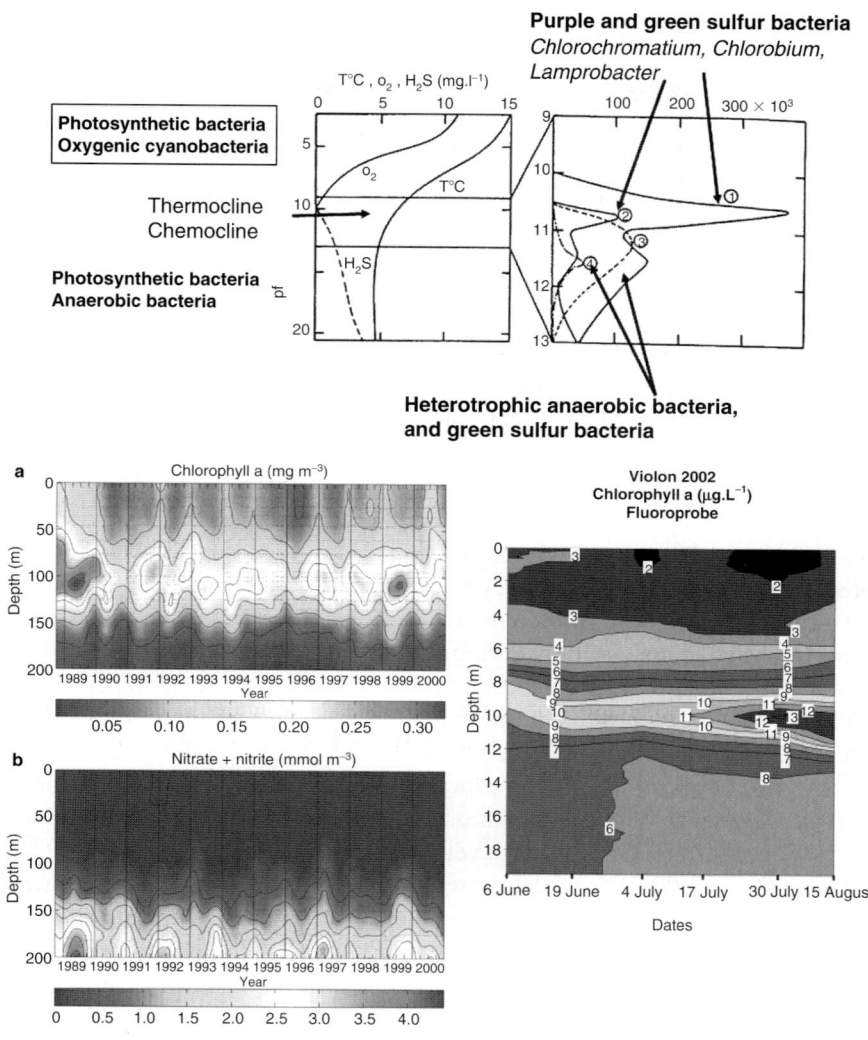

Figure 8-6. (A) Deep layers of green and purple photosynthetic bacteria (adapted from Servais et al., 1995); (B) deep layers of chlorophyll *a* and phytoplankton functional groups in marine systems (Huisman et al., 2006); (C) deep layers of chlorophyll *a* and spectral groups in oligo-mesotrophic lakes.

1999; Rohrlack et al., 1999; Kurmayer and Jüttner, 1999; Ghadouani et al., 2003, 2004). In meso-eutrophic lakes, they are attracted by non-toxic strains of *Planktothrix agardhii* developing in metalimnetic DCM but do not move toward toxic strains (Lauren-Määta et al., 1997). In north temperate oligotrophic lakes, some copepods species which are glacial cold-water relicts showed habitat selection restricted to relatively deep lakes (>30 m) offering a hypolimnetic refuge characterized by low temperatures (less than −10°C) and well-oxygenated waters during summer (Kasprzak et al., 2005). Small-scale horizontal patchiness in the distribution of cladocerans between microhabitats (plants, rocks, sediments, open-water) in the littoral zone of lakes is a common phenomenon reported by DiFonzo and Campbell (1988) and Sakuma and Hanazato (2002). Chydoridae differ in their habitat preference depending on the animal species/genus; *Alona* spp. selected plant habitat rich in epiphytic algae and oxygen and avoid hypoxic water while *Chydorus sphaericus* was found in lake water or at the sediment surface.

As seen in freshwaters, one major feature of fine-scale patchiness of marine zooplankton is their DVM behaviour. DVM by marine herbivorous copepods is commonly observed in pelagic zones of oceans and seas (Atkinson et al., 1996; Steele and Henderson, 1998; Pearre, 2003; Bonnet et al., 2005; Smith and Madhupratap, 2005) and in coral reefs (Yahel et al., 2005). In the Sea of Japan or East Sea, the vertical distributions of zooplankton are modulated by DVM patterns of specific copepods species (Ashjian et al., 2005). In the Arabian Sea, very high biomass of DVM move from the surface waters at night to the suboxic waters during the day at depth varying from 150 to 400 m (Smith and Madhupratap, 2005). In the Black Sea, some copepods as *Calanus* do DVM to stay during night in the phytoplankton-rich layer, grazing 10–14% of primary production (Besiktepe et al., 1998). DVM behaviour in marine zooplankton (copepods, chaetognaths) is a very flexible pattern, depending of predation risk and vulnerability based on their body size, colour (carotenoid pigment levels) and morphology (Hays et al., 1994; Irigoien et al., 2004). In contrast with temperate oceanic habitats where DVM is observed for only few species of copepods or euphausiids, the migrating community of coral reef is highly diverse and many different taxa ascended to the water column in a highly synchronized manner (Yahel et al., 2005). Small zooplankton emerge first at dusk time, whereas larger zooplankton emerge into the water column only at dark (>1 h after sunset).

4.5.2 Processes

Fine-scale horizontal patchiness of phytoplankton during surface algal blooms in marine coastal zones are linked to hydrological and advective physical forcing by coastal currents and retentive gyres which tend to favour reseeding of resting cysts of planktonic dinoflagellates *Alexandrium fundyense* within gulfs and bays (McGillicuddy et al., 2005), In lakes, small-scale phytoplankton patches during intensive blooms of *Ceratium* are linked to current gyres in the central zone and wind sheltering effect by vegetation in the littoral zone. There is a very good agreement between location and size of current gyres and phytoplankton patches. Inside the gyre, the low-flow velocity and vertical turbulence allow *Ceratium* to form distinct vertical layers with high density close to the water surface, according to the light gradient (Schernewski et al., 2005).

Deep microbial layers in marine and freshwaters have varying characteristics that suggest multiple processes contributing to their formation and maintenance (Gould, 1987; Cowles, 2003). Firstly, physical and chemical water stratification associated to a strong density gradient is a prerequisite for the formation of deep microbial layers. In oceans, DCM formation is strongly dependent on water-column properties (frontal characteristics, nutrient gradients, and euphotic depth) and historical wind stress (Franks and Walstad, 1997; Basterretxea et al., 2002). These deep layers of high chlorophyll reflect a trade-off of phytoplankton communities exposed to two opposing resource gradients: light supplied from above and nutrients supplied from below (Fig. 8-6A) (Huisman et al., 2006). Deep layers of bacteria are closely linked to DCM formation in oceanic provinces and gulfs. Bacteria exhibits population maxima in the upper half of the subsurface chlorophyll maximum as bacterial distribution and production are linked to phytoplankton as the major source of labile bacterial substrates through their production and exudation (Jochem, 2001; Lie et al., 1992; Jiao and Ni, 1997; Robarts et al., 1996; Pedrós-Alió et al., 1999). In meromictic lakes, the extreme water stability and salinity gradient promotes the formation and maintenance of deep layers of bacteria and algae in the chemocline at the bottom of the mixolimnion (Miracle et al., 1993). In Lake Baikal, subsurface maxima of bacterial abundance were recorded at the same depth as the maximum of chlorophyll biomass caused by a concentration of phytoflagellates at the lower of the thermocline (15 m) which were sustained by nutrient supply from the hypolimnion (Nakano et al., 2003).

In dimictic lakes, strong vertical physical and chemical gradients enable DCM to form below the thermocline where light, temperature, and turbulence are low, but nutrients relatively rich. The formation and persistence of DCM in lakes is directly tied to their trophic status as indicated by the nutrient levels and water light irradiance (Moll and Stoermer, 1982; Pick et al., 1984). Most DCM layers were reported for oligotrophic and mesotrophic lakes. As lakes move toward increasing eutrophy, the environment available for subsurface phytoplankton growth is restricted by shading from epilimnetic algae growth, and DCM do not developed. Secondly, mechanistic models for the formation of DCM in oceans and lakes include several biotic processes: phytoplankton sinking, adaptation to low-light photosynthesis, buoyancy, respiration, grazing, and mixing (Gould, 1987; Figueroa et al., 1998). These factors interact to produce a constantly changing vertical profile of cell abundances and chlorophyll biomass. Most of the DCM layers are controlled by differential sinking rates of algal cells (sinking diatoms versus motile dinoflagellates) generated in the epilimnion, or by *in situ* production in the metalimnion of algae adapted to low-light environment. Below the euphotic zone at photosynthetic active radiation (PAR) corresponding to 0.1–1.3% of surface irradiance (Ryan et al., 2005), algae in the DCM layer have greater photosynthetic efficiency and higher chlorophyll content than algae at the surface (Karlson et al., 1996; Gross et al., 1997). Planktonic cyanobacteria forming DCM in the metalimnion of meso-eutrophic lakes are all adapted to low-irradiance level but species composition can vary with water column stability: *Limnothrix redekei*, *Planktothrix agardhii*, and *P. rubescens* prefer turbulent waters, whereas *Microcystis* prefer water column stability (Wojciechowska et al., 2004). Using a hydrodynamic model, Hedger et al. (2004) simulate the vertical distribution of phytoplankton species having different properties: the motile large dinoflagellate *Ceratium hirundinella* and the buoyant cyanobacteria *Microcystis*. They showed that the interaction of the phytoplankton properties with the wind-forced circulation was instrumental in the determining of spatial distribution of phytoplankton. Buoyant cyanobacteria blooming in surface layers became entrained in surface drift currents, and accumulate at downwelling areas. In contrast, surface-avoiding dinoflagellates that sank from the surface, were pushed upwind and entrained in subsurface return currents, and accumulated in deep layers at upwelling areas. Nutrient availability also controls the vertical position of DCM in aquatic systems. In oceans, as major inorganic nutrients (nitrogen, phosphorus, silica, iron) are limited in the surface waters (Holm-Hansen and Hewes, 2004; Holm-Hansen et al., 2005), algal populations in DCM are utilizing nutrients brought up from depth, and diffusing up across the pycnocline (Gould, 1987). In oligotrophic lakes, experimental fertilization or inflow of cold nutrient-rich waters in the metalimnion coming from stream inflows or hypolimnion waters induce the formation and maintenance of DCM during summer stratification (Wurtsbaugh et al., 2001; Sawatzky et al., 2006). The vertical position of *Cryptomonas* in DBM of lakes and reservoirs depends of both the nutrient and light conditions. *Cryptomonas* diurnal migration is largely regulated by daily light conditions (photon irradiance and gradient), whereas *Cryptomonas* abundance is strongly influenced by the nutrient supply levels in the hypolimnion, most importantly by TP and ammonia (Knapp et al., 2003b). Phycoerythrin-rich picocyanobacteria (*Synechococcus*) are abundant in DCM of deep oligotrophic clear lakes, while picocyanin-rich cyanobacteria (*Planktothrix rubescens* and *P. agardhii*) are dominant in DCM of meso-eutrophic lakes (Camacho et al. 2003b; Humbert et al., 2001). The higher phycoerythrin content in coccoid picocyanobacteria of DCM in oligotrophic lakes allow these cyanobacteria to efficiently harvest the predominant yellow-green light available at the metalimnion, where nutrient availability during stratification is higher than in the epilimnion (Perez et al., 2002). Another contributing factor to the development of a DCM is algal mixotrophic metabolism. In oceans, mixotrophic flagellates as *Chrysochromulina* benefit from bacteria accumulating in the pycnocline, and have advantage over obligate autotrophs in light limited situations (Karlson et al., 1996). In freshwaters, Chrysomonads (*Dinobryon, Ochromonas*) forming deep chlorophyll maxima in the metalimnion of lakes are sustained by mixotrophic metabolism made of photosynthesis and bacteria phagotrophic feeding (Bird and Kalff, 1989). Finally, zooplankton grazing may also contribute to DCM formation. Zooplankton grazing in the epilimnion along with a downward movement of nutrients via sedimentation could contribute to a nutrient-enriched DCM especially in absence of large grazers (Pilati and Wurtsbaugh, 2003). Furthermore, losses by grazing at the DCM level can be partially offset by nutrient recycling and consequent increase in algal growth rates. Depth-differential nutrient deficiencies, as well as zooplankton grazing and nutrient recycling interact to maintain the DCM in high mountain lakes (Sawatzky et al., 2006).

In some lakes, migration from one plant to the other constitutes an important factor controlling habitat selection and small-scale patchiness of chydorid cladocerans (Sakuma et al., 2004). The presence of macrophytes induces higher zooplankton biomass and different species composition of zooplankton from that at the pelagic and other littoral sites without vegetation (Chang and Hanazato, 2005). Furthermore, the species composition and the spatial structure (length of plant stems and plant density) of macrophytes can also influence the fine-scale spatial distribution of rotifer and cladoceran communities in shallow lakes (Kuczynska-Kippen and Nagengast, 2006). Thus the development of macrophyte beds is a key factor establishing the heterogeneity of zooplankton species composition in the littoral zone of lakes. In a small South Dakota impoundment, cyclopoids and daphnids were more abundant in the vegetated area, while calanoids were more abundant in the non-vegetated area (Olson et al., 2004). The littoral zone of shallow lakes represents an ecotone, and the macrophytes beds act as a daytime refuge for cladocerans as observed for diel horizontal migration of pelagic zooplankton.

Relative changes in light intensity at dawn and dusk can be considered the primary stimuli inducing endogenous migratory rhythms in marine zooplankton (Ringelberg, 1995). Even moonlight and lunar eclipse can influence marine DVM patterns. For instance, euphausiids can perceive moonlight and stay in surface waters during lunar eclipse instead of sinking after their arrival to the surface (Tarling et al., 1999). The phototactic DVM behaviour is enhanced by cues from predators and seems to be inhibited by low food concentration (Ringelberg, 1995). As in other marine habitats, DVM in coral reefs is a light-dependent diel behaviour developed for avoiding intense daytime predation by the highly abundant visual planktivorous fish (Yahel et al., 2005). A third category of causal factors that may influence zooplankton DVM behaviour and amplitude in marine systems, as seen in freshwaters, are the temperature and oxygen vertical gradients and the presence of chlorophyll layers (Ringelberg, 1995). In the Sea of Japan or East Sea, the vertical distribution of both small copepods and algal fluorescence were associated with the vertical structure of the water column and, consequently, changed with the changing depth of the pycnocline (Ashjian et al., 2005). In the Baltic Sea, the vertical distribution of *Oithona similis* is determined by the salinity stratification in the central basin forcing the marine cyclopoid copepod to accumulate in the deep halocline layer, limited from above by low salinity and from below by low oxygen concentrations (Hansen et al., 2004). In the coastal zone of northern Chile, the DVM pattern of the diaptomids *Eucalanus inermis* is related to the strong vertical oxygen gradient associated with an intense OMZ. Most developmental stages of the calanoid population remained near the base of the oxycline (30–80 m) and within the upper zone of the OMZ (30–200 m). This strategy of movement may resust in a better utilization of food resources since the strong physical and chemical gradients at the base of the oxycline and upper OMZ boundary might serve as a site of bacterial and algal accumulation (Hidalgo et al., 2005). For instance, the marine copepod *Acartia tonsa* has higher egg production with patchy food distribution in vertically structured layers than with homogeneous food distribution (Saiz et al., 1993). Feeding on small particles and being the major food for micronekton (fish larvae), marine copepods form an important link in the food chain, coupling protists to higher trophic levels. A recent study on the control of marine copepod DVM using *Calanus finmarchicus* as a model organism (Thorisson, 2006), gave strong support to the multiple forces hypothesis and proposed both physical (light avoidance) and biological (active food searching, buoyancy) complex mechanisms to explain the diel and seasonal vertical migrations of herbivorous copepods in boreal and polar waters. The hypothesis is based on the following assumptions. Light avoidance is assumed to operate solely while the copepods are satiated and stayed below the euphotic zone to avoid visual predators hunting in surface waters. Hungry copepods are assumed to react to food smell by increased upward swimming, whereas satiated copepods maintain low activity and remain in deep zones. High lipid content is assumed to affect the buoyancy of the copepods during the season according to egg production and lipid storage during periods of algal blooms at spring autumn and late winter.

4.6 Microscale and nanoscale patterns and processes of microbial patchiness

4.6.1 Patterns

Micro- (1 mm–10 cm) and nanoscale (1 μm–1 mm) patchiness of bacteria has been recently detected in marine and freshwaters systems. Spatial heterogeneity in marine bacterial abundances was observed in natural samples at the centimetre scale with changes in abundance of up to 3.5-fold across distance of 10 cm (Mitchell and Fuhrman, 1989). Microscale distribution patterns showed patches of high bacterial density ($>10^7$ cells·ml^{-1}) scattered within a matrix of low bacterial density (10^5 cells·ml^{-1}) (Duarte and Vaqué, 1992). Recently, microscale sampling techniques coupled with flow cytometry (FCM) analysis help detecting the existence of microscale heterogeneity in the abundance and metabolic activity of marine bacterial communities at the millimetre scale with 2–16-fold variation across distances of 9–32 mm (Seymour et al., 2000, 2004). Using phylogenetic analysis and experimental protocol, Long and Azam (2001) showed that bacterial species richness and composition also vary at the millimetre scale in the marine pelagic environment. In freshwaters, rapid freezing of small samples of water (spatial information preservation method (SIP)) allow detecting nanoscale patchiness of bacteria in lake water at scales of tens to hundreds of micrometres (Krembs et al., 1998a, b).

In marine waters, there is increasing evidence that persistent small-scale features on the 10 cm scale exist over a wide range of marine environments. An example of these small-scale features is thin layers of plankton measured with *in vivo* fluorescence as recently documented for protected bays and the coastal oceans (Waters and Mitchell, 2002; Waters et al., 2003; Seuront and Schmitt, 2005a, b; Seymour et al., 2005). These features occur over surprisingly large horizontal scales (>1 km) and frequently persisted over weeks (Cowles et al., 1998). They can consist of a combination of bacterioplankton and phytoplankton which persist within specific density gradients in well-stratified marine waters; they may be beneficial for zooplankton grazers if composed for edible algae or harmful if formed by toxic algal species (Cowles, 2003).

Microscale patchiness of zooplankton is well documented both in marine and freshwaters. Microscale patchiness of marine zooplankton due to swarm formation, swimming behaviour and grazing of krill, mysids, and copepods (O'Brien, 1988; Price, 1989; Tiselius, 1992) have important implications for zooplankton trophodynamics, resources patchiness, sexual encounter and mating, and predation vulnerability (Seuront et al., 2004a–c). In small lakes, aggregation and schooling behaviour is common in littoral crustaceans and is also very important for food searching, bisexual reproduction and predator avoidance. At the microscale (1–100 cm), cladocerans such as *Polyphemus pediculus* form vertical schools scattered over shallow waters as patches composed of feeding and mating groups (Butorina, 1986). In alpine lakes, the males and gravid females of the copepod *Arctodiaptomus alpinus* forms dense swarms in scattered patches of the macrophytes *Ranonculus eradicatus* in the shallow zone while ovigerous females were more concentrated in the lake center to avoid predation (Schabetsberger and Jersabek, 2004). Mate tracking or mate-searching behaviour in calanoid copepod has been explored in laboratory by video recording. Males of freshwater calanoids increased its speed after detecting females while marine calanoid have a hopping behaviour associated with mecanoreception. The reactive distance ranges from 2 to 34 mm in marine calanoids and 12–30 mm in freshwater calanoids. Males do not swim directly to the females but use chemoreception to capture the female from below (Nihongi et al., 2004).

4.6.2 Processes

Microscale distribution patterns of subpopulations of bacteria in marine coastal habitats reflect physiologically distinct bacterial populations of dissimilar activity levels, perhaps influenced by different microscale features in the distribution of organic substrates (Long and Azam, 2001; Seymour et al., 2004). Marine bacteria aggregate into microscale patches or "hot spots" in response to favourable environmental conditions such as around the dissolved organic matter diffusing from phytoplankton cells, organic detritus particles, and discrete microscale nutrient patches (Seymour et al., 2000). At the microscale dimension, seawater is an organic matter continuum of tangled polymers and embedded particles including transparent exopolymeric particles (TEP), proteinaceous coomassie stained particles (CSP) and organic sub-micrometre particles. The potentially gel-like nature of this organic matter continuum and the bulk effects of phytoplankton-exuded polysaccharides, may act to increase water viscosity and influence turbulent drag by elastic effects, which may play an important role in the aggregation of microbes, via processes of polymer bridging. Oceanic turbulence could influence the dynamics of the polymer gel field, and seston particles, and in turn affect levels of microscale patchiness of marine bacteria in the sea. However, before an understanding of these processes can be obtained, focussed research into the relationship between aquatic microorganisms, turbulence and the polymer matrix in the oceans and freshwaters, and a greater understanding of polymer dynamics within turbulent flow is required (Seymour et al., 2000).

For the marine phytoplankton, there is evidence that turbulent processes can generate and control microscale patchiness rather than randomness (Seuront et al., 1999). The extreme similarity observed between microscale intermittent distributions of temperature, salinity, and phytoplankton biomass suggests a passive behaviour of phytoplankton cells in the turbulent flows. Very layers of plankton in marine systems are associated with physical features such as thermoclines, pycnoclines, or oxyclines; they are under the control of the same physical (water stability and circulation, nutrient inflow across the pycnocline) and biological (low-light adaptation, buoyancy, grazing) forces as described for the fine-scale DCM in marine systems (Cowles, 2003). The presence of these thin plankton layers poses a new set of challenges for the accurate representation of trophic processes. First, if the sampling method is too coarse, a local maximum can easily escape detection and the average amount of prey is miscalculated (Cowles et al., 1998). This problem is not trivial, since water column averages have often shown to be below maintenance levels for zooplankton which may be sustained by the deep plankton layers. The second major challenge in understanding rate processes related to thin layers is that predator–prey interactions may change quantitatively due to the presence of dense and spatially confined layers of food. These changes may occur because of behavioural changes of plankton such as aggregation in or avoidance of these layers. Hence conventional functional response models that derive from steady-state experiments may be inadequate when dealing with spatially limited, highly aggregated preys (Bochdansky and Bollens, 2004).

Swarms and schools of zooplankton are a feeding and mating unions, and vision is the main channel of communication of crustaceans. In freshwaters, the size, shape, and density of schools depend on illumination of water, water body type, and meteorological conditions (Butorina, 1986). The optimization of successful sexual encounters in copepods is the ultimate driving forces behind this segregation of both sexes (Schabetsberger and Jersabek, 2004; Nihongi et al., 2004). Food searching and predators avoidance are also biological forces explaining microscale spatial distribution of rotifers and daphnids. Experimental studies showed that swimming behaviour of rotifers at microscale extent is dependent of the food environment and nutritive status of females. The responses of *Brachionus calyciflorus* to the presence of food consisted of an increased rate of turns and a decreasing swimming speed (Charoy, 1995). Video recording of swimming behaviour of *Daphnia* clones in presence of infochemicals (kairomones) excreted by their main invertebrate (*Chaoborus*) and fish (young-of-the-year (YOY) perch (*Perca fluviatilis*)) predators indicated that microscale vertical distribution of *Daphnia* was strongly clone dependent. Some clones respond in a particular trait (swimming speed, trajectory length, and vertical distribution) to the presence of *Chaoborus* infochemicals, whereas others react only to *Perca* infochemicals (Weber and Van Noordwijk, 2002). Searching for food or mate also induces zooplankton patchiness at microscales in marine systems. For instance, krill have some means of detecting and remaining within algal patches using active swimming behaviour. Krill density was an order of magnitude higher in algal patches and their swimming speed doubled. Sinking bouts were almost entirely eliminated inside the algal patch. The ability to detect and remain within algal patches must be considered as an explanation of euphausiid small-scale patchiness (Price, 1989). In marine copepods, mate finding behaviour is mediated by pheromone clouds of female copepods, as seen with three-dimensional video analysis (Kiørboe et al., 2005; Bagøien and Kiørboe, 2005; Kiørboe and Bagøien, 2005). Females produce a short and spatially patchy pheromone cloud or trail, which induces males to swarm and adopt a rapid zigzag or trail swimming behaviour around the females.

5. IMPORTANCE OF ABIOTIC AND BIOTIC COUPLING FOR MICROBIAL PATCHINESS ACROSS THE MULTISCALE CONTINUUM

In any ecological system, factors that regulate the abundance of species vary with spatial scale (Wiens, 1989; Ricklefs, 1990; Levin, 1992). The relative importance of abiotic and biotic processes in driving aquatic microbial patchiness may vary across spatial scales and plankton food webs. One of the major challenges for the disciplines of oceanography and limnology is to measure the relative strength of these processes in natural ecosystems, examine the interactions among them, and combine this information in an effort to explain the multiscale patterns of microorganisms' distribution, abundance, and function. Recent attention to processes driving plankton patchiness stems from the likely influence of patchiness on species interaction, the modelling of population dynamics and assessment of community function.

The spatial heterogeneity observed in plankton microbial communities in marine and freshwaters has a multiplicity of origins. Two classical models are found in the literature: (1) *the environmental abiotic control model*, where abiotic factors related to hydrodynamics, thermal, and chemical conditions of the water column, are deemed responsible for the observed spatial patterns of microbial organisms; and (2) *the environmental biotic control model*, where the links among microorganisms in the food web (competition and predation) are considered to be the primary factors structuring microbial communities. These models have often been viewed as competing or mutually exclusive hypotheses. However, the hypothesis of multiple coupling of abiotic and biotic processes has recently been proposed to better explain plankton spatial structures in freshwater and marine systems (Pinel-Alloul, 1995). Here, we discuss the importance of abiotic and biotic coupling for driving patchiness for bacteria, phytoplankton and zooplankton across the multiscale continuum in aquatic systems.

5.1 Bacterioplankton

Spatial patchiness of microbial organisms (bacteria, picocyanobacteria) in marine and freshwater systems results from the environmental variation in physical, chemical, and biological factors. Different forcing mechanisms and ecological interactions dominated at different spatial scales (Fig. 8-7). Our multiscale survey shows that abiotic forces are the major drivers of the spatial structuring of bacterial microorganisms at every scale, even on the nanoscale extent. Physical advective forcing generated by water circulation, major hydrographic and topographic fronts, cyclonic gyres, eddies and upwellings, river inflows and tidal currents play a predominant role in shaping bacterial community attributes on the mega- and mesoscale extents (Table 8-1). This physical forcing constitutes a first step in structuring the abiotic environment, such as sea temperature, salinity gradients, and nutrient fluxes. Even on large-scale extents in freshwaters, variation in bacterial communities among lakes and rivers is mainly driven by abiotic processes operating at the regional and landscape scales, such as lake and river topography and morphology, trophic gradient, water retention time, water chemistry and nutrients (Table 8-2). On coarse-scale and fine-scale extents, water vertical stratification and the associated gradients in light, nutrients, and anoxia within lakes control the vertical distribution of bacterioplankton, whereas longitudinal and transversal discontinuities in water residence time, light, and nutrients conditions influence bacterial distribution patterns in rivers. Even on the microscale extent, abiotic processes associated with the polymerization of the organic matter gel and the micro-patches of nutrients and substrates are the first determinants of bacterial distribution (Table 8-1). Secondly, bacteria are also under a biotic control at each spatial scale via the coupling between bacterial and phytoplankton communities, which both are under bottom-up control by nutrients and organic substrates (Tables 8-1, 8-2). The biotic interactions within the microbial food web have strong influence on the maintenance of the spatial structuring of microbial plankton in aquatic systems, except in rivers and humic lakes where the bacterial/algal coupling is weak because allochthonous sources of carbon are more important for sustaining bacterial activity than autochthonous sources of carbon. Biotic control by top-down factors such as grazing is only important on fine-scale extents in stratified lakes where deep bacterial layers offer dense patches of resources for mixotrophic algae and zooplankton (Table 8-2). Overall, on the largest scale extents, physical and bottom-up forcing (nutrients) shape the global distribution of bacterial communities (Li and Harrison, 2001). However, at smaller scales in local environments, both bottom-up (nutrients) and top-down (bacterivores) processes control bacterial community abundance and diversity (Schafer et al., 2001). However, bacterial abundance increases primarily with primary productivity is marine and freshwater systems (Gasol and Duarte, 2000; Gasol et al., 2002), and top-down effects of bacterivores are, in general, less important than bottom-up effects of nutrient and trophic status in determining the overall abundance and diversity of bacteria in aquatic systems (Horner-Devine et al., 2003). Often the patterns of patchiness observed in bacterial community at local scales may be interpreted as of little ecological importance, and averaged out, especially in studies conducted over regional and global scales. However, this fine-scale and microscale variability will often have an ecological relevance that is equivalent to, or greater than, the changes observed at larger scales. For instance, microscale intermittency of bacteria and edible algae density will have a definite effect on the grazing efficiency and growth and survival of planktonic consumers, subsequently influencing the flow of carbon through the microbial loop (Azam et al., 1983; Overman et al., 1999).

5.2 Phytoplankton

As seen for the bacterioplankton, phytoplankton patchiness in marine and freshwater systems is mostly under a strong abiotic control by physical advective forces and bottom-up processes (Fig. 8-8). However, biotic forcing due to phytoplankton specific adaptations to low-light photosynthesis, its mixotrophic metabolism, and top-down effects of zooplankton grazing are more important forces than for the bacterioplankton. Oceanic circulation and fronts, mesoscale eddies and gyres, upwelling and tidal currents, as well as freshwater inflows in coastal zones shape the mega-scale and mesoscale distribution of surface chlorophyll in oceans (Table 8-1). At coarse-scale extents, phytoplankton distribution is generally associated with the salinity

Spatial Heterogeneity of Planktonic Microbes

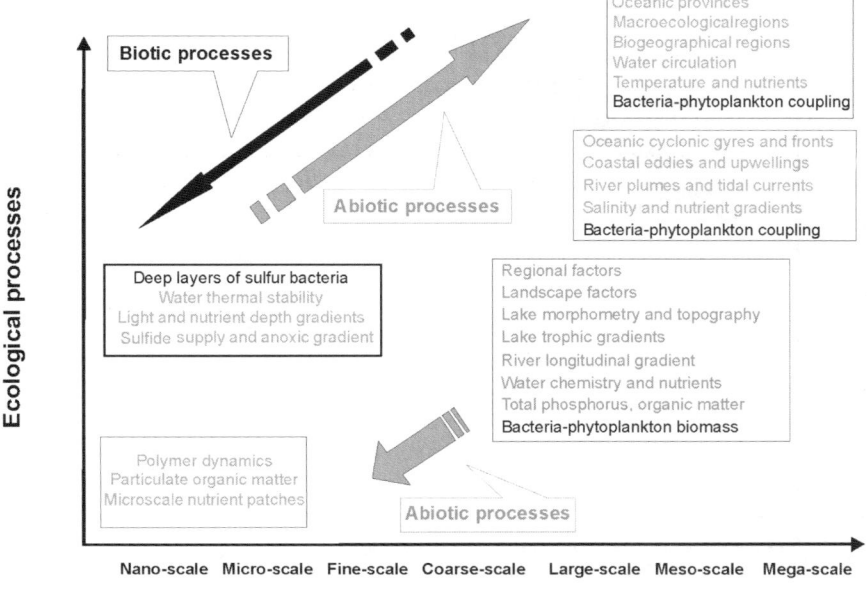

Figure 8-7. Generative processes driving bacterioplankton patchiness along the multiscale-continuum in marine and freshwater systems.

and nutrients gradients induced by river plumes, retentive gyres and coastal currents. Secondly, patchiness patterns in phytoplankton abundance are also influenced by *in situ* algal growth, and zooplankton grazing (Table 8-1). Important algal blooms of toxic algae may occur in coastal zones under a strong bottom-up forcing by nutrient retention and resuspension of resting cysts in retentive gyres when zooplankton grazing is weak (Townsend et al., 2005). On large-scale extents in lakes, phytoplankton patchiness is mainly under abiotic control by landscape factors related to lake topography and morphometry, water residence time and chemistry (Table 8-2). Total phosphorus is the main factor explaining among-lake variation in chlorophyll biomass, although biotic factors such as zooplankton grazing can drive the phytoplankton dynamics (Beisner et al., 2003). Physical forcing by wind-induced circulation, water stability, and inflow of rivers shapes lakewide spatial distribution of phytoplankton. In rivers, large-scale variation in phytoplankton biomass is also under abiotic control and related to trophic gradients (total phosphorus), river channel discontinuities and water retention time, even though biotic factors such as macrophyte development may affect transversal distribution of phytoplankton within rivers (Table 8-2). On fine-scale

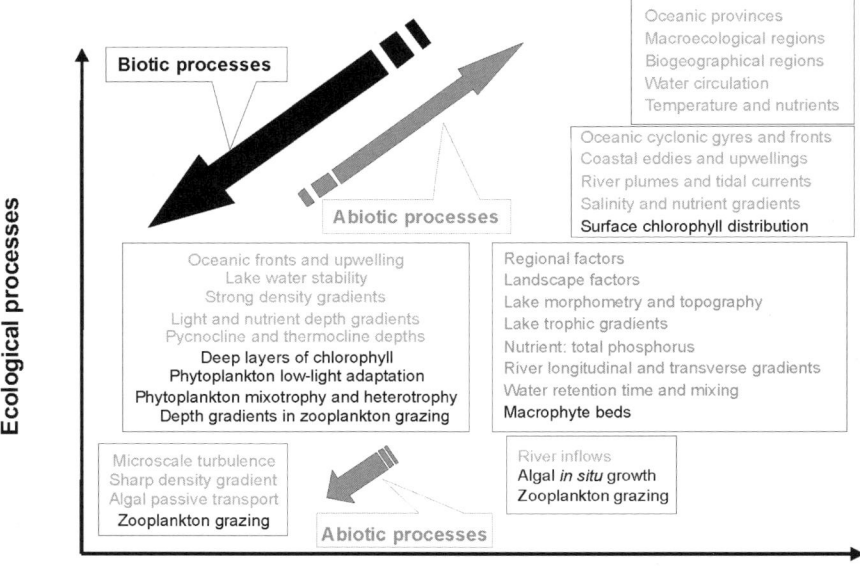

Figure 8-8. Generative processes driving phytoplankton patchiness along the multiscale-continuum in marine and freshwater systems.

extents in marine and freshwater systems, DCM were attributed to the effects of both abiotic and biotic factors (Tables 8-1 and 8-2). Strong stability of the water column associated to sharp vertical gradients in light, temperature, and nutrients are prerequisite conditions for the formation of DCM, whereas phytoplankton metabolic adaptations to low-light environment (chlorophyll content, buoyancy, motility, and mixotrophy) and depth-differential zooplankton grazing are very important biotic factors for the maintenance and persistence of DCM. As seen for bacterioplankton at microscale extent, abiotic forcing by small-scale turbulence is also of major influence for microscale phytoplankton patchiness, even though zooplankton may graze on phytoplankton micro-patches.

5.3 Zooplankton

Zooplankton organisms were long considered passive members of patches that were the product of large-scale physical processes. In oceans, it was stated that the structure of zooplankton assemblages was influenced by local hydrodynamic features, which through their action on surface water temperature, salinity, stratification, and mixing conditions, lead to spatial differentiation of the phytoplankton and zooplankton communities (physical–biological coupling) (Harvey et al., 2001). This prevailing viewpoint has shifted to accepting that biological processes also contribute strongly to zooplankton patchiness (Folt and Burns, 1999), compared to more passive plankton microorganisms (bacterio- and phytoplankton). Physical mechanisms alone proved insufficient to explain many spatial patterns found in zooplankton (Wiafe and Frid, 1996). In general, the environmental processes that generate and maintain the spatial patterns of zooplankton are of two types: (1) *physical processes* mainly generated by climatic and hydrodynamic regimes associated with bottom topography, such as currents, eddies, turbulent mixing, internal waves, and tidal and regional wind forcing are the dominant drivers of large-scale distribution patterns, and (2) *biological processes* due to zooplankton behaviour such as DVM, food searching, predator avoidance, swimming capability, swarming and mating behaviour are the dominant drivers of fine-scale distribution patterns. New studies about microscale patchiness (Tiselius, 1992, 1998) have altered our perception of the behavioural capacity and flexibility of zooplankton. Food search, predation reduction, and reproductive facilitation have been proposed as the major benefits of aggregation at fine and microscales (Ritz, 1994). Overall, all physical and biological processes take place at preferential spatial scales, and their relative importance for controlling zooplankton patchiness is scale-, species- and size-dependent (Avois-Jacquet, 2002).

Our multiscale survey supports the multiple control of zooplankton patchiness by biotic and abiotic forces in marine and freshwater systems (Fig. 8-9). Compared to generally passive members of planktonic communities (bacteria and algae), zooplankton is the community for which biotic factors have a greater contribution to spatial patchiness than abiotic factors, at least for scale of extent of <100 km (Tables 8-1 and 8-2). In oceans and estuaries, physical forcing by water circulation patterns is still the major force, although bottom-up coupling with primary producers, plankton diel migration, and food-web interactions have important contributions for structuring zooplankton patchiness. In lakes, local and regional processes interact to influence the composition of zooplankton communities in lacustrine habitats at large-scale in landscapes. At local sites, biotic and abiotic factors shown to correlate with patterns of zooplankton distribution include lake trophic gradient as expressed by TP concentration, water transparency, and acidity, chemistry, and predator assemblages (Table 8-2). Regional-scale processes include landscape position of lakes, colonization history, and dispersal mechanisms, which will influence large-scale trends in zooplankton distribution (Cottenie et al., 2003). Besides among-lake spatial variation at regional scale, factors operating at local scale within lakes, such as spatial heterogeneity among limnetic strata, including local effects of vertical gradients in light, oxygen, and nutrients, and species migratory behaviour and habitat preference can exert an important influence on zooplankton community patchiness. While zooplankton biomass is predictable at regional scale from lake trophic gradient, the size structure of zooplankton communities is independent of lake trophic gradient but changed art local scale in relation to fish and invertebrate predation pressure (Masson et al., 2004; Masson and Pinel-Alloul, 1998). Overall, studies on zooplankton DVM patterns in lakes support the model of multiple driving force of spatial patchiness of zooplankton at coarse and fine scales. Ecological plasticity in crustacean DVM and feeding patterns results from interactive effects of multiple abiotic and biotic forces. Bottom-up control by food resources and top-down control by fish predators govern zooplankton DVM and DHM in both marine and freshwater systems. The model suggests an endogenous control of zooplankton migration behaviour by the relative change in light intensity at dusk and dawn, and an exogenous control by the selective predation by fish or invertebrates. Other factors such as food resources, water transparency, and vertical gradients in temperature and oxygen also regulate migration patterns. Food distribution and temperature vertical profiles modulate the actual depth at which the animals stop migrating, whereas fish and UVR likely determine the timing of migration and its synchronization with rapid light changes at dawn and dusk. The most complex model of interactions between abiotic and biotic factors for the control of marine copepod DVM has been presented recently by Thorisson (2006). It also gives strong support to the multiple forces hypothesis and proposes that both physical (light avoidance) and biological (active food searching, buoyancy) mechanisms explain the diel and seasonal vertical migrations of herbivorous copepods in boreal and polar waters. Light avoidance is assumed to operate solely while the copepods are satiated and stayed below the euphotic zone to avoid visual predators hunting in surface waters. Hungry copepods are assumed to react to food smell from algae-rich surface waters by increased upward swimming, whereas satiated copepods maintain low activity and remain in deep zones. High-lipid content can also affect buoyancy of the copepods during the periods of egg production and lipid storage associated to algal blooms at spring, autumn and late winter. Finally, microscale distribution of marine and freshwater zooplankton is totally governed by biotic factors related to food- or mate-searching behaviour (Tables 8-1 and 8-2). Overall, spatial patchiness of zooplankton microorganisms which possess important dispersal and chemoreception

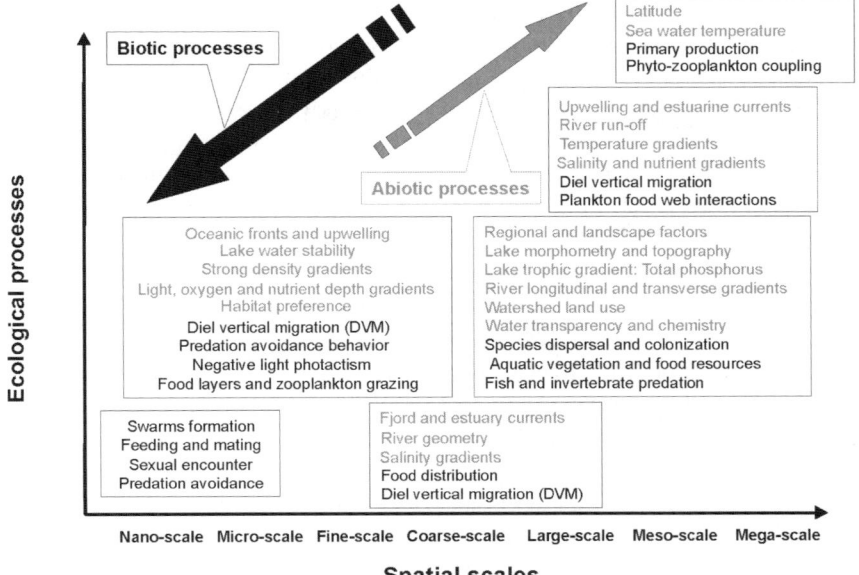

Figure 8-9. Generative processes driving zooplankton patchiness along the multiscale-continuum in marine and freshwater systems.

capacities is governed more by biotic than abiotic processes at regional (dispersal, colonization) and local (habitat preference, food searching, mate-searching) scales. Abiotic factors appear prevalent only on very large-scale extent of several hundreds or thousands of kilometres (mega- and meso-scales in oceans) because zooplankton distribution is closely linked to algal food resources which spatial distribution is governed by physical forcing.

6. ADVANCED SAMPLING AND ANALYTICAL TECHNOLOGIES

Historically, the study of spatial distribution of microorganisms in aquatic systems has always relied on advances in technology for the simultaneous sampling and measurements of the organisms' distribution and their immediate environment. This challenging adventure has seen its first day in the early 1930s when Alistar Hardy designed and built the first mark of the CPR, coinciding with the launch of the most ambitious plankton survey in the history of aquatic sciences (Hardy, 1935; Hardy, 1939). After 70 years since the first trial, the CPR survey collected samples from over 5 million nautical miles resulting in the largest database of plankton distribution in the world (Reid et al., 2003). While the CPR survey experienced throughout these years a large variety of ships and consequently towing speeds, the machine itself remains relatively unchanged in comparison to the original prototype (Batten et al., 2003b). In the meantime, technology continued to evolve and various sensors designed to measure physical, chemical, and biological variables were fitted onto the CPR as they became available resulting in a whole series of sister instruments such as flowmeters and sensors for measuring temperature, conductivity, depth, radiant energy, chlorophyll fluorescence, and trubidity (Reid et al., 2003; Brande et al., 2003). In this sense, the CPR could be considered a precursor to today's seagliders which are long-range autonomous underwater vehicles fitted with a series of sensors, capable of gliding through thousands of kilometres of seawater (Eriksen et al., 2001; Rudnick et al., 2004). Wireless technology allows the instrument to keep in touch with a land base taking instructions and sending data on a near real time fashion (Eriksen et al., 2001). In this section, we will review some of the emerging technologies currently applied in aquatic systems for the study of plankton patchiness. We will also discuss the future directions and how various technologies could be combined and also supported by advanced numerical models to better describe plankton patchiness and the driving forces behind it. A special focus of this review is to examine how the various technologies used in the study of the distribution of microorganism in water can successfully resolve the relevant spatial scales across the continuum.

6.1 Continuous Plankton Recorder used in large-scale marine and freshwater surveys

CPR could certainly be considered more than a sampling apparatus, but rather a concept which was motivated by Hardy's strong belief that by sampling fixed stations, which are miles apart, investigators may not be able to describe the true distributions of planktonic organisms. As he wrote in one of the earliest description of CPR, published as an appendix in Kemp (1926), "at one point one may strike a swarm of Copepods, or between two others miss an important zone of Diatoms". This demonstrates one of the earliest awareness that the uneven distribution of plankton in the ocean was the rule rather than the exception as it was commonly believed. The first deployment of the CPR started an era of investigation and documentation of plankton patchiness at ocean-basin scale which lasted for decades (Reid et al., 2003). Detailed descriptions of CPR and other comparable instruments such as the Longhurst–Hardy Plankton Recorder (LHPR), the Fast Continuous Plankton Recorder (FCPR), and the Undulating Oceanographic Recorder (UOR) could be found in the abundant literature that originated from what is now called the Sir Alister Hardy Foundation for Ocean Science (Longhurst and William, 1976; Aiken, 1981; Warner and Hays, 1994), and especially in a recent issue of Progress in Oceanography (e.g., Batten et al., 2003b; Reid et al., 2003; Richardson et al., 2004). For the purpose of this review we will focus only on some aspects of the technology in the context of the spatial distribution of plankton organisms in aquatic systems.

While CPR first mark was limited to the collection of planktonic organisms of a certain size (mesh size ~270 µm), later versions of CPR such as CPER were equipped with a suite of sensors designed to measure a number of physical and chemical variables such as sea temperature and conductivity, depth, PAR, and phytoplankton colour and chlorophyll fluorescence (Table 8-3). CPRs were designed to operate in towing mode and were typically towed behind volunteer commercial vessels also known as ship of opportunity. The CPRs were towed at a subsurface depth of ~10 m and collected planktonic samples continuously for about 800 km in a typical sample length (Batten et al., 2003a). This sampling regime has allowed the collection of about 4 millions of nautical miles from which 190,000 samples were analysed for phytoplankton color index (PCI) and zooplankton abundance representing a sampling rate of, roughly, a sample point per ~40 km (Reid et al., 2003). The speed of the silk advancement through the mechanism can be adjusted reaching sometime a spatial resolution of a sample per ~20 km (Richardson et al., 2006). Section of the silk band equivalent to a given mileage (between 20 and 40 km) is typically cut and PCI visually assessed. The sample is then taxonomically analysed for phytoplankton and zooplankton species, usually to the species level (Richardson et al., 2006). The abundances are then expressed in cubic metres of what would be interpreted as an integrated sample in a running average fashion over the distance of the sample, typically hundreds of nautical miles. The analyses of the continuous samples generates a set of semiquantitative estimates of large zooplankton species, which have been shown to consistently underestimate absolute abundances (Richardson et al., 2004, 2006). CPR surveys suffer some caveats due to under-sampling of some fast-moving and delicate components of the zooplankton (Hunt and Hosie, 2003) and to decreasing volume of filtered water with increasing speed of ships (Jonas et al., 2004). However, many of these issues of consistency and comparability with other plankton samplers have been investigated throughout the decades of the CPR survey and corrected with calibration factors (Batten et al., 2003a; Richardson et al., 2004). CPR surveys provide pan-oceanic data on geographic distribution, species composition, seasonal cycles of abundance, and long-term changes during the last 70 years on zooplankton populations in the North Atlantic ocean (Beare et al., 2003), the Southern Ocean (Hunt and Hosie, 2005), and the North Sea (Vezzulli and Reid, 2003). CPR data was used successfully to resolve spatial and temporal variability at the scale sufficient to address ocean basin issues such as algal blooms and eutrophication, plankton biodiversity and invasion, fisheries management, climate and hydrographic changes (Brander et al., 2003; Richardson and Schoeman, 2004; Leterme et al., 2005).

In freshwaters, a light prototype of the oceanographic CPR (2.22 m long with a frame diameter of 0.46 m, a mouth area of 0.024 m^2, and a net bag 1.17 m long with a 0.51 m long filter net of 200 µm mesh size) equipped with a digital flow meter, and temperature and pressure probes was used to assess vertical and horizontal distribution of crustacean zooplankton, and its driving forces in a large alpine lake (Masson et al., 2001). This study was a first attempt at coupling high-frequency zooplankton recording and hydroacoustic fish survey to assess simultaneously the large-scale distribution patterns of YOY fish and their zooplankton prey over a diel cycle (day, dusk, and night sampling). It showed that spatio-temporal distribution of crustaceans was shaped by predation loss to YOY perch and by antipredator behaviour (DVM, DHM) of zooplankton.

6.2 Remote sensing imagery used on large-scale surveys in marine and freshwater systems

Remote sensing imagery analysis with satellite or airbone-calibrated multispectral scanners has been widely used since several decades for the quantitative detection of the spatial distribution of Chl-*a* in aquatic systems. In oceans, the dynamics and spatial patterns of sea temperature, ocean colour, phytoplankton distribution, and associated changes in taxa (diatoms versus mixed algae) can be revealed by remote sensing at megascale extent (Platt et al., 2005). Recently, the successful launch of the ocean colour program Sea-Viewing Wide Field-of-View (SeaWiFS) by NASA in 1997 coincided with the start of a new and unprecedented era for the study of phytoplankton patchiness in oceans at the global scale (Lavender and Groom, 1999; Hooker and McClain, 2000; Werdell and Bailey, 2005). The images produced through SeaWiFS program became quickly very popular among the oceanographic communities and their use in the study of mesoscale oceanographic features is now part of everyday routine in oceanographic institutions. In the last 10 years, SeaWiFS data have been used in a large number of studies ranging from the description of oceanographic features such as large-scale algal blooms (Lin et al., 2003; Tang et al., 2003; Tomlinson et al., 2004; Zeichen and Robinson, 2004; Perez et al., 2005; Hu et al., 2005) to very detailed process-based studies of iron-induced phytoplankton blooms in combination with *in situ* experiments (Abraham et al., 2000; Boyd et al., 2004). The objective of the program were to record a 5-year period of high resolution and calibrated ocean colour data and to make it available to the oceanographic community to aid in the understanding of the ocean biological productivity at the global scale (Hooker and McClain, 2000; Ruddick et al., 2000). SeaWiFS provides global coverage every 2 days of ocean surface colour reflectance of visible light which is processed through a data processing system SeaAPS (Lavender and Groom, 1999) and converted into layers of equivalent surface chlorophyll concentrations in milligram per cubic metre. SeaWiFS data appear in a variety of ways in more than 600 publications dealing with various aspects of the ocean. The data collected since the start of the program have contributed to the understanding of the temporal and spatial dynamics of large-scale patchiness and their implications in the functioning of ocean processes. The data collection with SeaWiFS came to an end in 2004, but all the data collected between August 1, 1997 and December 23, 2004 will still be available to the scientific communities as announced by NASA. Other global programs (NASA's New Millenium Program (NMP)) will be implemented as a replacement of SeaWiFS and would include NASA's Earth Observing Satellite (EOS), Moderate Resolution Imaging Spectroradiometer (MODIS), and the European Medium Resolution Imaging Spectrometer (MERIS) among others (Hooker and McClain, 2000). The first hyperspectral sensor in space (Hyperion and Advanced Land Imager (ALI)) has sufficient spectral resolution (30 m) to enable mapping from space the phycocyanin and phycoerythrin pigments of cyanobacterial blooms in complexe estuarine and coastal systems (Kutser, 2004), and colored dissolved organic matter (CDOM) in lake ecosystems (Kutser et al., 2005).

Airbone remote sensing using the calibrated airborne multispectral scanner (CAMS) or the airborne imaging spectrophotometer for applications (AISA) were also used in freshwater systems to evaluate and monitor lake water quality at the scale of landscapes. It performs well in retrieving water quality variables (total suspended solids, turbidity, Secchi transparency, and absorption coefficient of aquatic dissolved organic mater, Chl-*a* and phaeophytin *a*) in various lake ecosystems (Kallio et al., 2001) and wetlands (Cózar et al., 2005).

Table 8-3. New technologies applied to assess microbial patchiness at multiple scales for the bacterioplankton, phytoplankton, and zooplankton communities in marine and freshwater systems.

Technology	Variable measured/sampled	Spatial scale	Depth range (m)	Data processing
Flow cytometry (FCM)	Bacteria	Macro to small	subsurface	Sample processing[a]
Airborne remote sensing (CAMS, AISA)	Chl-a	Large to small	Surface	Multispectral scanner imaging spectro photometer[b–d]
Ocean colour (SeaWiFS)	Chl-a	Macro to global	surface	Extensive data processing and calibration (SeaAPS)[b–d]
Fluorescence	Chl-a	Micro to small	0–200	Processing and calibration[e,f]
Spectrofluorescence	Chl-a + other pigments	Micro to small	0–100	Dedicated software[g,h]
Continuous Plankton Recorder (CPR)	Plankton	Small to meso	0–10	Sample processing[i,j]
Continuous Plankton and Environmental Recorder (CPER)	Plankton, Chl-a, T, Cond., PAR, depth	Small to meso	0–10	Sample processing[i,k]
Longhurst–Hardy plankton recorder (LHPR)	Plankton	Small to meso	0–10	Sample Processing[i,l]
Fast Continuous Plankton Recorder (FCPR)	Plankton	Small to meso	0–10	Sample processing[i,k]
Undulating oceanographic recorder (UOR)	Plankton, Chl-a, T, Cond., PAR, depth	Small to meso	0–120	Sample processing[i,k]
Bioacoustic techniques	Zooplankton	Large to meso	0–120	Dedicated software[l–r]
Video plankton recorder (VPR)	Plankton	Micro	0–130	Dedicated software
Digital in-line holographic microscopy (DIHM)	Plankton and bacteria size and tracks	Micro	0–15 (prototype)	Numerical reconstruction algorithm[s]
Optical plankton counter (OPC)	Particle size	Micro to meso	0–1000	Dedicated software[t]

continued

Technology	Variable measured/sampled	Spatial scale	Depth range (m)	Data processing
Laser optical plankton counter (LOPC)	Particle size and shape	Micro to meso	0–660	Dedicated software[u]
Long-range autonomous underwater vehicle: seaglider	Chl-*a*, T, Cond., PAR, depth	Micro to macro	0–200	Dedicated software; near real-time wireless transmission[v]

[a]Ducklow (2000); [b]Kallio et al. (2001); [c]Cózar et al. (2005); [d]Hooker and McClain (2000); [e]Lorenzen (1966); [f]Berman (1972); [g]Beutler et al. (2002); [h]Ghadouani and Smith (2005); [i]Reid et al. (2003); [j]Hardy (1936), (1939); [k]Aiken (1981); [l]Longhurst and William (1976); [m]Megard et al. (1977); [n]Hembre and Megard (2003); [o]Lorke et al. (2004); [p]Mair et al. (2005); [q]Sutor et al. (2005); [r]Davis et al. (2005); [s]Garcia-Sucerquia et al. (2006); [t]Herman (1988), (1992); [u]Herman et al. (2004); [v]Eriksen et al. (2001)

6.3 Flow cytometry, fluorometry, and spectrofluorometry used on small-scale surveys in marine and freshwater systems

Sampling strategies and analysis have been developed to get an accurate estimation of microbial processes not only at large scales but also at small scales. Indeed it has been recognized for several years that large volume samples may lead to underestimates of productivity due to the inadequate sampling of discrete microzones of plankton patchiness, and that even if one is interested in large-scale patterns and processes, samples should be taken at smaller intervals or distances, otherwise a misleading picture may be obtained. The ability to collect and process small samples is critical for monitoring microbial populations in spatially heterogeneous and temporally dynamic systems such as coastal oceans, lakes, and rivers. The understanding of microbial dynamics at small scales is vital not only to understand and model the ecology of aquatic microorganisms but for understanding the functioning and phenomenology of entire aquatic ecosystems (Seymour et al., 2005).

With the development of sensitive optics, practical laser systems, and higher fluorescence with blue-light-excited cyanine dyes and DNA staining, the flow cytometric detection and enumeration of bacteria is becoming an attractive alternative to epifluorescence microscopy for assessing small-scale spatial variation of bacterial communities in marine and freshwater systems (Ducklow, 2000; Gasol and del Giorgio, 2000; Jochem, 2001; Legendre et al., 2001). Discrete subpopulations of heterotrophic bacteria can be identified according to variations in green fluorescence and side scatter while autotrophic picocyanobacteria (*Synechococcus*) can be clearly defined according to higher levels of orange and red fluorescence. FCM technology was used to describe spatial heterogeneity and temporal dynamics of coastal planktonic microbial community using 10 s (microscale) or 30 min (small-scale) sampling intervals. New sampling techniques were recently developed to assess micro- and nanoscale patchiness of bacteria. In oceans, a new pneumatically sampling device allows to collect water samples with a 4.5 mm resolution (Seymour et al., 2000, 2005). In freshwaters, a sampling method using rapid freezing of small samples of water (Spatial information preservation method (SIP)) enable to demonstrate that plankton patchiness exists on scales of tens to hundreds of micrometres (Krembs et al., 1998a, b). Futhermore, novel molecular tools such as genetic fingerprinting with a high degree of phylogenetic resolution (automated ribosomal intergenic spacer analysis (ARISA), 5′ nuclease assays) were applied for estimating the relative spatial diversity in the expression of functionally important genes that may serve as proxy of metabolic activities of bacteria in marine and freshwater ecosystems (Fisher and Triplett, 1999; Suzuki et al., 2001; Yannarell and Triplett, 2004). Bacterial operational taxonomic units (OTUs) generated can be used as a surrogate of predominant biodiversity units (Horner-Devine et al., 2003; Reche et al., 2005). Recently, it has been shown that the combination of analytical flow cytometry (AFC) and denaturing gradient gel electrophoresis (DGGE) gives much more information about the diversity and dynamics of microbial communities (virus, bacteria, and phytoplankton) than either method on its own (Goddard et al., 2005).

During the past decades, underwater equipments for *in situ* investigations of the distribution of chlorophyll fluorescence in aquatic systems have been developed based on applications of fluorometric and spectrofluorometric technologies (Grunwald and Kühl, 2004). The phenomenon of chlorophyll fluorescence was discovered back in the mid-nineteenth century, however, it is until a century later that its use to quantify phytoplankton and photosynthetic bacteria in aquatic systems was attempted (Lorenzen, 1966). Lorenzen used a modified model III turner fluorometer to produce a very tight relationship between fluorescence and chlorophyll concentrations. This method was presented as suitable for continuous measurement of chlorophyll along a ship route where water was regularly pumped through and its *in vivo* fluorescence measured on board. It is not too long after Lorenzen's paper that Strickland (1968) published a short note drawing attention to the technical problems which caused non-linearity in the relationship between chlorophyll and *in vivo* fluorescence. Other pitfalls of the method have since been pointed out and are now known to all the limnological and oceanographic communities (Kruskopf and Flynn, 2006). It is now well accepted that the interpretation and the use of *in vivo* fluorescence must undergo a series of calibrations especially when it is used a proxy for biomass of photosynthetic organisms. However, the use of *in vivo* fluorescence in limnological and oceanographical investigations is now considered as routine as the use of Secchi disk measurement. This popularity was probably aided by the arrival of the first underwater fluorometers which allowed high-resolution measurement of *in vivo* fluorescence and very detailed description of vertical and horizontal distribution of chlorophyll (Wolk et al., 2001). With the recent advances in sensor technology, fluorescence sensors are now one of many sensors available as options to add to a wide variety of commercially available multiprobe instruments. Despite the known problems, *in vivo* fluorescence presents a number of advantages that makes it very useful in the study of plankton patchiness especially when it is combined with the other direct measurement such as algal counts. One of these features is certainly the ability of this technique to provide high-resolution measurements (~10 cm) and its strong correlation with measurement derived from more traditional techniques (Wolk et al., 2001).

Recently, spectrofluorescence techniques have been introduced and are believed to have resolved some of the common problem with what we can call now traditional *in vivo* fluorescence. The main difference is the ability to measure fluorescence at a number of wavelengths as opposed to usual 660 nm used by the common submersible fluorometers. Another important feature is the extensive calibration which resulted in the development of an algorithm capable of distinguishing between a number of algal groups based on their specific spectral signature (Beutler et al., 2002). Essentially, this instrument measures Chl-*a* fluorescence after a 0.1 ms excitation by 5 light-emitting diodes (450, 525, 570, 590, and 610 nm). The fluorescence is then transformed through an algorithm and total biomass of phytoplankton expressed in equivalent microgram Chl-*a* L^{-1}. It is possible to distinguish up to four spectral algal groups based on their distinct fluorescence patterns following the multiwavelength excitation, which are determined by the composition of their antenna pigments contained in the light harvesting complex of photosystem II (Dau, 1994; Beutler et al., 2003). The green group corresponds to phytoplankton species containing Chl-*a*, *b*, and xanthophylls, mainly chlorophytes and euglenophytes in freshwater lakes. The cyanobacteria group includes all those cyanobacteria (usually the majority) with phycocyanin as the main phycobilin pigment. The diatom group contains algae with Chl-*a*, *c*, and xanthophylls (i.e., fucoxanthin or peridinin), which in freshwater plankton are mainly diatoms, chrysophytes, and dinoflagellates. The mixed group, usually dominated by cryptophytes and some phycoerythrin-containing cyanobacterial species, has a combination of Chl-*a*, *c*, and phycoerythrin as the dominant phycobilin pigment (Beutler et al., 2003). The biomass of each of these groups is given in equivalent microgram Chl-a L^{-1}. However, phytoplankton group taxonomic reference should not be taken literally but rather as a fluorescence spectral designation as defined in Beutler et al. (2003). A research team from the Max Planck Institute (Plön, Germany) used several algal species representative of each algal group to produce a set of fingerprints used to capture the signature of each algal group (Beutler et al., 2002, 2003). This calibration procedure resulted in the generation of algorithm for each spectral group which are made available with the instrument and used in conjunction with a dedicated software (FluoroProbe 1.62). The fluoroprobe is also equipped with depth-pressure probe and a thermistor for measuring temperature with an accuracy of 0.05°C according to the manufacturer. While the application of this new technology to aquatic systems is still in its infancy, its early validations have demonstrated a great potential (Leboulanger et al., 2002; Ghadouani and Smith, 2005).

6.4 Acoustic, optical, and non-optical imaging techniques used at coarse and small scales

Bioacoustic techniques are commonly used in fisheries to detect large schools of fish; however, it is only since a decade that this technology has been successfully used to detect patches of smaller organisms such as zooplankton and large colonies of phytoplankton (Greene et al., 1998). It was shown that high-frequency (192 kHz) sonar systems could be used to detect patches formed by small zooplankton organisms measuring approximately 1 mm (Megard et al., 1997). Recently, this method was successfully used to study the distribution of daphnids in a small lake (Hembre and Megard, 2003). This technology can provide valuable information on the size and shape of patches and their temporal and spatial variability (Hembre and Megard, 2003); however, little information is provided on the actual composition of the patches. It is therefore necessary to complement the data by direct measurements such as net hauls. The composition of the patches have been successfully verified to be the target species (i.e., *Daphnia*) in most of the published studies (Megard et al., 1997; Hembre and Megard, 2003); however, this technology would require extensive on-site validation as the composition of plankton and non-plankton particles varies significantly from one system to the other, especially in high detritus environments. The use of multifrequency (38, 120, 200, and 420 kHz) acoustic technology in studies of zooplankton patchiness is now commonplace, and allows to describe fine-scale spatial distribution of small copepods as well as larger zooplankters such as *Mysis relicta*, *Chaoborus*, euphausiids, chaetognaths, and polychaetes (Gal et al., 1999; Lorke et al., 2004; Mair et al., 2005; Sutor et al., 2005).

The Optical Plankton Counter (OPC) is a novel technology designed firstly to assess zooplankton distribution in oceans (Herman, 1988), and later applied in limnology (Stockwell and Sprules, 1995; Sprules et al., 1998). The laboratory version OPC-1L (Focal Technologies, Halifax, Nova Scotia, Canada) was designed in the late 1980s with the aim to electronically count biological particles of a certain size based on the Coulter principle (Herman, 1988). It was shown that the OPC was able to detect most of the small and large zooplankton ranging in size from 250 µm up to 5 mm in equivalent spherical diameter (ESD) (Herman, 1992; Wieland et al., 1997). There was a general good agreement between the estimates of zooplankton abundance or biomass based on OPC analysis and those based on other sampling systems (LHPR or cantilever net plankton) and analytical methods (taxonomic-derived biomass using microscope counting and organic biomass of fractionated seston) despite the variability in zooplankton shape and composition among aquatic systems and the presence of interfering algal particles (Grant et al., 2000; Woodd-Walker et al., 2000; Baumgartner, 2003; Patoine et al., 2006). The measurements were found to be accurate even in water with high proportion of detritus (Zhang et al., 2000). The OPC has the ability not only to count the particles but also to provide information about the size of each particle (Herman, 1992) and the size distribution of crustacean zooplankton communites in oceanic and lacustrine systems (Beaulieu et al., 1999; Patoine et al. 2002). Overall, the OPC provides a potentially useful tool to monitor zooplankton size structure and biomass over large spatial scales in oceans (Edvardsen et al., 2002; Nogueira et al., 2004) and lake ecosystems (Sprules et al. 1998; Patoine et al., 2002, 2006). However, the effectiveness of the OPC to estimate zooplankton biomass depends of rigourous calibration studies and applications of specific ellipse models to convert ESD data of OPC into biovolume based on zooplankton shape and composition (cladoceran versus copepod dominance) (Mustard and Anderson, 2005; Patoine et al., 2006). The underwater version of the OPC (OPC-1T and OPC-2T) mounted on a towned vehicle along with a fluorometer and sensors unit for conductivity, temperature, and depth has the ability to perform a large number of *in situ* measurements of both the zooplankton distribution at a fine temporal and spatial resolution and the environmental heterogeneity in oceans (Herman et al., 1993) and large lakes (Stockwell and Sprules, 1995). The last generation of OPC uses a laser beam and is consequently called Laser-OPC or LOPC (Herman et al., 2004). This new instrument is meant to address some of the technical issues encountered in the use of OPC. LOPC operates with a significant advance relative to previous models that is its ability to provide more detailed information about the shape of the particles which reflect the light beam. The feature opens new perspectives for its use in aquatic systems as it could allow some level of fast taxonomical resolution in combination with an image analysis system equipped with a shape recognition algorithm. While providing useful information at similar spatial scale, bioacoustics, OPC, and LOPC technologies still require extensive calibration effort by means of a more direct observation technique.

Advances in high-resolution video technology during the last decade could provide a valuable support to bioacoustics or OPC methods as it would help determine the composition of the counted particles and provide near real-time validation of the data, especially when these instruments are deployed from ship on long-range cruises. Since the first development of video recording of zooplankton and fish larvae (Bergström et al., 1992; Lenz et al., 1995; Tiselius, 1998), digital imagery for the study of mesozooplankton distribution at sea in real time has been successfully applied. The newly developed fast-two digital video plankton recorder (VPR) systems mounted on towed platform could prove useful in combination with these technologies (Davis et al., 2004; Davis et al., 2005). It allows automatically identifying and classifying plankton to major taxa and some dominant species while determining their spatial distribution in real time. The results of the first deployment of this new VPR are described in details in Davis et al. (2005).

6.5 Emerging technologies and future directions

It is clear that new technology has always been at the forefront of investigations of the spatial distribution of microorganisms in aquatic systems, starting from the smallest bacteria and up to the largest zooplankton species. Table 8-3 presents a selection of instrumentation used in these investigations. While the list is not exhaustive it clearly shows that with few exceptions, every technique only addressed a narrow range of the much wider spatial domain which act as a host of various expression of spatial distribution and patchiness. It is clear that it is not possible to address a continuum of spatial scales when the investigation is limited to the use one or two techniques, which is usually the case. One of the challenges facing the exciting field of research is to widen the perspective of the investigation by addressing a continuum of scale while assessing the spatial heterogeneity in both the microbial organisms and their environment. A much broader perspective along a continuum of scale is needed to this research to be able to unravel the driving forces of spatial patchiness. As stated by Simon A. Levin (1992) in his seminal address to the Ecological Society of America, there is no single natural scale at which ecological phenomena, especially patchiness, should be studied. The key to a better understanding of patchiness and its driving forces is obviously dependent on how well we know how information is passed from one scale to the other (Levin, 1992), as most of what we observe at a given scale is only an expression of processes operating at a continuum of smaller scales. Any significant development in our understanding of this phenomenon will have to be guided by our ability to perform simultaneously high-resolution measurement of the physics, the chemistry, and the biology along a continuum spatial and temporal scale. This means that we need to combine some of the currently available technologies in a way that would guarantee some coherence between the data collected. Limnology and oceanography have developed a critical dependency on new instrumentation, which is able to collect a number of variables that are transmitted by satellite in real time in some cases and near real time in most of the cases. It is to be noted that the variables measured are more often physical and chemical variables while the biology is yet to catch up in this area.

Legendre et al. (2001) present a new perspective where the application of FCM could help advancing our ability to collect biological data at similar resolution as the physical data. It is expected that FCM and other cytometric approaches will improve the abibility of biological oceanography and limnology to address the major environment challenges in aquatic systems. Moreover, the new advancement in spectroflurorometry and optical plankton counting would have the potential to increase our ability to collect the required biological data to address patchiness questions. These instruments are becoming smaller in size and their mounting on moored buoys, drifter floats, or even long-range gliders is possible in the near future. This would allow us to collect biological data able to provide taxonomical information directly and without direct observation. *In situ* optics and hydrographic devices such as high-resolution bio-optical sensor capable of resolving centimetre scales of fluorescence, temperature, turbidity, and light irradiance are crucial for measuring microbial patchiness and determining their generative processes. Recently, automated underwater vehicle (AUV) equipped with a series of bio-optical sensors and a submersible microscope with hologrammetry and digital imaging technologies are under development for detecting microorganisms' distribution in waters without chemical treatment and was used for detecting the three-dimensional distribution of freshwater algal red tide (Ishikawa et al., 2005). The high-resolution sensor mounted on a free-fall profiler "Turbulence Ocean Microstructure Acquisition Profiler" (TurboMAP) is specifically designed to record simultaneously biological and physical properties of marine water column, i.e., shear, velocity, temperature, conductivity, *in vivo* fluorescence and turbidity (Wolk et al., 2002, 2003). The Shadowed Image Particle Profiling and Evaluation Recorder (SIPPER) coupled with an OPC in a towed platform served to describe spatial variation in the abundance, biomass, taxonomic composition, and size distribution of marine mesoplankton (Remsen et al., 2004). The most novel technology of digital in-line holographic microscopy (DIHM) was developed recently as a new tools that allows real three-dimensional observation and four-dimensional tracking of aquatic microorganisms as small as bacteria, protists, and rotifers (Garcia-Sucerquia et al., 2006).

7. STATISTICAL METHODS TO ASSESS AND MODEL MICROBIAL FUNCTIONAL HETEROGENEITY

The spatial heterogeneity of plankton distribution rose acute methodological problems in quantitative ecology since two decades (Angeli et al., 2006). Statistical methods of spatial analysis (descriptive or inferential) have been under development since the 1980s and have offered many approaches to demonstrate the existence of spatial structures, describe and map the spatial patterns, and analyse the influence of environmental abiotic and biotic factors (Legendre and Fortin, 1989; Legendre and Legendre, 1998). Starting around two decades ago, various statistical methods are widely used by oceanographers and limnologists to describe spatial patterns of biological data and model their environmental control (Table 8-4). They include: (1) univariate spatial correlograms and multivariate Mantel correlograms, (2) geostatistical methods such as univariate variogram and multivariate clustering with spatial contiguity constraint, and methods of kriging for mapping spatial distribution patterns, (3) multivariate analysis as principal component analysis, cluster analysis, indicator value method, and non-metric multidimensional scaling (MDS), (4) spectral analysis and multifractal analysis, and (5) variance partioning and canonical analysis which include space as an explanatory variable for modelling the effects of environmental and spatial factors and their interactions. Multiscaling approaches such as spectral analysis, wavelet analysis, multifractal analysis, and principal coordinates of neighbour matrices were recently developed to assess spatial patchiness across multiple scales (Seuront et al., 1999; Borcard et al. 2004; Franks, 2005; Keitt and Urban, 2005).

Here, we review these statistical methods and present how they were used to describe and model spatial patterns of aquatic microorganisms. However, no mathematical information is given, and readers must refer to specialized books (Legendre and Legendre, 1998). Recently, Dungan et al. (2002), Dale et al. (2002), and Perry et al. (2002) presented a balanced view of scale in spatial statistical analysis and concepts; they showed mathematical relationships among methods for spatial analysis and gave guidelines for selecting statistical methods for quantifying spatial pattern in ecological data. The recent review of Beaugrand et al. (2003a) has emphasized how important statistical analyses have been, and are likely to continue to be, in the interpretation of CPR data for assessing zooplankton mesoscale distribution. These statistical methods improved the sorting of information in the CPR database collected from 1958 to 2000 on marine zooplankton, and help to evaluate relationships between biological and environmental data

However, statistical methods used for assessment of spatial patchiness cannot be conducted without taking into account a number of theoretical and practical considerations. Generally, data collected on the spatial distributions of microbial organisms are spatially autocorrelated and thus not suitable for traditional statistical analysis (Dutilleul, 1998). Autocorrelation is a very general property of ecological variables, and indeed, of all variable observed across geographical space (spatial autocorrelation) or along time series (temporal autocorrelation) (Legendre, 1993). Spatial autocorrelation which comes either form the physical forcing of environmental variables or from community processes, present a problem for statistical testing because autocorrelated variables violate the assumption of independence of most standard statistical procedures. Thus, many of the basic statistical methods used in ecological studies are impaired by autocorrelated data (Legendre and Fortin, 1989). Recently, several methods have been developed to perform valid statistical tests such as t-test, correlation analysis, and ANOVAs in the presence of spatial autocorrelation (Dutilleul and Pinel-Alloul, 1996; Legendre and Legendre 1998).

7.1 Spatial analysis and geostatistical approaches

Spatial and geostatistical analyses require biological data that are labeled with the spatial coordinates at which measurements were collected. The spatial structure of microbial community can be illustrated by spatial univariate or multivariate correlograms using either Moran's I and Geary's c spatial autocorrelation coefficients. In case of multivariate data (such as species composition), we can use Mantel correlogram (Legendre and Fortin, 1989). Correlograms well performed to describe lakewide distribution of microbial communities (bacteria, Chl-a, and zooplankton) and environmental factors (water temperature and stability, nutrients, fish) in Lake Geneva (Pinel-Alloul et al., 1999) (Fig. 8-4).

Table 8-4. Statistical methods used to assess spatial distribution of planktonic microorganisms and its generative processes in marine and freshwater studies.

Statistical methods	Systems	Microbial organisms	Scales	Spatial patterns
Correlograms Mantel correlograms	Northern Pacific, Continental Shelf[1]	Zooplankton	Meso	Cross-shelf, along-shore
	Lake Geneva[2]	Bacteria, Chl-*a* zooplankton	Large	Lakewide pattern
Geostatistical methods, slope of the semi-variogram	Reservoir, China[3]	Chl-*a*	Large	Horizontal patterns
Semi-variance Kriging method	North Atlantic[4–6]	Calanoid Copepods	Meso	Horizontal patterns
Geostatistical methods, variograms	Continental shelf[1]	Zooplankton	Meso	Horizontal patterns
	Tropical lagoon[7]	Planktonic ciliates	Small	Patchy distribution
	Gulf of St. Lawrence[8]	Northern Shrimp	Large	Horizontal patterns
	Lakes[9–11]	Zooplankton	Large	Lakewide patterns
	Lake Trasimeno (Italy)[12]	Phytoplankton Zooplankton	Small to large	Horizontal distribution, Lakewide scale
Multivariate analysis PCA Centred PCA Three-mode PCA	North Atlantic[13–15]	Zooplankton Calanoid copepods	Meso	Horizontal patterns
PCA	Lakes[16]	Zooplankton	Large	Landscape patterns
Cluster analysis	Lakes[17–19]	Zooplankton	Large	Landscape patterns
Indicator value method	North Atlantic[6]	Zooplankton Calanoid copepods	Meso	Horizontal patterns
Non-metric multidimensional scaling (MDS)	North Sea[20]	Zooplankton	Meso	Horizontal patterns
Canonical correspondence analysis CCA, partial CCA	Lakes[19]	Zooplankton	Large	Landscape patterns
Spectral analysis	Open Ocean[21]	Temperature, Chl-*a*, Zooplankton	Mega	Trans-atlantic distribution
Multifractal analysis	North Sea[22–25]	Zooplankton *Temora longicornis*	Meso	Horizontal patterns

continued

Statistical methods	Systems	Microbial organisms	Scales	Spatial patterns
		Copepods[26–28]	Fine	Swimming patterns
	Oceans[29–32]	Phytoplankton	Meso	Multiscale patterns
Scaling and multi-scaling methods Individual-based models (IBM)	Oceans[33,34]	Zooplankton	Meso	Individual spatial patterns
Spatial analysis by distance indices (SADIE)[35] Nearest- and multiple-neighbour statistics[36]		Phytoplankton		
Principal coordinates of neighbour matrices PCNM analysis	Reef lagoon (Guadaloupe)[37]	Zooplankton	Large	Costal transects
	Thau marine Lagoon (France)[38]	Chlorophyll	Large	Whole lagoon

[1]Mackas (1984); [2]Pinel-Alloul et al. (1999); [3]Zhao and Cai (2004); [4]Beaugrand and Ibanez, (2002); [5]Beaugrand (1999); [6]Beaugrand et al. (2002b); [7]Bulit et al. (2003); [8]Simard et al. (1992); [9]Pinel-Alloul and Pont (1991); [10]Masson et al. (2001); [11]Thackeray et al. (2004); [12]Ludovisi et al. (2005); [13]Reid and Beaugrand (2002); [14]Beaugrand et al. (2001); [15]Beaugrand et al. (2000b); [16]Masson et al. (2004); [17, 18]Pinel-Alloul et al. (1990a, b); [19]Pinel-Alloul et al. (1995); [20]Lindley and Williams (1994); [21]Piontkovski et al. (1997); [22]Seuront and Lagadeuc (2001); [23,24]Seuront and Schmitt (2005a, b); [25]Pascual et al. (1995); [26]Schmitt and Seuront (2005); [27, 28]Seuront et al. (2004a, b); [29]Lovejoy et al. (2001); [30]Seuront et al. (1996); [31]Seuront et al. (1999); [32]Denman and Abbott (1994); [33]Seuront and Scrutton (2004); [34]Souissi et al. (2005); [35]Xu and Madden (2003); [36]Strutton et al. (1997); [37]Borcard et al. (2004); [38]Avois-Jacquet (2002)

Empirical variograms are used as exploratory graphical tools constructed to detect and describe spatial patterns, and identify differences in spatial structure. Variograms can be used to measure the fractal dimension of environmental gradients. However, classical experimental semi-variograms are highly sensitive to irregular distribution of observations or sampling sites inside the spatial domain (Wackernagel, 1995). As sampling with CPR or other plankton sampler is irregular, unbiased estimates of the experimental semi-variogram are difficult to obtain. Sen (1989) proposed to calculate cumulative semi-variograms that were applied to CPR data by Beaugrand and Ibanez (2002). Selection of appropriate interpolation methods for spatial representation of plankton data is a key stage in making spatial and temporal comparisons of biological variables. Kriging method is the most commonly used because it has the advantage that it takes into consideration spatial scales of change in ecological variability. The diversity of calanoid copepods in the North Atlantic based on CPR sampling has recently been mapped using this procedure of spatial interpolation (Beaugrand, 1999; Beaugrand et al., 2002b). The use of the kriging procedure is limited, however by three main problems. First, there is the interpretation of the variogram and its approximation using theoretical model. Secondly, both geometric and zonal anisotropy must be corrected. Thirdly, it is difficult to use kriging for rare species because of the high proportion of zeros in the matrices. In practice, it is hard to verify all these parameters when a large number of maps is produced, so another method, called inverse squared distance, has been applied (Beaugrand et al., 2000a; Planque and Batten, 2000) to map the mean spatial distribution of some key species of calanoid copepods in the North Atlantic Ocean. In this case, semi-variograms were modelled using a spherical model. Geostatistical analysis (semi-variogram) and fractal analysis (slope of the semi-variogram) was used to evaluate whole-lake spatial distribution of Chl-a in a Chinese reservoir (Zhao and Cai, 2004). Geostatistical techniques were also used to examine plankton ciliates patches in a tropical coastal lagoon (Bulit et al., 2003), shrimp distribution in the St. Lawrence Golf (Simard et al., 1992) and zooplankton distribution in the continental shelf of Northern Pacific (Mackas, 1984) and macrozooplankton in lakes (Pinel-Alloul and Pont, 1991; Masson et al., 2001; Thackeray et al., 2004).

7.2 Multivariate analysis

Multivariate analysis help to distinguish plankton assemblages associated to specific environmental features. An overview of the multivariate methods applied to CPR data from marine zooplankton has been presented by Beaugrand et al. (2003a). They include standardized principal component analysis (PCA), centred PCA, three-mode PCA, cluster analysis, indicator value method, and non-metric multidimensional scaling (MDS).

Standardised Principal Component Analysis (PCA) helped to illustrate the spatial distribution of marine zooplankton in the North Sea around United Kingdom, and separates different species associations related to northern and southern oceanic, northern and southern intermediate and neritic habitats (Reid and Beaugrand, 2002). The spatial pattern was in part explained by effect of temperature and salinity gradients. Centred PCA was used to identify the spatial patterns in diversity (in terms of number of taxa per CPR sample) of calanoid copepods (Beaugrand et al., 2001). The three-mode PCA was applied to CPR data to distinguish species association of phytoplankton and zooplankton communities along the South Atlantic route from the channel of the North Sea near England to the Bay of Biscay (Spain) in the

North Atlantic (Beaugrand et al., 2000b). In freshwaters, PCA analysis served to study the effects of abiotic and biotic factors on the multiscale variation in zooplankton biomass in a lake district (Masson et al., 2004).

Cluster analysis techniques are powerful multivariate tools that is used to group objects and descriptors. Cluster analysis and ordination methods were applied to investigate distribution of zooplankton species and enabled visual inspection of species and environmental descriptors in the projection of the multidimensional space reduced to two dimensions (Pinel-Alloul et al., 1990a, b; Pinel-Alloul et al., 1995).

The indicator-value method (Dufrêne and Legendre, 1997) has been applied to identify calanoid copepod species association (Beaugrand et al., 2002b) in the North Atlantic. This enables to distinguish six different associations: warm-temperate oceanic species assemblage, Bay of Biscay and southern European shelf-edge assemblage, temperate neritic, and pseudo-oceanic species assemblage, cold-temperate, subarctic and arctic species assemblage, subtropical and warm-temperate species assemblage. Four modulating factors have been identified: (1) temperature, (2) hydrodynamics, (3) stratification, and (4) seasonal variability. These factors are often linked, but they can act at different scales, and their contribution can vary geographically. Moreover, this study clearly detected the influence of warm currents on diversity and hence the functional characteristics of ecotones west of Europe and the Gulf stream extension.

MDS is a non-parametric ordination method that aims to project multidimensional space into a reduced number of dimensions, generally two. MDS offers some advantages over current ordination analysis as principal coordinates, reciprocal averaging, and correspondence analysis by its less direct approach – first constructing a dissimilarity matrix to suit the particular form of the data and then allowing a general monotonic transformation to distance. MDS has the advantage of better handle missing data, replication and data of non-uniform reliability. The MDS method was used to distinguish zooplankton species association in the North Sea (Lindley and Williams, 1994).

Methods of canonical correspondence analysis (CCA and partial CCA; ter Braak 1986, 1988; Palmer, 1993) are useful to explaining a spatial typology base on biological data using several sets of environmental factors, and to estimate the values of environmental factors from species abundance (indicator species), and to demonstrate the agreement between the typologies resulting from faunistic and environmental data. Co-inertia analysis has been shown as an alternative to CCA (Franquet et al., 1995). Mantel tests also allow comparing spatial patterns of biological and environmental variables, and testing if there is a significant relationship between the two spatial structures (Legendre and Fortin, 1989). Partial mantel tests can be computed to include space coordinates as explanatory covariates, and testing the biological–environmental relationships after controlling for the influence of the space. Finally, a method has been proposed to partition the variation of species abundance into independent components: pure spatial, pure environmental, spatial component of environmental influence, and undetermined (Borcard et al., 1992). It has been used to evaluate the relative importance of abiotic and biotic environmental factors, and space on freshwater zooplankton distribution over large-scale extent (Pinel-Alloul et al., 1995). To evaluate the effects of local (small-scale) and global (large-scale) spatial patchiness in the spatial distribution of plankton communities (phyto and zoo) in a lake, Ludovisi et al. (2005) used the decomposition procedure proposed by Thioulouse et al. (1995) and multivariate geostatistical approaches.

7.3 Spectral analysis and multiscale approach

Spectral analysis (spatial spectrum and wavelet analysis) has been used since the 1970s to analyse spatial patterns in plankton communities (Platt and Denman, 1975; Franks, 2005; Keitt and Urban, 2005; Currie and Roff, 2006). The advantage of this method is that it allows analysis of anisotropic data, which are frequent in ecology. Its main disadvantage is that, like spectral analysis for time series, it requires a large database. In spectral analysis, the periodogram assumes that the spatial pattern results from a combination of repeatable patterns. The periodogram and its R-spectrum and Θ-spectrum are very sensitive to repeatability in the data. The slope of spatial spectrum (k-spectrum) varied form -2.5 for flat distribution to -0.6 for distribution with high numbers of spikes (patchy distribution) (Franks, 2005). In open oceans, the power spectra of temperature, Chl-a and zooplankton biomass in the surface layer were similar, varying within the range of -3 to -2 in a band of wavelengths from 200 km to 10 km (Piontkoski et al., 1997). Wavelet analysis is an approach to analysing spatial data, related to spectral analysis. Wavelet transform and wavelet-coefficient regression efficiently characterize scale-specific patterns (Keitt and Urban, 2005). Wavelet transform have become the preferred representation in which to analyse pattern and scale in environmental data. Wavelet transform decompose a pattern into a hierarchy of different scales. Scale-specific analysis using wavelets hold great promise for ecological application concerned with multivariate, multiscaled patterns.

However, the most interesting method for detecting multiscale patterns of spatial distribution recently developed is the principal coordinates of neighbour matrices (PCNM) (Borcard and Legendre, 2002; Borcard et al., 2004). The PCNM method, based on the close neighbourhood relationships among the sampling sites, can be used to detect and quantify the spatial patterns over a wider range of scales. Continuous recording of all variables is not necessary, and the method can be used with irregularly spaced data. PCNM analysis achieves a spectral decomposition of the spatial relationships among the sampling sites, creating variables that correspond to all the spatial scales that can be perceived in a given data set. The significant PCNM variables can be directly interpreted in terms of spatial scales, or included in a procedure of variance decomposition with respect to spatial and environmental components. PCNM analysis was used to describe the multiscale spatial variation of zooplankton biomass inhabiting a coastal reef lagoon (Guadeloupe) and to test hypotheses about the biological and physical processes causing zooplankton patchiness (Avois-Jacquet, 2002).

7.4 Multifractal analysis

There is a great deal of interest in the potential usefulness in applying the concepts of fractals to the study of spatial structures in plankton ecology (Seuront et al., 1996, 1999; Seuront and Schmitt, 2005a, b). Contrary to basic analysis techniques such as power spectral analysis, universal multifractals allow the description of the whole statistics of a given spatial field with only three parameters (spectral exponent ß, first moment scaling H, universal multifractal parameter C). There is now more and more experimental evidence of the intermittent and multifractal nature of plankton distribution in marine waters (Pascual et al., 1995; Seuront et al., 1996, 1999; Seuront and Lagadeuc, 2001; Lovejoy et al., 2001; Seuront and Schmitt, 2005b). Theoretical considerations based upon detailed (multifractal) descriptions of the intermittency of turbulent kinetic energy dissipation rates and phytoplankton and zooplankton distribution at small scales suggest that taking into account intermittency in turbulence for critical processes such as predator–prey encounter rates, nutrient fluxes around phytoplankton cells, phytoplankton coagulation, and the relative size of phytoplankton aggregates and vertical fluxes has consequences far from being negligible (Seuront, 2001; Seuront et al., 2001). Seuront et al. (1999) used universal multifractals to describe phytoplankton distribution in turbulent coastal waters of the English Channel in the North Sea. The phytoplankton exhibited a very specific heterogeneous distribution over smaller (>10 m) and larger scales (>500 m) dominated by turbulent processes whereas, the pattern was obviously dominated by biological processes over intermediate scales (10–500 m). Universal multifractal analysis can be regarded as a way to delineate the relative contribution of biological and physical processes to the spatial pattern in plankton communities, a major issue in marine ecology.

Multifractal analysis was used to characterize spatial patterns of distribution of a calanoid copepod (*Temora longicornis*) along the coastal area of the North Sea (Seuront and Lagadeuc, 2001). The fractal dimension of the spatial structure was D = 1.79, which correspond to patchiness at scales ranging from 92 m and 120 km. The fractal dimension estimated from the distribution of *T. longicornis* is very similar to that estimated for the oceanic copepod *Neocalanus cristatus* abundance transects from the subarctic Pacific (D = 1.80), over a similar range of scales between 10 m and >100 km. Fractal dimensions of phytoplankton distributions range from 1.61–1.67 for *in vivo* fluorescence (Seuront et al., 1999) and from 0.98 to 1.69 for satellite images of sea surface Chl-*a* patterns (Denman and Abbott, 1994). It is not new to state that turbulence is the underlying dominant influence on oceanic distributions. Variables of primary physical origin, such as temperature, have spectra expected of a passive tracer in a turbulent system. Lovejoy et al. (2001) showed that phytoplankton must be considered as an active rather than passive scalar. Large-scale phytoplankton distribution is determined both by passive advection and growth. The fact that copepod's fractal dimension was higher than that of intermittent turbulence and phytoplankton distribution suggests that copepod behaviours such as diel migration, phototaxis, rheotaxis, social and predation pressure behaviour relevant at the different spatial scales induced larger fractal dimensions (i.e., flatter power spectrum and weaker scale dependence) in comparison with the phytoplankton under a stronger control of physical forces. Furthermore, zooplankton distributions are less conservative and sparser than those of phytoplankton. Fractal dimensions also confirms that, generally large life forms show more extreme spatial scale aggregations than do smaller and less mobile organisms.

Multifractals, and in particular universal multifractals, lead to a very precise characterization of the overall statistical structure of any given intermittent spatial pattern. Thus they appear to be an efficient descriptive tool which should also allow modelling of the multiscale detailed variability of intermittently fluctuating biological fields as well as one of the properties of their surrounding environment (Seuront et al., 1999). Precise knowledge of the distribution of pelagic organisms is indeed of fundamental importance in understanding the trophic relationships between planktonic organisms and also the related fluxes of matter. Low fractal dimension means a smooth and predictable distribution of food particles which can be gathered in a small number of patches, whereas high fractal dimension means rough, fragmented, space filling, and less predictable distribution of food. Therefore, when a predator can remotely detect its food in the surrounding, prey distributions with a low dimension should be more efficient. In contrast, when a predator has no detection ability, prey distribution with high fractal dimension should be relatively better, because available food quantity and encounter rate become proportional to the searched volume as fractal dimension increases. Moreover, the very complex patchy structure associated with a multifractal distribution may also change the food signal, usually considered as homogeneously distributed in time and space in models of predator–prey encounter rates. Indeed, planktonic animals have been shown to remain within patches when feeding, or exhibit fine-scale movements in areas of higher food concentration. Thus, encounter rates might be very different when organisms feed within patches (intensive search) as opposed to the search of new patches (extensive search). Thus foraging models will probably incorporate switching between feeding and searching behaviours as scaled to organism size, in order to simulate these complex physical–biological relationships effectively.

Multifractal analysis was also applied to describe three-dimensional copepod displacements using two-dimensional orthogonally focused and synchronized CCD cameras (Schmitt and Seuront, 2005), and zooplankton swimming behaviour under different intensities of turbulence (Seuront et al., 2004a, b). A simple, smooth, quasi-linear path is characterized by a low fractal dimension ($D = 1$) while that of a more-tortuous path has a higher fractal dimension reaching a value of $D = 2$.

Scaling and multiscaling approaches such as individual-based models (IBM) are based on life history and demographic parameters of zooplankton populations (mainly marine copepods) and can be used to evaluate the consequence of individual behaviour and spatial heterogeneity on the emerging properties at the population scale (Seuront and Scrutton, 2004; Souissi et al., 2005). The SADIE methodology (spatial analysis by distance indices) has recently been developed to quantify spatial patterns, particularly in determining the randomness of observed patterns, estimating patches and gaps, and quantifying correlation between species (Xu and Madden, 2003). Finally, Scheuerell (2004) has developed new statistical methods for gauging the degree of aggregation among organisms which can distribute in three dimensions in aquatic environments. The nearest neighbour approach (nearest- and multiple-neighbour statistics) is based upon the probability of observing at least one neighbour within a give n radius (Ripley's K statistics). The multi-neighbour statistical analysis utilizes information on all of the measured distance among individuals. Near-neighbour algorithm was used to compare the spatial structure of chlorophyll, a non-conservative tracer, with that of a conservative tracer, salinity in eastern Antarctica (Strutton et al., 1997).

8. CONCLUSIONS

The concept of spatial heterogeneity at multiple scales is very promising as the integrative basis for assessing spatial patterns in aquatic microbial communities and their driving forces. Our review shows that the spatial heterogeneity observed in plankton microbial communities in marine and freshwaters has a multiplicity of patterns and origins. In aquatic systems, plankton heterogeneity arises as the result of vertical and horizontal structuring of marine and freshwater habitats. Plankton patchiness occurs along a hierarchical continuum of spatial scales from mega-scale to micro- and nanoscale patterns, in both marine and freshwater systems. Plankton patchiness patterns are driven by many abiotic processes interacting with many biotic processes, and the relative influence of abiotic and biotic processes varies along the scale continuum. The relative importance of these processes corresponds to a gradation in effects over scales, the physical effects predominating at broad spatial scales while biological effects predominate at finer scales. Our multiscale survey supports the model of multiple driving forces for planktonic communities. However, abiotic forces are the most important drivers of patchiness for the bacterial and algal components whereas biotic forces greatly influence zooplankton patchiness.

Abiotic factors such as physical (water circulation) and bottom-up forcing (nutrients) shape the large-scale distribution of marine and freshwater bacterioplankton, whereas at smaller scales, both bottom-up (nutrients) and top-down (bacterivores) processes control bacterial community abundance and diversity. Bacterial abundance increases primarily with primary productivity and top-down effects of bacterivores are, in general, less important than bottom-up effects of nutrient and trophic status in determining the overall abundance and diversity of bacteria in aquatic systems. Phytoplankton large-scale patchiness in marine and freshwater systems is also mostly under a strong abiotic control by physical advective forces, landscape factors, and bottom-up processes. On fine-scale, deep algal layers are driven by both abiotic (water stability, light, temperature, and nutrient gradients) and biotic (chlorophyll-content, buoyancy, motility, and mixotrophy, zooplankton grazing) factors. However, biotic forcing due to phytoplankton specific adaptations to low-light photosynthesis, mixotrophic metabolism, and top-down effects of zooplankton grazing is more important than for the bacterioplankton. Spatial patchiness of zooplankton organisms which possess important dispersal and chemoreception capacities is governed more by biotic than abiotic processes at regional (dispersal, colonization) and local (habitat preference, food searching, mate-searching) scales. Abiotic factors appear prevalent only on very large-scale extent of several hundreds or thousands of kilometres (mega- and mesoscales in oceans) because zooplankton distribution is closely linked to algal food resources which spatial distribution is governed by physical forcing.

A large series of new technologies has been developed during the last two decades for investigate *in situ* spatial distribution of microorganisms in aquatic systems, starting from the smallest bacteria and up to the largest zooplankton species. Every technology addressed specific scales and types of microorganisms. It is still difficult to address a continuum of spatial scales when the investigation is limited to the use one or two techniques. One of the challenges facing the exciting field of research is to address a much broader range of scales simultaneously using complementary technologies advancing our ability to collect biological data at similar resolution as the physical data. New sampling strategy combining *in vivo* flow cytometry, fluorometry or spectrofluorometry with CPR or OPC would have the potential to increase our ability to collect the required biological data to address patchiness questions. *In situ* optics and hydrographic devices such as high-resolution bio-optical sensor capable of resolving centimetre scales of fluorescence, temperature, turbidity, and light irradiance, and AUV equipped with a series of bio-optical sensors and a submersible microscope with hologrammetry and digital imaging technologies are crucial for measuring microbial patchiness and determining their generative processes.

Various statistical methods are now used by oceanographers and limnologists to describe spatial patterns of microorganisms in aquatic systems. They include geostatistical and spatial analysis (correlograms, variograms, krigeage), multivariate statistics (clustering, canonical analysis, variance partitioning), spectral and multifractal analyses. However, few methods allow assessing spatial patterns with a multiscale perspective. The most innovative method for detecting multiscale patterns of spatial distribution is the PCNM.

REFERENCES

Abbott, M. R., K. L. Denman, T. M. Powell, P. J. Richerson, R. C. Richards, and C. R. Goldman, 1984, Mixing and the dynamics of the deep chlorophyll maximum in Lake Tahoe, *Limnol. Oceanogr.* **29**:862–878.

Abraham, E. R., C. S. Law, P. W. Boyd, S. J. Lavender, M. T. Maldonado, and A. R. Bowie, 2000, Importance of stirring in the development of an iron-fertilized phytoplankton bloom, *Nature* **407**:727–730.

Adler, M., F. Gervais, and U. Siedel, 2000, Phytoplankton species composition in the chemocline of mesotrophic lakes, *Arch. Hydrobiol.* **55**:513–530.

Adrian, R., and T. Schipolowski, 2003, Bacterial and protozoan mass accumulation in the deep chlorophyll maximum of a mesotrophic lake, *Arch. Hydrobiol.* **157**:27–46.

Aiken, J., 1981, The undulating oceanographic recorder mark 2, *J. Plankton Res.* **3**:551–560.

Alonso, C., V. Rocco, and J. P. Barriga, 2004, Surface avoidance by freshwater zooplankton: field evidence on the role of ultraviolet radiation, *Limnol. Oceanogr.* **49**:225–232.

Amblard, C., J. F. Carrias, G. Bourdier, and N. Maurin, 1995, The microbial loop in a humic lake: seasonal and vertical variations in the structure of the different communities, *Hydrobiologia* **300/301**:71–84.

Amon, R. M. W., and R. Benner, 1998, Seasonal patterns of bacterial abundance and production in the Mississipi River plume and their importance for the fate of enhanced primary production, *Microb. Ecol.* **35**:289–300.

Angeli, N., B. Pinel-Alloul, G. Balvay, and I. Ménard, 1995, Diel patterns of feeding and vertical migration in daphnids and diaptomids during the clear water phase in Lake Geneva (France), *Hydrobiologia* **300/301**:163–184.

Angeli, N., D. Gerdeaux, and J. Guillard, 2006, Analyse géostatistique des répartitions horizontales printanières de la biomasse zooplanctonique et des variables physico-chimiques dans un petit lac. *Rev. Sciences de l'Eau*, **19**:285–294.

Armengol, X., A. Esparcia, and M. R. Miracle, 1998, Rotifer vertical distribution in a strongly stratified lake: a multivariate analysis, *Hydrobiologia* **387/388**:161–170.

Arnott, S. E., and M. J. Vanni, 1993, Zooplankton assemblages in fishless bog lakes: influence of biotic and abiotic factors, *Ecology* **74**:2361–2380.

Ashjian, C. J., C. S. Davis, S. M. Gallaper, and P. Alatalo, 2005, Characterization of the zooplankton community, size composition, and distribution in relation to hydrography in the Japan/East Sea, *Deep Sea Res. II* **52**:1363–1392.

Atkinson, A., P. Ward, and E. J. Murphy, 1996, Diel periodicity of subantarctic copepods: relationship between vertical migration, gut fullness and gut evacuation rate, *J. Plankton Res.* **18**:1387–1405.

Attayde, J. L., and R. L. Bozelli, 1998, Assessing the indicator properties of zooplankton assemblages to disturbance gradients by canonical correspondence analysis, *Can. J. Fish. Aquat. Sci.* **55**:1789–1797.

Avois-Jacquet, C., 2002, Variabilité spatiale multi-échelle du zooplancton dans un lagon récifal côtier (Multiscale spatial variability of zooplankton in a coastal reef lagoon). Ph.D. Thesis. Université Pierre et Marie Curie et Université de Montréal, Canada, 302 p.

Azam, F., T. Fenchel, J. G. Field, J. S. Gray, L. A. Meyer-Reil, and F. Thingstad, 1983, The ecological role of water-column microbes in the sea, *Mar. Ecol. Prog. Ser.* **10**:257–263.

Azovsky, A. I., 2000, Concept of scale in marine ecology: linking the words to the world? *Web Ecol.* **1**:28–34.

Bagøien, E., and T. Kiørboe, 2005, Blind dating – mate finding in planktonic copepods. I. tracking the pheromone trail of *Centropagus typicus*, *Mar. Ecol. Prog. Ser.* **300**:105–115.

Barberio, R. P., and C. M. McNair, 1996, The dynamics of vertical chlorophyll distribution in an oligo-mesotrophic lakes, *J. Plankton Res.* **18**:225–237.

Barbiero, R. P., and M. L. Tuchman, 2004, The deep chlorophyll maximum in Lake Superior, *J. Great Lakes Res.* **30**:256–268.

Basterretxea, G., E. D. Barton, P. Tett, P. Sangra, E. Navarro-Perez, and J. Aristegui, 2002, Eddy and deep chlorophyll maximum response to wind-shear in the lee of Gran Canaria, *Deep Sea Res. I* **49**:1087–1101.

Basu, B. K., and F. R. Pick, 1996, Factors regulating phytoplankton and zooplankton biomass in temperate rivers, *Limnol. Oceanogr.* **41**:1572–1577.

Basu, B. K., and F. R. Pick, 1997a, Factors related to heterotrophic bacterial and flagellate abundance in temperate rivers, *Aquat. Microbiol Ecol.* **12**:123–129.

Basu, B. K., and F. R. Pick, 1997b, Phytoplankton and zooplankton development in a lowland, temperate river, *J. Plankton Res.* **19**:237–253.

Basu, B. K., J. Kalff, and B. Pinel-Alloul, 2000a, Mid-summer plankton development along a large temperate river: The St. Lawrence River, *Can. J. Fish. Aquat. Sci.* **57**:7–15.

Basu, B. K., J. Kalff, and B. Pinel-Alloul, 2000b, The influence of macrophyte beds on plankton communities and their export from fluvial lakes in the St. Lawrence River, *Freshw. Biol.* **45**:373–382.

Batten, S. D., and W. R. Crawford, 2005, The influence of coastal origin eddies on oceanic plankton distributions in the eastern Gulf of Alaska, *Deep Sea Res.* **52**:991–1009.

Batten, S. D., R. Clark, J. Flinkman, G. Hays, E. John, A. W. G. John, T. Jonas, J. A. Lindley, D. P. Stevens, and A. Walne, 2003a, CPR sampling: the technical background, materials and methods, consistency and comparability, *Prog. Oceanogr.* **58**:193–215.

Batten, S. D., A. W. Walne, M. Edwards, and S. B. Groom, 2003b, Phytoplankton biomass from continuous plankton recorder data: an assessment of the phytoplankton colour index, *J. Plankton Res.* **25**:697–702.

Baumgartner, M. F., 2003, Comparisons of *Calanus finmarchicus* fifth copepodite abundance estimates from nets and an optical plankton counter, *J. Plankton Res.* **25**:855–868.

Beare, D. J., S. D. Batten, M. Edwards, E. McKenzie, P. C. Reid, and D. G. Reid, 2003, Summarizing spatial and temporal information in CPR data, *Prog. Oceanogr.* **58**:217–233.

Beaulieu, S. E., M. M. Mullin, V. T. Tang, S. M. Pyne, A. L. King, and B. S. Twining, 1999, Using an optical plankton counter to determine the size distributions of preserved zooplankton samples, *J. Plankton Res.* **21**:1939–1956.

Beaugrand, G., 1999, Le programme Continuous Plankton Recorder (CPR) et son application à l'étude des changements spatio-temporels de la biodiversité pélagique en Atlantique nord et en mer du Nord, *Océanis* **25**:417–433.

Beaugrand, G., 2004a, Continuous plankton records: plankton atlas of North Atlantic Ocean (1958–1999). I. Introduction and methodology, *Mar. Ecol. Progress Ser. Suppl.* **2004**:3–10.

Beaugrand, G., 2004b, Continuous plankton records: plankton atlas of North Atlantic Ocean (1958–1999), II biogeographical charts, *Mar. Ecol. Prog. Ser. Suppl.* **2004**:11–75.

Beaugrand, G., 2004c, The North Sea regime shift: evidence causes, mechanism and consequences, *Prog. Oceanogr.* **60**:245–262.

Beaugrand, G., 2005, Monitoring pelagic ecosystems using plankton indicators, *ICES J. Mar. Sci.* **62**:333–338.

Beaugrand, G., and F. Ibañez, 2002, Spatial dependence of pelagic diversity in the North Atlantic ocean, *Mar. Ecol. Prog. Ser.* **232**:197–211.

Beaugrand, G., and F. Ibañez, 2004, Monitoring marine plankton ecosystems, II, Long term changes in the North Sea calanoid copepods in relation to hydro-climatic variability, *Mar. Ecol. Prog. Ser.* **284**:35–47.

Beaugrand, G., F. Ibañez, and J. A. Lindley, 2001, Geographical distribution and seasonal and diel changes in the diversity of calanoid copepod in the North Atlantic and North Sea, *Mar. Ecol. Prog. Ser.* **219**:205–219.

Beaugrand, G., P. C. Reid, F. Ibañez, and P. Planque 2000a, Biodiversity of North Atlantic and North Sea calanoid copepods, *Mar. Ecol. Prog. Ser.* **204**:299–303.

Beaugrand, G., F. Ibañez, and P. C. Reid, 2000b, Long-term and seasonal fluctuations of plankton in relation to hydroclimatic features in the English Channel, Celtic Sea and Bay of Biscay, *Mar. Ecol. Prog. Ser.* **200**:93–102.

Beaugrand, G., P. C. Reid, F. Ibañez, J. A. Lindley, and M. Edwards, 2002a, Reorganization of North Atlantic marine copepod biodiversity and climate, *Science* **296**:1692–1694.

Beaugrand, G., F. Ibañez, J. A. Lindley, and P. C. Reid, 2002b, Diversity of calanoid copepods in the North Atlantic and adjacent seas: species associations and biogeography, *Mar. Ecol. Prog. Ser.* **232**:179–195.

Beaugrand, G., F. Ibañez, and J. A. Lindley, 2003a, An overview of statistical methods applied to CPR data, *Prog. Oceanogr.* **58**:235–262.

Beaugrand, G., K. M. Brander, J. A. Lindley, S. Souissi, and P. Reid, 2003b, Plankton effect on cod recruitment in the North Sea, *Nature* **426**:661–664.

Bergström, B. I., A. Gustavsson, and J.-O. Strömberg, 1992, Determination of abundance of gelatinous plankton with a remotely operated vehicle (ROV*)*, *Arch. Hydrobiol. Beih. Ergebn. Limnol.* **36**:59–65.

Bertoni, R., R. Piscia, and C. Callieri, 2004, Horizontal heterogeneity of seston, organic carbon and picoplankton in the photic zone of Lago Maggiore, Northern Italy, *J. Limnol.* **63**:244–249.

Beisner, B. E., C. L. Dent, and S. R. Carpenter, 2003, Variability of lakes on the landscape: roles of phosphorus, food webs, and dissolved organic carbon, *Ecology* **84**:1563–1575.

Benfield, M. C., C. J. Schwehm, R. G. Fredericks, G. Squyres, S. F. Keenan, and M. V. Trevorrow, 2003, Measurement of zooplankton distributions with a high-resolution digital camera system, Chapter 2, in: *Handbook of Scaling Methods in Aquatic Ecology: Measurements, Analysis, Simulation*, L. Seuront, and P. G. Strutton, eds., CRC Press LLC, Boca Raton, FL, pp. 17–31.

Besiktepe, S., A. E. Kideys, and M. Unsal, 1998, *In situ* grazing pressure and diel vertical migration of female *Calanus euxinus* in the Black Sea, *Hydrobiologia* **363**:323–332.

Beutler, M., K. H. Wiltshire, B. Meyer, C. Moldaenke, C. Lüring, M. Meyerhöfer, U.-P. Hansen, and H. Dau, 2002, A fluorometric method for the differentiation of algal populations *in vivo* and *in situ*, *Photosynth. Res.* **72**:39–53.

Beutler, M., K. H. Wiltshire, M. Arp, J. Kruse, C. Reineke, C. Moldaenke, and U.-P. Hansen, 2003, A reduced model of the fluorescence from the cyanobacterial photosynthetic apparatus designed for the *in situ* detection of cyanobacteria, *Biochim. Biophys. Acta* **1604**:33–46.

Biddanda, B., and R. Benner 1997, Major contribution from mesopelagic plankton to heterotrophic metabolism in the upper ocean. *Deep Sea Res. I* **44**:2069–2085.

Bird, D., and J. Kalff, 1984, Empirical relationships between bacterial abundance and chlorophyll concentration in fresh and marine waters, *Can. J. Fish. Aquat. Sci.* **41**:1015–1023.

Bird, D. F., and J. Kalff, 1989, Phagotrophic sustenance of a metalimnetic phytoplankton peak, *Limnol. Oceanogr.* **34**:155–162.

Bochdansky, A. B., and S. M. Bollens, 2004, Relevant scales in zooplankton ecology: distribution, feeding, and reproduction of the copepod *Acartia hudsonica* in response to thin layers of the diatom *Steletonema costatum*, *Limnol. Oceanogr.* **49**:625–636.

Bode, A., B. Casaa, E. Fernandez, E. Marañon, P. Serret, and M. Varela, 1996, Phytoplankton biomass and production in shelf waters of NW Spain: spatial and seasonal variability in relation to upwelling, *Hydrobiologia* **341**:225–234.

Bonnet, D., A. Richardson, R. Harris, A. Hirst, G. Beaugrand, M. Edwards, S. Ceballos, R. Diekman, A. Lopez-Urrutia, L. Valdes, F. Carlotti, J. C. Molinero, H. Weikert, W. Greve, D. Lucic, A. Albaina, N. Daly Yahia, S. Fonda Umani, A. Miranda, A. dos Santos, K. Cook, S. Robinson, and M. L. Fernandez de Puelles, 2005, An overview of *Calanus helgolandicus* ecology in European waters, *Prog. Oceanogr.* **65**:1–53.

Borcard, D., and P. Legendre, 2002, All-scale spatial analysis of ecological data by means of principal coordinates of neighbour matrices, *Ecol. Model.* **153**:51–68.

Borcard, D., P. Legendre, and P. Drapeau, 1992, Partialling out the spatial component of ecological variation, *Ecology* **73**:1045–1055.

Borcard, D., P. Legendre, C. Avois-Jacquet, and H. Tuomisto, 2004, Dissecting the spatial structure of ecological data and multiple scales, *Ecology* **85**:1826–1832.

Boyd, P. W., C. S. Law, C. S. Wong, Y. Nojiri, A. Tsuda, M. Levasseur, S. Takeda, R. Rivkin, P. J. Harrison, R. Strzepek, J. Gower, R. M. McKay, E. R. Abraham, M. Arychuk, J. Barwell-Clarke, W. Crawford, D. Crawford, M. Hale, K. Harada, K. Johnson, H. Kiyosawa, I. Kudo, A. Marchetti, W. Miller, J. Needoba, J. Nishioka, N. Ogawa, J. Page, M. Robert, H. Saito, A. Sastri, N. Sherry, T. Soutar, N. Sutherland, Y. Taira, F. Whitney, S.K.E. Wong, and T. Yoshimura, 2004, The decline and fate of an iron-induced subarctic phytoplankton bloom, *Nature* **428**:549–553.

Brancelj, A., and A. Blejec, 1994, Diurnal vertical migration of *Daphnia hyalina* Leydig 1860 (Crustacea: Cladocera) in Lake Bled (Slovenia) in relation to temperature and predation, *Hydrobiologia* **284**:125–136.

Brander, K. M., R. R. Dickson, and M. Edwards, 2003, Use of Continuous Plankton Recorder information in support of marine mangement: applications in fisheries, environmental protection, and in the study of ecosystem response to environmental change, *Prog. Oceanogr.* **58**:175–191.

Bulit, C., C. Díaz-Avalos, M. Signoret, and D. J. S. Montagnes, 2003, Spatial structure of planktonic ciliate patches in a tropical coastal lagoon: an application of geostatistical methods, *Aquat. Microb. Ecol.* **30**:185–196.

Burks, R. L., D. M. Lodge, E. Jeppesen, and Lauridsen T. L., 2002, Diel horizontal migration of zooplankton: costs and benefits of inhabiting the littoral, *Freshw. Biol.* **47**:343–365.

Butorina L. G., 1986, On the problem of aggregations of planktonic crustaceans (*Polyphemus pediculus* (L.), Cladocera), *Arch. Hydrobiol.* **105**:355–386.

Camacho, A., E. Vicente, L. J. Garcia-Gil, M. R. Miracle, M. D. Sendra, X. Vila, and C. M. Borrego, 2002, Factors determining changes in the abundance and distribution of micro-, nano- and picoplanktonic phototrophs in Lake El Tobar (central Spain), *Verh. Internat. Verein Limnol.* **28**:613–619.

Camacho, A., W. A. Wurtsbaugh, M. R. Miracle, X. Armengol, and E. Vicente, 2003a, Nitrogen limitation of phytoplankton in a Spanish karst lake with a deep chlorophyll maximum: a nutrient enrichment bioassay approach, *J. Plankton Res.* **25**:397–404.

Camacho, A., M. R. Miracle, and E. Vicente, 2003b, Which factors determine the abundance and distribution of picocyanobacteria in inland waters? A comparison among different types of lakes and ponds, *Arch. Hydrobiol.* **157**:321–338.

Chang, K.-H., and T. Hanazato, 2004, Diel vertical migrations of invertebrate predators (*Leptodora kindtii, Thermocyclops taihokuensis,* and *Mesocyclops* sp.) in a shallow, eutrophic lake, *Hydrobiologia* **528**:249–259.

Chang, K.-H., and T. Hanazato, 2005, Heterogeneous distribution of zooplankton in a shallow eutrophic lake: species composition and diversity of the zooplankton community associated with habitat structure in the littoral area, *Verh. Internat. Verein. Limnol.* **29**:922–926.

Chapin, B. R. K., F. DeNoyelles, Jr., D.W. Graham, and V. H. Smith, 2004, A deep maximum of green sulphur bacteria (*Chloromatium aggregatum*) in a strongly stratified reservoir, *Freshw. Biol.* **49**:1337–1354.

Charlson, R. J., J. E. Lovelock, M. O. Andreae, and S. G. Warren, 1987, Ocean phytoplankton, atmospheric sulphur, cloud albedo and climate, *Nature* **326**:665–661.

Charoy, C., 1995, Modification of the swimming behaviour of *Brachionus calyciflorus* (Pallas) according to food environment and individual nutritive state, *Hydrobiologia,* **313/314**: 197–204.

Chételat, J., F. R. Pick, and P. B. Hamilton, 2006, Potamoplankton size structure and taxonomic composition: influence of river size and nutrient concentrations, *Limnol. Oceanogr.* **51**:681–689.

Chin-Leo, G., and R. Benner, 1992, Enhanced bacterioplankton production and respiration at intermediate salinities in the Mississipi River plume, *Mar. Ecol. Prog. Ser.* **87**:87–103.

Cho, B. C., and F. Azam, 1988, Major role of bacteria in biogeochemical fluxes in the ocean's interior, *Nature* **332**:441–443.

Cho, B. C., and F. Azam, 1990, Biogeochemical significance of bacterial biomass in the ocean's euphotic zone, *Mar. Ecol. Prog. Ser.* **63**:253–259.

Clark, D. R., K. V. Aazem, and G. C. Hays, 2001, Zooplankton abundance and community structure over a 4000 km transect in the north-east Atlantic, *J. Plankton Res.* **23**:365–372.

Cohen, G. M., and J. B. Shurin, 2003, Scale-dependence and mechanisms of dispersal in freshwater zooplankton, *Oikos* **103**:603–617.

Cole, J. J., Findlay S., and M. L. Pace, 1988, Bacterial production in fresh and saltwater: a cross-system overview, *Mar. Ecol. Progr. Ser.* **43**:1–10.

Cole, J. J., M. L. Pace, N. F. Caraco, and Steinhar G. S., 1993, Bacterial biomass and cell size distributions in lakes: More and larger cells in anoxic waters, *Limnol. Oceanogr.* **38**:1627–1632.

Cottenie, K., 2005. Integrating environmental and spatial processes in ecological community dynamics, *Ecol. Lett.* **8**:1175–1182.

Cottenie, K., and L. De Meester, 2003, Connectivity and cladoceran species richness in a small interconnected pond system, *Oikos* **48**:823–832.

Cottenie, K., and L. De Meester, 2004, Metacommunity structure: synergy of biotic interactions as selective agents and dispersal as fuel, *Ecology* **85**:114–119.

Cottenie, K., E. Michels, N. Nuytten, and L. De Meester, 2003, Zooplankton metacommunities structure: regional vs local process in highly interconnected ponds, *Ecology* **84**:991–1000.

Cowles, T. J., 2003, Planktonic layers: physical and biological interactions on the small scale, Chapter 3, in: *Handbook of Scaling Methods in Aquatic Ecology: Measurements, Analysis, Simulation*, L. Seuront, and P. G. Strutton, eds., CRC Press LLC, Boca Raton, FL, pp. 31.

Cowles, T. J., R. A. Desiderio, and M.-E. Carr, 1998, Small-scale plankton structure: persistence and trophic consequences, *Oceanography* **11**:4–9.

Coyle, K. O., 2005, Zooplankton distribution, abundance and biomass relative to water masses in eastern and central Aleutian Island passes, *Fish. Oceanogr.* **14**:77–92.

Cózar, A., C. M. Garcia, J. A. Gálvez, S. A. Loiselle, L. Bracchini, and A. Cognetta, 2005, Remote sensing imagery analysis of the lacustrine system of Ibera wetland (Argentina), *Ecol. Model.* **186**:29–41.

Crawford W. R., P. J. Brickley, T. D. Peterson, and A. C. Thomas, 2005, Impacts of Haida eddies on chlorophyll distribution in the Eastern Gulf of Alaska, *Deep Sea Res. II* **52**:975–989.
Cullen, J. J., 1982, The deep chlorophyll maximum: comparing vertical profiles of chlorophyll a, *Can. J. Fish. Aquat. Sci.* **39**:791–803.
Currie, W. J. S., and J. C. Roff, 2006, Plankton are not passive tracers: plankton in a turbulent environment, *J. Geophys. Res.* **111**:C05S07.
Dale, M. R. T., P. Dixon, M.-J. Fortin, P. Legendre, D. E. Myers, and M. S. Rosenberg, 2002, Conceptual and mathematical relationships among methods for spatial analysis, *Ecography* **25**:558–577.
Dau, H., 1994, Short-term adaptation of plants to changing light intensities and its relation to Photosystem II photochemistry and fluorescence emission, *J. Photochem. Photobiol. B.* **26**:3–27.
Davis, C. S., Q. Hu, S. M. Gallager, X. Tang, and C. J. Ashjian, 2004, Real-time observation of taxa-specific plankton distributions: an optical sampling method, *Mar. Ecol. Prog. Ser.* **284**:77–96.
Davis, C. S., F. T. Thwaites, S. M. Gallager, and Q. Hu, 2005, A three-axis fast-tow digital Video Plankton Recorder for rapid surveys of plankton taxa and hydrography, *Limnol. Oceanogr. Meth.* **3**:59–74.
Dejen, E., J. Vijverberg, L. A. J. Nagelkerle, and F. A. Sibbling, 2004, Temporal and spatial distribution of microcrustacean zooplankton in relation to turbidity and other environmental factors in a large tropical lake (L. Tana, Ethiopa), *Hydrobiologia* **513**:39–49.
De Meester, L., S. Maas, K. Dierckens, and H. J. Dumont, 1993, Habitat selection and patchiness in *Scapholeberis*: horizontal distribution and migration of *S. mucronata* in a small pond, *J. Plankton Res.* **15**:1129–1139.
DeMott, W.R., 1999, Foraging strategies and growth inhibition in five daphnids feeding on mixtures of a toxic cyanobacterium and a green algae, *Freshw. Biol.* **42**:263–274.
Denman, K. L., and M. R. Abbott, 1994, Time scales of pattern evolution from cross-spectrum analysis of advanced high resolution radiometer and coastal zone color scanner imagery, *J. Geophys. Res.* **99**:7433–7442.
Denman, K. L., and T. Platt, 1976, The variance spectrum of phytoplankton in a turbulent ocean, *J. Mar. Res.* **34**:593–601.
Descroix, A., M. Harvey, S. Roy, and P. S. Galbraith, 2005, Macrozooplankton community patterns driven by water circulation in the St. Lawrence marine system, Canada, *Mar. Ecol. Prog. Ser.* **302**:103–119.
DiFonzo, C. D., and J. M. Campbell, 1988, Spatial partitioning of microhabitats in littoral cladoceran communities, *J. Freshw. Ecol.* **4**:303–313
Dodson, S., 1990, Predicting diel vertical migration of zooplankton, *Limnol. Oceanogr.* **35**:195–200.
Dodson, S., I. Richard, A. Lillie, and S. Will-Wolf, 2005, Land use, water chemistry, aquatic vegetation, and zooplankton community structure of shallow lakes, *Ecol. Appl.* **15**:1191–1198.
Downing J. A., 1991, Biological heterogeneity in aquatic ecosystems, Chapter 9, in: *Heterogeneity in Ecological Systems. Ecol. Studies 86*, J. Kolosa, and S. T. A. Pickett, eds., Springer, New York, pp. 160–180.
Downing, J. A., M. Pérusse, and Y. Frenette, 1987, Effect of interreplicate variance on zooplankton sampling design and data analysis, *Limnol. Oceanogr.* **32**:673–680.

Duarte, C. M., and D. Vaqué, 1992, Scale dependence of bacterioplankton patchiness, *Mar. Ecol. Prog. Ser.* **84**:95–100.

Ducklow, H., 2000, Bacterial production and biomass in the oceans, in: D. Kirchman, ed., *Microbial Ecology of the Oceans*, Wiley-Liss, New York, pp. 85–120.

Dûfrêne, M., and P. Legendre, 1997, Species assemblages and indicator species: the need for a flexible asymmetrical approach, *Ecol. Monogr.* **67**:345–366.

Dungan, J. L., J. N. Perry, M. R. T. Dale, P. Legendre, S. Citron-Pousty, M.-J. Fortin, A. Jakomulska, M. Mitri, and M. S. Rosenberg, 2002, A balanced view of scale in spatial statistical analysis, *Ecography* **25**:626–640.

Dutilleul, P., 1998, Incorporating scale in ecological experiments – data analysis, in: *Ecological Scale*, D. L. Peterson, and V. T. Parker, eds., Columbia University Press, New York, pp. 387–425.

Dutilleul, P., and B. Pinel-Alloul, 1996, A doubly multivariate model for statistical analysis of spatio-temporal environmental data, *Environmetrics* **7**:551–565.

Ediger, D., and A. Yilmaz, 1996, Characteristics of deep chlorophyll maximum in the northeastern Mediterranean with respect to environmental conditions, *J. Mar. Syst.* **9**:291–203.

Edvardsen, A., M. Zhou, K. S. Tande, and Y. Zhu, 2002, Zooplankton population dynamics: measuring in situ growth and mortality rates using an Optical Plankton Counter, *Mar. Ecol. Prog. Ser.* **227**:205–219.

Engelhardt, C., A. Krüger, A. Sukhodolov, and A. Nicklish, 2004, A study of phytoplankton spatial distributions, flow structure and characteristics of mixing in a river reach with groynes, *J. Plankton Res.* **26**:1351–1366.

Eriksen, C. C., T. J. Osse, R. D. Light, T. Wen, T. W. Lehman, P. L. Sabin, J. W. Ballard, and A. M. Chiodi, 2001, Seaglider: a long-range autonomous underwater vehicle for oceanographic research, *IEEE J. Ocean. Engin.* **26**:424–436.

Ernst, B., B. Hitzfeld, and D. Dietrich, 2001, Presence of *Planktothrix* sp. and cyanobacterial toxins in Lake Ammersee, Germany, and their impact on whitefish (*Coregonus lavaretus* L.), *Environ. Toxicol.* **16**:483–488.

Easton, J., and M. Gophen, 2003, Diel variation in the vertical distribution of fish and plankton in Lake Kinneret: a 24-h study of ecological overlap, *Hydrobiologia* **491**:91–100.

Falkenhaug, T., E. Nordby, H. Svendsen, and K. Tande, 1995, Impact of advective processes on displacement of zooplankton biomass in a North Norwegian fjord system: a comparison between spring and autumn, in: *Ecology of Fjords and Coastal Waters*, H. R. Skjoldal, C. Hopkins, K. E. Erikstad, and H. P. Leinass, eds., Elsevier Science, Amsterdam, The Netherlands, pp. 195–217.

Falkenhaug, T., K. Tande, and A. Timonin, 1997, Spatio-temporal patterns in the copepod community in Malangen, Northern Norway, *J. Plankton Res.* **19**:449–468.

Fahnenstiel, G. L., and J. M. Glim, 1983, Subsurface chlorophyll maximum and associated *Cyclotella* pulse in Lake Superior, *Int. Rev. Ges. Hydrobiol.* **68**:605–616.

Fee, E. J., 1976, The vertical and seasonal distribution of chlorophyll in lakes of the Experimental Lakes Area, Northwest Ontario: implications for primary production-estimates, *Limnol. Oceanogr.* **21**:767–783.

Fennel, K., and E. Boss, 2003, Subsurface maxima of phytoplankton and chlorophyll: Steady-state solutions from a simple model, *Limnol. Oceanogr.* **48**:1521–1534.

Fernandez, E., J. Cabal, J. L. Acuña, A. Bode, A. Botas, and C. Garcia-Soto, 1993, Plankton distribution across a slope current-induced front in the southern Bay of Biscay, *J. Plankton Res.* **15**:619–641.

Feuillade, J., 1994, The cyanobacterium (blue-green alga) *Oscillatoria rubescens* D. C., *Arch. Hydrobiol. Beih. Ergebn. Limnologie* **41**:77–93.

Figueroa, F. L., F. X. Niell, Figueiras F. G., and M. L. Villarino, 1998, Diel migration of phytoplankton and spectral light field in the Ria de Vigo (NW Spain), *Mar. Biol.* **130**:491–499.

Fisher, M. M., and E. W. Triplett, 1999, Automated approach for ribosomal intergenic spacer analysis of microbial diversity and its application to freshwater bacterial communities, *Appl. Environ. Microbiol.* **65**:4630–4636.

Fietz, S., G. Kobanova, L. Izmesteva, and A. Nicklisch, 2005, Regional, vertical and seasonal distribution of phytoplankton and photosynthetic pigments in Lake Baikal, *J. Plankton Res.* **27**:793–810.

Findlay, S., M. L. Pace, D. Lints, J. J. Cole, N. F. Caraco, and B. Peierls, 1991, Weak coupling of bacterial and algal production in a heterotrophic ecosystem: The Hudson River Estuary, *Limnol. Oceanogr.* **36**:268–278.

Folt, C. L., and C.W. Burns, 1999, Biological drivers of zooplankton patchiness, *Trends Ecol. Evol.* **14**:300–305.

Forbes, A. E., and J. M. Chase, 2002, The role of habitat connectivity and landscape geometry in experimental zooplankton metacommunities, *Oikos* **96**:433–440.

Franks, P. J. S., 2005, Plankton patchiness, turbulent transport and spatial spectra, *Mar. Ecol. Prog. Ser.* **294**:295–309.

Franks, P. J., and L. J. Walstad, 1997, Phytoplankton patches at fronts: a model of formation and response to wind events, *J. Mar. Res.* **55**:1–29.

Franklin, R. B., and A. L. Mills, 2003, Multi-scale variation in spatial heterogeneity for microbial community structure in an eastern Virginia agricultural field, *FEMS Microbiol. Ecol.* **44**:335–346.

Franquet, E., S. Dolédec, and D. Chessel, 1995, Using multivariate analyses for separating spatial and temporal effects with species-environment relationships, *Hydrobiologia* **300/301**:425–431.

Friedrich U., M. Schallenberg, and C. Holliger, 1999, Pelagic bacteria-particle interactions and community-specific growth rates in four lakes along a trophic gradient, *Microb. Ecol.* **37**:49–61.

Froneman, P. W., 2004, Zooplankton community structure and biomass in a southern African temporarily open/closed estuary, *Estuar. Coast. Shelf Sci.* **60**:125–132.

Gal, G., L. G. Rudstam, and C. H. Greene, 1999, Acoustic characterization of *Mysis relicta*, *Limnol. Oceanogr.* **44**:371–381.

Garcia-Sucerquia, J., W. Xu, S. K. Jericho, P. Klages, M. H. Jericho, and H. J. Kreuzer, 2006, Digital in-line holographic microscopy, *Appl. Opt.* **45**:836–850.

Garçon, V. C., A. Oschlies, S. C. Doney, D. McGillicuddy, and J. Waniek., 2001, The role of mesoscale variability on plankton dynamics in the North Atlantic, *Deep Sea Res. II* **48**:2199–2226.

Gasol, J. M., and C. M. Duarte, 2000, Comparative analysis in aquatic microbial ecology: how far do they go? *FEMS Microbiol. Ecol.* **31**:99

Gasol J. M., and P. A. del Giorgio, 2000, Using flow cytometry for counting natural planktonic bacteria and understanding the structure of planktonic bacterial communities, *Sci. Mar.* **64**:197–224.

Gasol, J. M., C. Pedrós-Alió, and D. Vaqué, 2002, Regulation of bacterial assemblages in oligotrophic plankton systems: results from experimental and empirical approaches, *A. Van Leeuw. J. Microbiol.* **81**:435–452.

Gerhardt, A., L. J. de Bisthoven, and S. Schmidt, 2006, Automated recording of vertical negative phototactic behaviour in Daphnia magna Straus (Crustacea), *Hydrobiologia* **559**:433–441.

Gervais, F., 1997, Diel vertical migration of *Cryptomonas* and *Chromatium* in the deep chlorophyll maximum of a eutrophic lake, *J. Plankton Res.* **19**:533–550.

Gervais, F., 1998, Ecology of cryptophytes coexisting near a freshwater chemocline, *Freshw. Biol.* **39**:61–78.

Ghadouani, A., and R. E. H. Smith, 2005, Phytoplankton distribution in Lake Erie as assessed by a new *in situ* spectrofluorometric technique, *J. Great Lakes Res.* **31**:154–167.

Ghadouani, A., B. Pinel-Alloul, and E. E. Prepas, 2003, Effects of experimentally induced cyanobacterial blooms on crustacean zooplankton communities, *Freshw. Biol.* **48**:363–381.

Ghadouani, A., B. Pinel-Alloul, K. Plath, W. Lampert, and G. A. Codd, 2004, Effects of *Microcystis aeruginosa* and purified microcystin-LR on the feeding behaviour of *Daphnia pulicaria*, *Limnol. Oceanogr.* **49**:267–280.

Ghan, D., K. D. Hyatt, and J. D. McPhail, 1998a, Benefits and costs of vertical migration by the freshwater copepod *Skistodiaptomus oregonensis*: testing hypotheses through population comparison, *Can. J. Fish. Aquat. Sci.* **55**:1338–1349.

Ghan, D., J. D. McPhail, and K. D. Hyatt, 1998b, The temporal-spatial pattern of vertical migration by the freshwater copepod *Skistodiaptomus oregonensis* relative to predation risk, *Can. J. Fish. Aquat. Sci.* **55**:1350–1363.

Giller, P. S., A. G. Hildrew, and D. Rafaelli, eds., 1994, *Aquatic Ecology: Scale, Pattern and Process*, Proceedings of British Ecological Society and American Society of Limnology and Oceanography Symposium, 1992, Blackwell Scientific Publications, Oxford, pp. 649.

Gliwicz, Z. M., 1986, Predation and the evolution of vertical migration in zooplankton, *Nature* **320**:746–748.

Goddard, V. J., A. C. Baker, J. E. Davy, D. G. Adams, M. M. De Ville, S. J. Tackeray, S. C. Maberly, and W. H. Wilson, 2005, Temporal distribution of viruses, bacteria and phytoplankton throughout the water column in a freshwater hypereutrophic lake, *Aquat. Microb. Ecol.* **39**:211–223.

Gould R. W., 1987, The deep chlorophyll maximum in the world ocean: a review, *Biologist* **66**:4–13.

Grant, S., P. Ward, E. Murphy, D. Bone, and S. Abbott, 2000, Field comparison of an LHPR net sampling system and an optical plankton counter (OPC) in the Southern Ocean, *J. Plankton Res.* **22**:619–638.

Greene, C. H., P. H. Wiebe, C. Pelkie, M. C. Benfield, and J. M. Popp, 1998, Three-dimensional acoustic visualization of zooplankton patchiness, *Deep Sea Res. II* **45**:1201–1217.

Gross, H. P., W. A. Wurtsbaugh, C. Luecke, and P. Budy, 1997, Fertilization of an oligotrophic lake with a deep chlorophyll maximum: predicting the effect on primary production, *Can. J. Fish. Aquat. Sci.* **54**:1177–1189.

Grunwald, B., and M. Kühl, 2004, A system for imaging variable chlorophyll fluorescence of aquatic phototrophs, *Ophelia* **58**:79–89.

Gyllström, M., L-A. Hansson, E. Jeppesen, F. Garçia-Criado, K. Irvine, T. Kairesalo, R. Kornijow, M. R. Miracle, N. Nykänen, T. Nôges, S. Romo, D. Stephen, E. Van Donk, and B. Moss, 2005, The role of climate in shaping zooplankton communities of shallow lakes, *Limnol. Oceanogr.* **50**:2008–2021.

Hall, S. R., N. K. Pauliukonis, E. L. Mills, L. G. Rudstam, C. P. Schneider, S. J. Lary, and F. Arrhenius, 2003, A comparison of total phosphorus, chlorophyll a, and zooplankton in embayment, nearshore and offshore habitats of Lake Ontario, *J. Great. Lakes. Res.* **29**: 54–69.

Håkanson, L., and R. H. Peters, 1995, *Predictive Limnology: Methods for Predictive Modeling.* SPB Academic Publishing, The Netherlands, pp. 464.

Haney, J. F., J. J. Sasner, and M. Ikawa, 1995, Effects of products released by *Aphanizomenon flos-aquae* and purified saxitoxin on the movements of *Daphnia carinata* feeding appendages, *Limnol. Oceanogr.* **40**:263–272.

Hansell, D. A., and H. W Ducklow, 2003, Bacterioplankton distribution and production in the bathypelagic ocean: directly coupled to particulate organic carbon export? *Limnol. Oceanogr.* **48**:150–156.

Hansen, F. C., C. Möllmann, U. Schütz, and H.-H. Hinrichsen, 2004, Spatio-temporal distribution of *Oithona similis* in the Bronholm Basin (Central Baltic Sea), *J. Plankton Res.* **26**:1–10.

Hardy, A. C., 1935, The Continuous Plankton Recorder, a new method of survey. Rapports et Proces-verbaux des Reunions, *Conseil Inter. Explor. Mer.* **95**:36–47.

Hardy, A. C., 1939, Ecological investigations with the Continuous Plankton Recorder: object, plan and methods, *Hull Bull. Mar. Ecol.* **1**:1–57.

Harris, G. P., 1994, Pattern, process and prediction in aquatic ecology: a limnological view of some general ecological problems. *Freshw. Biol.* **32**:143–160.

Harvey, M., J.-C. Therriault, and N. Simard, 1997, Late-summer distribution of phytoplankton in relation to water mass characteristics in Hudson Bay and Hudson Strait (Canada), *Can. J. Fish. Aquat. Sci.* **54**:1937–1952.

Harvey, M., J.-C. Therriault, and N. Simard, 2001, Hydrodynamic control of late summer species composition and abundance of zooplankton in Hudson Bay and Hudson Strait (Canada), *J. Plankton Res.* **23**:481–496.

Haury, L. R., McGowan J. S., and P. Wiebe, 1978, Patterns and processes in the time-space scales of plankton distribution, in: *Spatial Patterns in Plankton Communities*, J. H. Steele, ed., Plenum Press, New York, pp. 277–327.

Havel, J. E., and J. B. Shurin, 2004, Mechanisms, effects and scales of dispersal in freshwater zooplankton, *Limnol. Oceanogr.* **49**:1229–1238.

Havel, J. E., and K. A. Medley, 2006, Biological invasions across spatial scales: intercontinental, regional and local dispersal of cladoceran zooplankton, *Biol. Inv.* **8**:459–473.

Havel, J. E., and W. Lampert, 2006, Habitat partitioning of native and exotic Daphnia in gradients of temperature and food: mesocosm experiments, *Freshw. Biol.* **51**:487–498.

Hays, G. C., C. A. Proctor, A. W. G. John, and A. J. Warner, 1994, Interspecific differences in the diel vertical migration of marine copepods: the implications of size, color and morphology, *Limnol. Oceanogr.* **39**:1621–1629.

Hays, G. C., D. R. Clark, A. W. Walne, and A. J. Warner, 2001, Large-scale patterns of zooplankton abundance in the NE Atlantic in June and July 1996, *Deep Sea Res. II* **48**:951–961.

Hedger, R. D., N. R. B. Olsen, D. G. George, T. J. Malthus, and P. M. Atkinson, 2004, Modelling spatial distributions of *Ceratium hirundinella* and *Microcystis* spp. in a small productive Bristish lake, *Hydrobiologia* **528**:217–227.

Hembre, L. K., and R. O. Megard, 2003, Seasonal and diel patchiness of a *Daphnia* population: An acoustic analysis, *Limnol. Oceanogr.* **48**:2221–2233.

Herman, A. W., 1988, Simultaneous measurement of zooplankton and light attenuance with a new optical plankton counter, *Cont. Shelf Res.* **8**:205–221.

Herman, A. W., 1992, Design and calibration of a new optical plankton counter capable of sizing small zooplankton, *Deep Sea Res.* **39**:395–415.

Herman, A. W., N. A. Cochrane, and D. D. Sameoto, 1993, Detection and abundance estimation of euphausiids using an optical counter, *Mar. Ecol. Prog. Ser.* **94**:165–173.

Herman, A. W., B. Beanlands, and E. F. Phillips, 2004. The next generation of Optical Plankton Counter: the Laser-OPC, *J. Plankton Res.* **26**:1135–1145.

Hessen, D. O., B. A. Faafe, and T. Andersen, 1995, Replacement of herbivore zooplankton species along gradients of ecosystem productivity and fish predation pressure, *Can. J. Fish. Aquat. Sci.* **52**:733–742.

Hicks, R. E., P. Aas, and C. Jankovich, 2004, Annual and offshore changes in bacterioplankton communities in the western arm of Lake Superior during 1989 and 1990, *J. Great Lakes Res.* **30**:196–213.

Hidalgo, P., R. Escribano, and C. E. Morales, 2005, Ontogenic vertical distribution and diel migration of the copepod *Eucalanus inermis* in the oxygen minimum zone off northern Chile (20–21°S), *J. Plankton Res.* **27**:519–529.

Hietala, J., C. Laurén-Määttä, and M. Walls, 1997, Sensitivity of *Daphnia* to toxic cyanobacteria: effects of genotype and temperature, *Freshw. Biol.* **37**:299–306.

Holm-Hansen, P., and C. D. Hewes, 2004, Deep-chlorophyll-a maxima (DCMs) in Antarctic waters. I Relationships between DCMs and the physical, chemical, and optical conditions in the upper water column, *Polar Biol.* **27**:699–710.

Holm-Hansen, O., M. Kahru, and C. D. Hewes, 2005, Deep chlorophyll *a* maxima (DCMs) in pelagic Antarctic waters, II. Relation to bathymetric features and dissolved iron concentrations, *Mar. Ecol. Prog. Ser.* **297**:71–81.

Horner-Devine, C., M. A. Leibold, V. H. Smith, and B. J. M. Bohannan, 2003, Bacterial diversity patterns along a gradient in primary production, *Ecol. Lett.* **6**:613–622.

Hooker, S. B., and C. R. McClain, 2000, The calibration and validation of SeaWiFS data, *Prog. Oceanogr.* **45**:427–465.

Hu, C. M., F. E. Muller-Karger, C. Taylor, K. L. Carder, C. Kelble, E. Johns, and C. A. Heil, 2005, Red tide detection and tracing using MODIS fluorescence data: a regional example in SW Florida coastal waters, *Remote Sens. Environ.* **97**:311–321.

Hudon, C., 2000, Phytoplankton assemblages in the St. Lawrence River, downstream of its confluence with the Ottawa River, Québec, Canada, *Can. J. Fish. Aquat. Sci.* **57**:16–30.

Hudon, C., S. Paquet, and V. Jarry, 1996, Downstream variations of phytoplankton in the St. Lawrence River, *Hydrobiologia* **337**:11–26.

Huisman, J., N. N. Pham Thi, D. M. Karl, and B. Sommeijer, 2006, Reduced mixing generates oscillations and chaos in the oceanic deep chlorophyll maxima, *Nature Lett.* **439/19**:322–325.

Humbert, J.-F., G. Paolini, and B. Le Berre, 2001, Monitoring a cyanobacterial bloom and its consequences for water quality, in: *Harmful Algal Bloom 2000, Intergovernmental Oceanographic Commission of UNESCO 2001*, G. Hallegraeff et al., eds., pp. 496–499.

Hunt, B. P. V., and G. W. Hosie, 2003, The Continuous Plankton Recorder in the Southern Ocean: a comparative analysis of zooplankton communities sampled by the CPR and vertical net hauls along 140°E, *J. Plankton Res.* **25**:1561–1579.

Hunt, B. P. V., and G. W. Hosie, 2005, Zonal structure of zooplankton communities in the Southern Ocean South of Australia: results from a 2150-km continuous plankton recorder transect, *Deep Sea Res. I* **52**:1241–1271.
Ignoffo, T. R., S. M. Bollens, and A. B. Bochdansky, 2005, The effects of thin layers on the vertical distribution of the rotifer *Brachionus plicatilis*, *J. Exp. Mar. Biol. Ecol.* **316**:167–181.
Irigoien, X., D. V. P. Conway, and R. P. Harris, 2004, Flexible diel vertical migration behaviour of zooplankton in the Irish Sea, *Mar. Ecol. Prog. Ser.* **267**:85–97.
Ishikawa, K., M. Kumagai, and R. F. Walker, 2005, Application of autonomous underwater vehicle and image analysis for detecting the three-dimensional distribution of freshwater red tide *Uroglena americana* (Chrysophyceae), *J. Plankton Res.* **27**:129–134.
Jack, J. D., and J. H. Thorp, 2002, Impacts of fish predation on an Ohio River zooplankton community, *J. Plankton Res.* **24**:119–127.
Jacoby, J. M., D. C. Collier, E. B. Welch, F. J. Hardy, and M. Crayton, 2000, Environmental factors associated with a toxic bloom of *Mycrocystis aeruginosa*, *Can. J. Fish. Aquat. Sci.* **57**:231–240.
Jann-Para, G., I. Schwob, and M. Feuillade, 2004, Occurrence of toxic *Planktothrix rubescens* blooms in Lake Nantua, France, *Toxicon* **43**:279–285.
Jensen, K. H., P. Larsson, and G. Högstedt, 2001, Detecting food search in *Daphnia* in the field, *Limnol. Oceanogr.* **46**:1013–1020.
Jiao, N. Z., and I. H. Ni, 1997, Spatial variation of size-fractionated chlorophyll, cyanobacteria and heterotrophic bacteria in the central and western Pacific, *Hydrobiologia* **352**:219–230.
Johannsson, O. E. E., L. Mills, and R. O'Gorman, 1991, Changes in the nearshore and offshore zooplankton communities in Lake Ontario: 1981–1988, *Can. J. Fish. Aquat. Sci.* **48**:1546–1557.
Jochem, F. J., 2001, Morphology and DNA content of bacterioplankton in the Northern Gulf of Mexico: analysis by epifluorescence microscopy and flow cytometry, *Aquat. Microbiol Ecol.* **25**:179–194.
Jonas, T. D., A. Walne, G. Beaugrand, J. Gregory, and G. C. Hays, 2004, The volume of water filtered by a Continuous Plankton Recorder sample: the effect of ship speed, *J. Plankton Res.* **26**:1499–1506.
Jones, J. R., and M. F. Knowlton, 2005, Chlorophyll response to nutrients and non-algal seston in Missouri reservoirs and oxbow lakes, *Lake Reserv. Manage.* **21**:361–371.
Jones, R. I., A. S. Fulcher, J. K. U. Jayakody, J. Laybourn-Parry, A. J. Shine, M. C. Walton, and J. M. Young, 1995, The horizontal distribution of plankton in a deep oligotrophic lake – Loch Ness, Scotland, *Freshw. Biol.* **33**:161–170.
Kalff, J., 2002, *Limnology*. Prentice Hall, NJ, pp. 592.
Kallio, K., T. Kutser, T. Hannonen, S. Koponen, J. Pulliainen, Vepsäläinen, and T. Pyhälahti, 2001, Retrieval of water quality from airborne imaging spectrometry of various lake types in different seasons, *Sci. Total Environ.* **268**:59–77.
Karlson, B., L. Edler, W. Granéli, E. Sahlsten, and M. Kuylenstierna, 1996, Subsurface chlorophyll maxima in the Skagerrak – processes and plankton community structure, *J. Sea Res.* **35**:139–158.
Kasprzak, P., C. Reese, R. Koschel, M. Schulz, L. Hambaryan, and J. Mathes, 2005, Habitat characteristics of *Eurytemora lacustris* (Poppe 1887) (Copepoda, Calanoida): the role of lake depth, temperature, oxygen concentration and light intensity, *Int. Rev. Hydrobiol.* **90**:292–309.

Keller, W., and M. Conlon, 1994, Crustacean zooplankton communities and lake morphometry in Precambrian Shield lakes, *Can. J. Fish. Aquat. Sci.* **51**:2424–2434.

Kemp, S., 1926, The discovery expedition, *Nature* **118**:628–632.

Kessler, K., and W. Lampert, 2004, Depth distribution of *Daphnia* in response to a deep-water algal maximum: the effect of body size and temperature gradient, *Freshw. Biol.* **49**:392–401.

Kettle, W. D., M. F. Moffett, and F. DeNoyelles, Jr., 1987, Vertical distribution of zooplankton in an experimentally acidified lake containing a metalimnetic phytoplankton peak, *Can. J. Fish. Aquat. Sci.* **44**:91–95.

Keitt, T. H., and D. L. Urban, 2005, Scale-specific inference using wavelets, *Ecology* **86**:2497–2504.

Kiørboe, T., and E. Bagøien, 2005, Mobility patterns and mate encounter rates in planktonic copepods, *Limnol. Oceanogr.* **50**:1999–2007.

Kiørboe, T., E. Bagøien, and U. Høgsbro Thygesen, 2005, Blind dating – mate finding in planktonic copepods. II. The pheromone cloud of *Pseudocalanus elongates*, *Mar. Ecol. Prog. Ser.* **300**:117–128.

Kirchman, D. L., 2000, *Microbial Ecology of the Oceans*. Wiley, NY, pp. 552.

Kling, G. W., G.W. Kipphut, M. M. Miller, and J. W. O'Brien, 2000, Integration of lakes and streams in a landscape perspective: the importance of material processing on spatial patterns and temporal coherence, *Freshw. Biol.* **43**:477–497.

Knapp, C. W., D. W. Graham, R. J. Steedman, and F. deNoyelles, Jr., 2003a, Deep chlorophyll maxima in small boreal forest lakes after experimental catchment and shoreline logging, *Boreal Environ. Res.* **8**:9–18.

Knapp, C. W., F. DeNoyelles, Jr., D. W. Graham, and S. Bergin., 2003b, Physical and chemical conditions surrounding the diurnal vertical migration of *Cryptomonas* spp. (Cryptophyceae) in a seasonally stratified Midwestern reservoir (USA), *J. Phycol.* **39**:855–861.

Kolosa, J., and S. T. A. Pickett, 1991, *Heterogeneity in Ecological Systems, Ecological Studies 86*. Springer, NY, pp. 332.

Kolasa J., and C. D. Rollo, 1991, Introduction: the heterogeneity of heterogeneity: a glossary, in: *Heterogeneity in Ecological Systems. Ecological Studies 86,* J. Kolosa, and S. T. A. Pickett, eds., Springer, NY, pp. 1–23.

Köhler, J., 1994, Origin and succession of phytoplankton in a river-lake system (Spree, Germany), *Hydrobiologia* **289**:73–83.

Kratz, T. K., and T. M. Frost, 2000, The ecological organisation of lake districts: general introduction, *Freshw. Biol.* **43**:297–299.

Kratz, T. K., K. E. Webster, C. J. Bowser, J. J. Magnuson, and B. J. Benson, 1997, The influence of landscape position on lakes in Northern Wisconsin, *Freshw. Biol.* **37**:209–217.

Krembs, C., A. R. Juhl, R. A. Long, and F. Azam, 1998a, Nanoscale patchiness of bacteria in lake water studied with the spatial information preservation method, *Limnol. Oceanogr.* **43**:307–314.

Krembs, C., A. R. Juhl, and J. R. Strickler, 1998b, The spatial information preservation method: sampling the nanoscale spatial distribution of microorganisms, *Limnol. Oceanogr.* **43**:298–306.

Kruskopf, M., and K. J. Flynn, 2006, Chlorophyll content and fluorescence responses cannot be used to gauge reliably phytoplankton biomass, nutrient status or growth rate, *New Phytol.* **169**:525–536.

Kuczynska-Kippen, N. M., and B. Nagengast, 2006, The influence of the habitat structure of hyfromacrophytes and differentiating habitat on the structure of rotifer and cladoceran communities, *Hydrobiologia* **559**:203–212.

Kurmayer, R., and F. Jüttner, 1999, Strategies for the co-existence of zooplankton with the toxic cyanobacterium *Planktothrix rubescens* in Lake Kürich. *J. Plankton Res.* **21**:659–683.

Kutser, T., 2004, Quantitative detection of chlorophyll in cyanobacterial blooms by satellite remote sensing, *Limnol. Oceanogr.* **49**:2179–2189.

Kutser, T., D. C. Pierson, K. Y. Kallio, A. Reinart, and S. Sobek, 2005, Mappling lake CDOM by satellite remote sensing, *Remote Sen. Environ.* **94**:535–540.

Lacroix G., and F. Lescher-Moutoué, 1995, Spatial patterns of planktonic microcrustaceans in a small shallow lake, *Hydrobiologia* **300/301**:205–217.

Lampert, W., 1989, The adaptive significance of diel vertical migration of zooplankton, *Funct. Ecol.* **3**:21–28.

Lampert, W., 1993, Ultimate causes of diel vertical migration of zooplankton: new evidence for the predator avoidance behaviour, *Arch. Hydrobiol. Beiheft Ergebnisse der Limnologie* **39**:79–88.

Laurén-Määtä, C., C. Hietala, and M. Walls, 1997, Responses of *Daphnia pulex* populations to toxic cyanobacteria,. *Freshw. Biol.* **37**:635–647.

Lauridsen T. L., E. Jeppesen, S. F. Mitchell, D. M. Lodge, and R. L. Burks, 1999, Horizontal distribution of zooplankton in lakes with contrasting fish densities and nutrient levels, *Hydrobiologia* **408/409**:241–250.

Lavender, S. J., and S. B. Groom, 1999, The SeaWiFS automatic data processing system (SeaAPS), *Int. J. Remote Sens.* **20**:1051–1056.

Lawrence, D., I. Valiela, and G. Tomasky, 2004, Estuarine calanoid copepod abundance in relation to salinity, and land-derived nitrogen loading in Waquoit Bay, MA, *Estuar. Coast. Shelf Sci.* **61**:547–557.

Leboulanger, C., U. Dorigo, S. Jacquet, B. LeBerre, G. Paolini, and J.-F. Humbert, 2002, Application of a submersible spectrofluorometer for rapid monitoring of a freshwater cyanobacterial blooms: a case study, *Aquat. Microb. Ecol.* **30**:83–89.

Leech, D. M., Padeletti A., and C. E. Williamson, 2005, Zooplankton behavorial responses to solar UV radiation vary within and among lakes, *J. Plankton Res.* **27**:461–471.

Legendre, P., 1993, Spatial autocorrelation: trouble or new paradigm? *Ecology* **74**:1659–1673.

Legendre, P., and M.-J. Fortin, 1989, Spatial pattern and ecological analysis, *Vegetatio* **80**:107–131.

Legendre, P., and L. Legendre, 1998, *Numerical Ecology*, 2nd Edition. Elsevier Science, Amsterdam, The Netherlands.

Legendre, L., and J. Michaud, 1998, Flux of biogenic carbons in oceans: size-dependent regulation by pelagic food webs, *Mar. Ecol. Prog. Ser.* **164**:1–11.

Legendre, L., C. Courties, and M. Trousellier, 2001, Flow cytometry in Oceanography 1989–1999: environmental challenges and research trends, *Cytometry* **44**:164–172.

Lenz, J., D. Schnack, D. Petersen, J. Kreikemeier, B. Hermann, S. Mees, and K. Wieland, 1995, The Ichtyoplankton Recorder: avideo recording system for in situ studies of small-scale plankton distribution patterns, *ICES J. Mar. Sci.* **52**:409–417.

Letarte, Y., and B. Pinel-Alloul, 1991, Relationships between bacterioplankton production and limnological variables: necessity of bacterial size considerations, *Limnol. Oceanogr.* **36**:1208–1216.

Letarte, Y., H. J. Hansen, M. Sondergaard, and B. Pinel-Alloul, 1992, Production and abundance of different bacterial size classes: relationships with primary production and chlorophyll concentration, *Arch. Hydrobiol.* **126**:15–16.

Leterme, S. C., M. Edwards, L. Seuront, M. J. Attrill, P. C. Reid, and A. W. G. John, 2005, Decadal basin-scale changes in diatoms, dinoflagellates, and phytoplankton color across the North Atlantic, *Limnol. Oceanogr.* **50**:1244–1253.

Levin, S. A., 1992, The problem of pattern and scale in ecology, *Ecology* **73**:1943 1967.

Li, W. K. W., and W. G. Harrison, 2001, Chlorophyll, bacteria and picophytoplankton in ecological provinces of the North Atlantic, *Deep Sea Res. II* **48**:2271–2293.

Li, W. K. W., P. M. Dickie, B. D. Irwin, and A.M. Wood, 1992, Biomass of bacteria, cyanobacteria, prochlorophytes and photosynthetic eukaryotes in the Sargasso Sea, *Deep Sea Res. I* **39**:501–519.

Li, W. K. W., E. J. H. Head, and W. G. Harrison, 2004, Macroecological limits of heterotrophic bacterial abundance in the ocean, *Deep Sea Res. I* **51**:1529–1540.

Liebold, M. A., and J. Norberg, 2004, Biodiversity in metacommunities: Plankton as complex adaptative systems? *Limnol. Oceanogr.* **49**:1278–1289.

Liebold, M. A., M. Holyoak, N. Mouquet, P. Amarasekare, J. M. Chase, M. F. Hoopes, R. D. Holt, J. B. Shurin, R. Law, D. Tilman, M. Loreau, and A. Gonzalez, 2004, The metacommunity concept: a framework for multiscale community ecology, *Ecol. Lett.* **7**:601–613.

Lin, Q., Y. Zhang, Y. Nie, and Y. Guan, 2003, Detection of harmful algal blooms over the Gulf of Bohai Sea in China at visible and near infrared (NIR) wavelengths of remote sensing. *J. Electromagnet. Wave.* **17**:861–871.

Lindahl, O., and L. Hernroth, 1988, Large-scale and long-term variations in the zooplankton community of the Gullmar Fjord, Sweden, in relation to advective processes, *Mar. Ecol. Prog. Ser.* **43**:161–171.

Lindholm, T., 1992, Ecological role of depth maxima of phytoplankton, *Archiv. Für Hydrobiol. Beih. Erge. Limnol.* **35**:33–45.

Lindley, J. A., and Williams, 1994, Relating plankton assemblages to environmental variables using intruments towed by ships-of-opportunity, *Mar. Ecol. Prog. Ser.* **107**:245–262.

Lindström, E. S., 2000, Bacterioplankton community composition in five lakes differing in trophic state and humic content, *Microb. Ecol.* **40**:104–113.

Lindström, E. S., 2001, Investigating influential factors on bacterioplankton community composition: results from a field study of five mesotrophic lakes, *Microb. Ecol.* **42**:598–605.

Lindström, E. S., 2006, External control of bacterial community structure in lakes, *Limnol. Oceanogr.* **51**:339–342.

Lo, W.-T., J.-S. Hwang, and Q.-C. Chen, 2004, Spatial distribution of copepods in surface waters of the southeastern Taiwan Strait, *Zool. Stud.* **43**:218–228.

Long, R. A., and F. Azam, 2001, Microscale patchiness of bacterioplankton assemblage richness in seawater, *Aquat. Microb. Ecol.* **26**:103–113.

Longhurst, A. R., 1995, Seasonal cycles of pelagic production and consumption. *Prog. Oceanogr.* **36**:77–168.

Longhurst, A. R., 1998, *Ecological Geography of the Sea.* Academic Press, San Diego, pp. 398.

Longhurst, A. R., and R. Williams, 1976, Improved filtration systems for multiple-series plankton samples and their deployment, *Deep Sea Res.* **23**:1067–1073.

Loreau, M., N. Mouquet, and R. Holt, 2003, Meta-ecosystems: a theoretical frame work for spatial ecosystem ecology, *Ecol. Lett.* **6**:673–679.

Lorenzen, C. J., 1966, A method for the continuous measurement of *in vivo* chlorophyll concentration, *Deep Sea Res.* **13**:223–227.

Lorke, A., D. F. McGinnis, P. Spaak, and A. Wüest, 2004, Acoustic observations of zooplankton in lakes using a Doppler current profiler, *Freshw. Biol.* **49**:1280–1292.

Louette, G., and L. De Meester, 2005, High dispersal capacity of cladoceran zooplankton in newly founded communities, *Ecology* **86**:353–359.

Lovejoy, S., W. J. S. Currie, Y. Tessier, M. R. Claereboudt, E. Bourget, J. C. Roff, and D. Schertzer, 2001, Universal multifractals and ocean patchiness: phytoplankton, physical fields and coastal heterogeneity, *J. Plankton Res.* **23**:117–141.

Ludovisi, A., M. Minozzo, P. Pandolfi, and M. Illuminata Taticchi, 2005, Modelling the horizontal spatial structure of planktonic community in Lake Trasimero (Umbria, Italy) using multivariate geostatistical methods, *Ecol. Model.* **181**:247–262.

Lunven, M., J. F. Guillaud, A. Youénou, M. P. Crassous, R. Berric, R. Le Gall, R. Kérouel, C. Labry, and A. Aminot, 2005, Nutrient and phytoplankton distribution in the Loire River plume (Bay of Biscay, France) resolved by a new Fine Scale Sampler, *Estuar. Coast. Shelf Sci.* **65**:94–108.

Mackas, D. L., 1984, Spatial autocorrelation of plankton community composition in a continental shelf ecosystem, *Limnol. Oceanogr.* **29**:451–471.

Mackas, D. L., K. L. Denman, and M. R. Abbott, 1985, Plankton patchiness: biology in the physical vernacular, *Bull Mar. Sci.* **37**:652–674.

Mair, A. M., P. G. Fernades, A. Lebourges-Dhaussy, and A. S. Brierley, 2005, An investigation into the zooplankton composition of a prominent 38-kHz scattering layer in the North Sea, *J. Plankton Res.* **27**:623–633.

Malone, B. J., and D. J. McQueen, 1983, Horizonal patchiness in zooplankton populations in two Ontario kettle lakes, *Hydrobiologia* **99**:101–124.

Maranger, R. J., M. L. Pace, P. A. del Giorgio, N. F. Caraco, and J. J. Cole, 2005, Longitudinal spatial patterns of bacterial production and respiration in a large river-estuary: implications for ecosystem carbon consumption, *Ecosystems* **8**:318–330.

Martin, A. P., 2003, Phytoplankton patchiness: the role of lateral stirring and mixing, *Prog. Oceanogr.* **57**:125–174.

Martin, J. L., F. H. Page, A. Hanke, P. M. Strain, and M. M. LeGresley, 2005, *Alexandrium fundyense* vertical distribution patterns during 1982, 2001 and 2002 in the offshore bay of Fundy, eastern Canada, *Deep Sea Res. II* **52**:2569–2592.

Masson, S., and B. Pinel-Alloul, 1998, Spatial distribution of zooplankton biomass size fractions in a bog lake: abiotic and (or) biotic regulation, *Can. J. Zool.* **76**:805–823.

Masson, S., and B. Pinel-Alloul, and V. Smith, 2000, Total phosphorus – chlorophyll *a* size fraction relationships in southern Québec lakes, *Limnol. Oceanogr.* **45**:732–740.

Masson, S., B. Pinel-Alloul, N. Angeli, and J. Guillard, 2001, Diel vertical and horizontal distribution of crustacean zooplankton and YOY fish in a sub alpine lake: an approach based on high frequency sampling, *J. Plankton Res.* **23**:1041–1060.

Masson, S., B. Pinel-Alloul, and P. Dutilleul, 2004, Spatial heterogeneity of zooplankton biomass and size structure in southern Québec lakes: variation among lakes and within lake among epi-, meta- and hypolimnion strata, *J. Plankton Res.* **26**:1441–1458.

McCauley, E., and J. Kalff, 1981, Empirical relationships between phytoplankton and zooplankton biomass in lakes, *Can. J. Fish. Aquat. Sci.* **38**:458–463.

McGillicuddy, D. J. Jr., D. M. Anderson, D. R. Lynch, and D.W. Townsend, 2005, Mechanisms regulating large-scale seasonal fluctuations in *Alexandrium fundyense* populations in the Gulf of Maine: Results from a physical-biological model, *Deep Sea Res. II* **52**:2698–2714.

McNaught, A. S., D. Pavlik, and D. W. Schindler, 2000, Patterns of zooplankton biodiversity in the mountain lakes of Banff National Park, Canada. *Verh. Internat. Verein. Limnol.* **27**:494–499.

Megard, R. O., M. M. Kuns, M. C. Whiteside, and J. A. Downing, 1997, Spatial distributions of zooplankton during coastal upwelling in western Lake Superior, *Limnol. Oceanogr.* **42**:827–840.

Mehner, T., F. Hölker, and P. Kasprzak, 2005, Spatial and temporal heterogeneity of trophic variables in a deep lake as reflected by repeated singular sampling, *Oikos* **108**:401–409.

Méthé, B. A., and J. P. Zehr, 1999, Diversity of bacterial community in Adirondack lakes: do species assemblages reflect lake water chemistry? *Hydrobiologia* **401**:77–96.

Mitchell, J. G., and J. A. Fuhrman, 1989, Centimeter scale vertical heterogeneity in bacteria and chlorophyll *a*, *Mar. Ecol. Prog. Ser.* **54**:141–148.

Michels, E., K. Cottenie, L. Neys, and L. De Meester, 2001a, Zooplankton on the move: first results on the quantification of dispersal of zooplankton in a set of interconnected ponds, *Hydrobiologia* **442**:117–126.

Michels, E., K. Cottenie, L. Neys, K. De Gelas, P. Coppin, and L. De Meester, 2001b, Geographical and genetic distances among zooplankton populations in a set of interconnected ponds: a plea for using GIS modelling of the effective geographical distance, *Mol. Ecol.* **10**:1929–1938.

Miracle, M. R., and M. T. Alfonso, 1993, Rotifer vertical distributions in a meromictic basin of Lake Banyoles (Spain), *Hydrobiologia* **255/256**:371–380.

Miracle, M. R., J. Armengol-Díaz, and M. J. Dasí, 1993, Extreme meromixis determines strong differential planktonic vertical distributions, *Verh. Internat. Verein. Limnol.* **25**:705–710.

Moll, R. A., and E. F. Stoermer, 1982, A hypothesis relating trophic status and subsurface chlorophyll maxima of lakes. *Arch. Hydrobiol.* **94**:425–440.

Monti, D., Legendre, L., J.-C. Therriault, and S. Demers, 1996, Horizontal distribution of sea-ice microalgae: environmental control and spatial processes (Southeastern Hudson Bay, Canada). *Mar. Ecol. Prog. Ser.* **133**:229–240.

Moss, B., I. Hooker, H. Balls, and K. Manson, 1989, Phytoplankton distribution in a temperate flooplain lke and river system. I. Hydrology, nutrient sources and phytoplankton biomass, *J. Plankton Res.* **11**:813–835.

Mustard, A. T., and T. R. Anderson, 2005, Use of spherical and spheroidal models for calculate zooplankton biovolume from particle equivalent spherical diameter as measured by an optical plankton counter, *Limnol. Oceanogr.-Meth.* **3**:183–189.

Nakano, S., O. Mitamura, M. Sugiyama, A. Maslennikov, Y. Nishibe, Y. Watanabe, and V. Drucker, 2003, Vertical planktonic structure in the central basin of Lake Baikal in summer 1999, with special reference to the microbial food web, *Limnology* **4**:155–160.

Nagata, T.G., H. Fukuda, R. Fuguda, and I. Koidke, 2000, Bacterioplankton distribution and production in deep Pacific waters: large-scale geographic variations and possible coupling with sinking particle fluxes, *Limnol. Oceanogr.* **45**:426–435.

Neill, W. E., 1994, Spatial and temporal scaling and the organization of limnetic communities, Chapter 7, in: *Aquatic Ecology: Scale, Pattern and Process*, P. S., A. Giller, G. Hildrew, and D. Rafaelli, eds., Proceedings of British Ecological Society and American Society of Limnology and Oceanography Symposium, 1992. Blackwell Scientific Publications, Oxford, pp. 189–230.

Nihongi, A., S. B. Lovern, and R. Strickler, 2004, Mate-searching behaviours in the freshwater calanoid copepod Leptodiaptomus ashlandi, *J. Marine Syst.* **49**:65–74.

Nogueira, E., G. González-Nuevo, A. Bode, M. Varela, X. A. G. Morán, and L. Valdés, 2004, Comparison of biomass and size spectra derived from optical plankton counter data and net samples: application to the assessment of mesoplankton distribution along the Northwest and North Iberian Shelf, *ICES J. Mar. Sci.* **61**:508–517.

Norberg, J., 2004, Biodiversity and ecosystem functioning: a complex adaptative systems approach, *Limnol. Oceanogr.* **49**:1269–1277.

O'Brien, D. P., 1988, Direct observations of clustering (schooling and swarming) behaviour in mysids (Crustacea: Mysidacea), *Mar. Ecol. Prog. Ser.* **42**:235–246.
Olson, N. W., S. K. Wilson, and D. W. Willis, 2004, Effect of spatial variation on zooplankton community assessment in fisheries studies. *Fisheries* **29**:17–22.
O'Neill, R. V., D. L. DeAngelis, J. B. Waide, and T. F. H. Allen, 1986, *A Hierarchical Concept of Ecosystems.* Princeton University Press, NJ.
Overmann, J., J. T. Beatty, and K. J. Hall, 1996, Purple sulphur bacteria controls the growth of aerobic heterotrophic bacterioplankton in a meromictic salt lake, *Appl. Environ. Microbiol.* **62**:3251–3258.
Overmann, J., K. J. Hall, and T. G. Northcote, 1999, Grazing of the copepod *Diaptomus connexus* on purple sulphur bacteria in a meromictic salt lake, *Environ. Microbiol.* **1**:213–221.
Pace, M. L., 1986, An empirical analysis of zooplankton community size structure across lake trophic gradients, *Limnol. Oceanogr.* **31**:45–55.
Pace, M. L., S. E. G. Findlay, and D. Lints, 1991, Variance in zooplankton samples: evaluation of a predictive model, *Can. J. Fish. Aquat. Sci.* **48**:146–151.
Paerl, H. W., 1996, A comparison of cyanobacterial blooms dynamics in freshwater, estuarine and marine environments. *Phycologia* **35**:25–35.
Palmer, M.W., 1993, Putting things in even better order: The advantages of canonical correspondence analysis, *Ecology* **74**:2215–2230.
Pascual, M., A. Asciotti, and H. Caswell, 1995, Intermittency in the plankton: a multifractal analysis of zooplankton biomass variability, *J. Plankton Res.* **17**:1209–1232.
Patalas, K., and A. Salki, 1992, Crustacean plankton in Lake Winnipeg: variation in space and time as a function of lake morphology, geology and climate, *Can. J. Fish. Aquat. Sci.* **49**:1035–1059.
Patalas, K., and A. Salki, 1993, Spatial variation of crustacean plankton in lakes of different size, *Can. J. Fish. Aquat. Sci.* **50**:2626–2640.
Patoine, A., B. Pinel-Alloul, and E. E. Prepas, 2002, Influence of catchment deforestation by logging and natural forest fires on crustacean community size structure in lakes of the Eastern Boreal Canadian forest, *J. Plankton Res.* **24**:601–616.
Patoine, A., B. Pinel-Alloul, G. Méthot, and M-J. Leblnac, 2006, Correspondence among methods of zooplankton biomass measurements in lakes: effect of community composition on optical plankton counter and size fractionated seston, *J. Plankton Res.* **28**:1–11.
Pearre, S. Jr., 2003, Eat and run? The hunger/satiation hypothesis in vertical migration: history, evidence and consequences, *Biol. Rev.* **78**:1–79.
Pedersen, S. A., M. H. Ribergaard, and C. S. Simonsen, 2005, Micro- and mesozooplankton in Southern Greenland waters in relation to environmental factors, *J. Mar. Syst.* **56**:85–112.
Pedrós-Alió, C., J. M. Gasol, and R. Guerrero, 1987, On the ecology of a *Cryptomonas phaseolus* population forming a metalimnetic bloom in Lake Cisó, Spain: Annual distribution and loss factors, *Limnol. Oceanogr.* **32**:285–298.
Pedrós-Alió C., Calderón-Paz J. I., Guixa-Boixereu N., Estrada M., and J. M. Gasol, 1999, Bacterioplankton and phytoplankton biomass and production during summer stratification in the northwestern Mediterranean Sea, *Deep Sea Res. I* **46**:985–1019.
Pérez, G. L., C. P. Queimaliños, and B. E. Modenutti, 2002, Light climate and plankton in the deep chlorophyll maxima in North Patagonian Andean lakes, *J. Plankton Res.* **24**:591–599.

Pérez, V., E. Frenadez, E. Marañón, P. Serret, and C. García-Soto, 2005, Seasonal and interannual variability of chlorophyll *a* and primary production in the Equatorial Atlantic: in situ and remote sensing observations, *J. Plankton Res.* **27**:189–197.

Perry, J. N., M. Leibhold, M. S. Rosenberg, J. Dungan, M. Miriti, A. Jakomuska, and S. Citron-Pousty, 2002, Illustrations and guidelines for selecting statistical methods for quantifying spatial patterns in ecological data, *Ecography* **25**:578–600.

Pick, F. R., C. Nalewajko, and D. R. S. Lean, 1984, The origin of a metalimnetic chrysophyte peak, *Limnol. Oceanogr.* **29**:125–134.

Pilati, A., and W. A. Wurtsbaugh, 2003, Importance of zooplankton for the persistence of a deep chlorophyll layer: a limnocorral experiment, *Limnol. Oceanogr.* **48**:249–260.

Pinca, S., and S. Dallot, 1995, Meso- and macrozooplankton composition patterns related to hydrodynamic structures in the Ligurian Sea (Trophos-2 experiment, April-June 1986), *Mar. Ecol. Progress Ser.* **126**:49–65.

Pinel-Alloul, B., 1995, Spatial heterogeneity as a multiscale characteristics of zooplankton community, *Hydrobiologia* **300/301**:17–42.

Pinel-Alloul, B., and D. Pont, 1991, Spatial distribution patterns in freshwater macrozooplankton: variation with scale, *Can. J. Zool.* **69**:1557–1570.

Pinel-Alloul, B., J. A. Downing, M. Pérusse, and G. Codin-Blumer, 1988, Spatial heterogeneity in freshwater zooplankton variation with body size, depth, and scale, *Ecology* **69**:1393–1400.

Pinel-Alloul, B., G. Méthot, G. Verreault, and Y. Vigneault, 1990a, Zooplankton species associations in Quebec lakes: variation with abiotic factors, including natural and anthropogenic acidification, *Can. J. Fish. Aquat. Sci.* **47**:110–121.

Pinel-Alloul, B., G. Méthot, G. Verreault, and Y. Vigneault, 1990b, Phytoplankton in Quebec lakes: variation with lake morphometry, and with natural and anthropogenic acidification, *Can. J. Fish. Aquat. Sci.* **47**:1047–1057.

Pinel-Alloul, B., T. Niyonsenga, and P. Legendre, 1995, Spatial and environmental components of freshwater zooplankton structure, *Ecoscience* **2**:1–19.

Pinel-Alloul, B., N. Bourbonnais, and P. Dutilleul, 1996, Among-lake and within-lake variations of autotrophic pico- and nanoplankton biomass in six Québec lakes, *Can. J. Fish. Aquat. Sci.* **53**:2433–2445.

Pinel-Alloul, B., C. Guay, N. Angeli, P. Legendre, P. Dutilleul, G. Balvay, D. Gerdeaux, and J. Guillard, 1999, Large-scale spatial heterogeneity of macrozooplankton in Lake Geneva, *Can. J. Fish. Aquat. Sci.* **56**:1437–1451.

Pinel-Alloul, B., N. Z. Malinsky-Rushansky, and G. Méthot, 2004, A short-term study of vertical and horizontal distribution of zooplankton during thermal stratification in Lake Kinneret, Israel, *Hydrobiologia* **526**:85–98.

Pinto-Coelho, R., B. Pinel-Alloul, G. Méthot, and K. Havens, 2005, Crustacean zooplankton in lakes and reservoirs of temperate and tropical regions: variation with trophic status, *Can. J. Fish. Aquat. Sci.* **62**:348–361.

Piontkovski, S. A., and R. Williams, 1995, Multiscale variability of tropical ocean zooplankton biomass, *ICES J. Mar. Sci.* **52**:643–656.

Piontkovski, S. A., R. Williams, W. T. Peterson, O. A. Yunev, N. I. Minkina, V. L. Vladimirov, and A. Blinkov, 1997, Spatial heterogeneity of the planktonic fields in the upper mixed layer of the open ocean, *Mar. Ecol. Prog. Ser.* **148**:145–154.

Planque, B., and S. D. Batten, 2000, *Calanus finmarchicus* in the North Atlantic: the year of *Calanus* in the context of interdecadal change. *ICES J Mar. Sci.* **57**:1528–1535.

Platt, T., and K. L. Denman, 1975, Spectral analysis in Ecology, *Ann. Rev. Ecol. Syst.* **6**:189–210.

Platt, T., and S. Sathyendranath, 1999, Spatial structure of pelagic ecosystem processes in the global ocean, *Ecosystems* **2**:384–394.
Platt, T., H. Bouman, E. Devred, C. Fuentes-Yaco, and S. Sathyendranath, 2005, Physical forcing and phytoplankton distributions, *Sci. Mar.* **69**:55–73.
Postel, L., H. Fock, and W. Hagen, 2000, Biomass and abundance, in: *ICES Zooplankton Methodology Manual*, R. Harris, P. Wiebe, J. Lenz, H. Rune Skojdal, and M. Huntley, eds., Academic Press, Boston, pp. 83–192.
Pothoven, S. A., G. L. Fahnensteil, and H. A. Vanderploeg, 2004, Spatial distribution, biomass and population dynamics of *Mysis relicta* in Lake Michigan, *Hydrobiologia* **422**:291–299.
Price, H. L., 1989, Swimming behavior of krill in responses to algal patches: a mesocosm study, *Limnol. Oceanogr.* **34**:649–659.
Priscu, J. C., and C. R. Goldman, 1983, Seasonal dynamics of the Deep-Chlorophyll Maximum in Castle Lake, California, *Can. J. Fish. Aquat. Sci.* **40**:208–214.
Raffaelli, D. G., A. G. Hildrew and Giller P. S., 1994, Scale, pattern and process in aquatic systems: concluding remarks, Chapter 21, in: *Aquatic Ecology: Scale, Pattern and Process*, P. S. Giller, A. G. Hildrew and Rafaelli, D. eds., Proceedings of British Ecological Society and American Society of Limnology and Oceanography Symposium, 1992, Blackwell Scientific Publications, Boca Raton, FL, pp. 601–606.
Rantajärvi, E., V. Gran, S. Hällfors, and R. Olsonen, 1998, Effects of environmental factors on the phytoplankton community in the Gulf of Findland - unattended high frequency measurements and multivariate analyses, *Hydrobiologia* **363**:127–139.
Reche, I., E. Pulido-Villena, R. Morales-Baquero, and E. O. Casamayor, 2005, Does ecosystem size determine aquatic bacterial richness? *Ecology* **86**:1715–1722.
Reid, P.C., J.M. Colebrook, J.B.L. Matthews, and J. Aiken, 2003, The Continuous Plankton Recorder: concepts and history, from plankton indicator to undulating recorders, *Prog. Oceanogr.* **58**:117–173.
Remsen, A., T. L. Hopkins, and S. Samson, 2004, What you see is not what you catch: a comparison of currently collected net, Optical Plankton Counter, and Shadowed Image Particle Profiling Evaluation Recorder data from the northeast Gulf of Mexico, *Deep Sea Res. I* **51**:129–151.
Ressler, P. H., R. D. Brodeur, W. T. Peterson, S. D. Pierce, P. M. Vance, A. Røstad, and J. A. Barth, 2005, The spatial distribution of euphausiid aggregations in the Northern California Current during August 2000, *Deep Sea Res. II* **52**:89–108.
Reynolds, C. S., 1994, The role of fluid motion in the dynamics of phytoplankton in lakes and rivers, Chapter 6, in: *Aquatic Ecology: Scale, Pattern and Process*, P. S. Giller, A. G. Hildrew and D. Rafaelli, eds., Proceedings of British Ecological Society and American Society of Limnology and Oceanography Symposium, 1992, Blackwell Scientific Publications, Boca Raton, FL, pp. 141–187.
Reynolds, C. S., and A. C. Petersen, 2000, The distribution of planktonic Cyanobacteria in Irish lakes in relation to their trophic states, *Hydrobiologia* **424**:91–99.
Richardson, A. J., and D. S. Schoeman, 2004, Climate impact on plankton ecosystems in the Northeast Atlantic, *Science* **305**:1609–1612.
Richardson, A. J., E. H. John, X. Irigoien, R. P. Harris, and G. C. Hays, 2004, How well does the continuous plankton recorder (CPR) sample zooplankton? A comparison with the Longhurst Hardy Plankton Recorder (LHPR) in the northeast Atlantic, *Deep Sea Res. I* **51**:1283–1294.

Richardson, A. J., A. W. Walne, A. W. G. John, T. D. Jonas, J. A. Lindley, D. W. Sims, D. Stevens, and M. Witt, 2006, Using continuous plankton recorder data, *Prog. Oceanogr.* **68**:27–74.

Ricklefs, R. E., 1990, Scaling pattern and process in marine ecosystems, in: *Large Marine Ecosystems*, K. Sherman, L. M. Alexander, and B. D. Gold, eds., AAAS, Washington, pp. 169–178.

Reid, P. C., and G. Beaugrand, 2002, Interregional biological responses in the North Atlantic to hydrometeorological forcing, in: *Changing States of the Large Marine Ecosystems of the North Atlantic*, K. Sherman, and H.-R. Skjoldal, eds., Elsevier Science, Amsterdam, The Netherlands, pp. 27–48.

Riemann, B., and M. Søndergaard, eds., 1986, *Carbon Dynamics in Eutrophic, Temperate Lakes*, Elsevier Science Publishers, Amsterdam, The Netherlands, pp. 248.

Ringelberg, J., 1995, Changes in the light intensity and diel vertical migration: comparison of marine and freshwater environments, *J. Mar. Biol. Assoc. UK* **75**:15–25.

Ringelberg, J., 1999, The photobehaviour of *Daphnia* spp. as a model to explain diel vertical migration in zooplankton, *Biol. Rev.* **74**:397–423.

Ringelberg, J., and E. Van Gool, 2003, On the combined analysis of proximate and ultimate aspects in diel vertical migration (DVM) research, *Hydrobiologia* **491**:85–90.

Ringelberg, J., J. G. Flik, D. Lindenaar, and K. Royackers, 1991, Diel vertical migrations of *Daphnia hyalina* (sensu latiori) in Lake Maarsseveen: Part I. Aspects of seasonal and daily timing, *Arch. Hydrobiol.* **121**:129–145.

Ritz, D. A., 1994, Social aggregation in pelagic invertebrates, *Adv. Mar. Biol.* **30**:155–216.

Rixen, T., M. V. S. Guptha, and V. Ittekkot, 2005, Deep ocean fluxes and their link to surface ocean processes and the biological pump, *Prog. Oceanogr.* **65**:240–259.

Robarts, R. D., T. Zohary, M. J. Waiser, and Y. Z. Yacobi, 1996, Bacterial abundance, biomass, and production in relation to phytoplankton biomass in the Levantine Basin of the south-eastern Mediterranean Sea, *Mar. Ecol. Prog. Ser.* **137**:273–281.

Rohrlack, T., M. Henning, and J.-G. Kohl, 1999, Mechanisms of the inhibitory effect of the cyanobacterium *Microcystis aeruginosa* on *Daphnia galeata*'s ingestion rate, *J. Plankton Res.* **21**:1489–1500.

Roman, M., X. Zhang, C. McGilliard, and W. Boicourt, 2005, Seasonal and annual variability in the spatial patterns of plankton biomass in Chesapeake Bay, *Limnol. Ocenaogr.* **50**:480–492.

Ruddick, K. G., F. Ovidio, and M. Rijkeboer, 2000, Atmospheric correction of SeaWiFS imagery for turbid coastal and inland waters, *Appl. Opt.* **39**:897–912.

Rudnick, D. L., R. E. Davis, C. C. Eriksen, D. M. Fratantoni, and M. J. Perry, 2004, Underwater gliders for ocean research, *Mar. Technol. Soc. J.* **38**:73–84.

Rusak, J. A., N. D. Yan, K. M. Somers, K. L. Cottingham, F. Micheli, S. R. Carpenter, T. M. Frost, M. J. Paterson, and McQueen D. J., 2002, Temporal, spatial and taxonomic patterns of crustacean zooplankton variability in unmanipulated north-temperate lakes, *Limnol. Oceanogr.* **47**:613–625.

Ryan, E. F., D. P. Hamilton, J. A. Hall, and U. V. Cassie Cooper, 2005, Lake phytoplankton composition and biomass along horizontal and vertical gradients, *Verh. Internat. Verein. Limnol.* **29**:1033–1036.

Saiz, E., P. Tiselius, P. R. Jonsson, P. Verity, and G. A. Paffenhöfer, 1993, Experimental records of the effects of food patchiness and predation on egg production of *Acartia tonsa*, *Limnol. Oceanogr.* **38**:280–289.

Sakuma, M., and T. Hanazato, 2002, Abundance of Chydoridae associated with plant surfaces, water column and bottom sediments in the macrophyte zone of a lake, *Verh. Internat. Verein. Limnol.* **28**:975–979.

Sakuma, M., T. Hanazato, A. Saji, and R. Nakazato, 2004, Migration from plant to plant: an important factor controlling densities of the epiphytic cladoceran *Alona* (Chydoridae, Anomopoda) on lake vegetation, *Limnology* **5**:17–23.

Sawatzky, C. L., W. A. Wurtsbaugh, and C. Luecke, 2006, The spatial and temporal dynamics of deep chlorophyll layers in high-mountain lakes: effects of nutrients, grazing and herbivore nutrient recycling as growth determinants, *J. Plankton Res.* **28**:65– 86.

Schabetsberger, R., and C. D. Jersabek, 2004, Shallow males, deep females: sex-biased differences in habitat distribution of the freshwater calanoid copepod *Arctodiaptomus alpinus*, *Ecography* **27**:506–520.

Schäfer, H., L. Bernard, C. Courties, P. Lebaron, P. Servais, Pukall, R. E. Stackebrandt, M. Troussellier, T. Guindulain, J. Vives-rego, and G. Muyzer, 2001, Microbial community dynamics in Mediterranean nutrient-enriched seawater mesocosms: changes in the genetic diversity of bacterial populations, *FEMS Microbiol. Ecol.* **34**:243–253.

Schernewski, G., V. Podsetchine, and T. Huttula, 2005, Effect of the flow field on small scale phytoplankton patchiness, *Nordic Hydrology* **36**:85–98.

Scheuerell, M. D., 2004, Quantifying aggregation and association in three-dimensional landscapes, *Ecology* **85**:2332–2340.

Schmitt, F. G., and L. Seuront, 2001, Multifractal random walk in copepod behaviour, *Physica A*: **301**:375–396.

Schultz, G. E., E. D. White III, and H. W. Ducklow, 2003, Bacterioplankton dynamics in the York River estuary: primary influence of temperature and freshwater inputs. *Aquat. Microb. Ecol.* **30**:135–148.

Seip, K. L., and C. S. Reynolds, 1995, Phytoplankton functional attributes along trophic gradient and season, *Limnol. Oceanogr.* **40**:589–597.

Selig, U., T. Hübener, R. Heerkloss, and H. Schubert, 2004, Vertical gradient of nutrients in two dimictic lakes – Influence of phototrophic sulphur bacteria on nutrient balance, *Aquat. Sci.* **66**:247–256.

Sen, Z., 1989, Cumulative semi-variogram models of regionalized variables, *Math. Geol.* **21**:891–903.

Seppälä, J., and M. Balode, 1999, Spatial distribution of phytoplankton in the Gulf of Riga during spring and summer stages, *J. Marine Systems* **23**:51–67.

Servais, P., P. Dufour, P. Caumette, A. Hirschler, and R. Matheron, 1995, L'activité bactérienne. Chapitre 7, in. *Limnologie Générale,* R. Pourriot and M. Meybeck, eds., Masson, Paris, pp 253–295.

Seuront, L., 2001, Microscale processes in the ocean: why are they so important for ecosystem functioning? *La Mer* **39**:1–8.

Seuront, L., and Y. Lagadeuc, 2001, Multiscale patchiness of the calanoid copepod *Temora longicornis* in a turbulent coastal sea, *J. Plankton Res.* **23**:1137–1145.

Seuront, L., and P. G. Strutton, 2003, *Handbook of Scaling Methods in Aquatic Ecology: Measurements, Analysis, Simulation,* CRC Press LLC, Boca Raton, FL, pp. 600.

Seuront, L., and F. G. Schmitt, 2005a, Multiscaling statistical procedures for the exploration of biophysical coupling in interminent turbulence, Part I. Theory, *Deep Sea Res. II* **52**:1308–1324.

Seuront, L., and F. G. Schmitt, 2005b, Multiscaling statistical procedures for the exploration of biophysical couplings in intermittent turbulence, Part II. Applications, *Deep Sea Res. II* **52**:1325–1343.

Seuront, L., F. Schmitt, Y. Lagadeuc, D. Schertzer, S. Lovejoy, and S. Frontier, 1996, Mulitfractal analysis of phytoplankton biomass and temperature in the ocean, *Geophys. Res. Lett.* **23**:3591–3594.

Seuront, L., F. Schmitt, Lagadeuc Y., D. Schertzer, and Lovejoy S., 1999, Universal multifractal analysis as a tool to characterize multiscale intermittent patterns: example of phytoplankton distribution in turbulent coastal waters, *J. Plankton Res.* **21**:877–922.

Seuront, L., F. G. Schmitt, and Y. Lagadeuc, 2001, Turbulence intermittency, small-scale phytoplankton patchiness and encounter rates in plankton: where do we go from here? *Deep Sea Res. I* **48**:1199–1215.

Seuront, L., H. Yamazaki, and S. Souissi, 2004a, Hydrodynamics disturbance and zooplankton swimming behaviour, *Zool. Stud.* **43**:376–387.

Seuront, L., J.-S. Hwang, L.-C. Tseng, F. G. Schmitt, S. Souissi, and C.-K. Wong, 2004b, Individual variability in the swimming behaviour of the sub-tropical copepod *Oncaea venusta* (Copepoda, Poecilostomatoida), *Mar. Ecol. Prog. Ser.* **283**:199–217.

Seymour, J. R., J. G. Mitchell, L. Pearson, and R. L. Waters, 2000, Heterogeneity in bacterioplankton abundance from 4.5 millimetre resolution sampling, *Aquat. Microb. Ecol.* **22**:143–153.

Seymour, J. R., J. G. Mitchell, and L. Seuront, 2004, Microscale heterogeneity in the activity of coastal bacterioplankton communities, *Aquat. Microb. Ecol.* **35**:1–16.

Seymour, J. R., L. Seuront, and J. G. Mitchell, 2005, Microscale and small-scale temporal dynamics of a dynamics of a coastal planktonic microbial community, *Mar. Ecol. Prog. Ser.* **300**:21–37.

Sherr, E. B., B. F. Sherr, and T. J. Cowles, 2001, Mesoscale variability in bacterial activity in the Northeast Pacific Ocean off Oregon, USA, *Aquat. Microb. Ecol.* **25**:21–30.

Shurin, J. B., and J. E. Havel, 2002, Hydrological connections and overland dispersal in an exotic freshwater crustacean, *Biol. Invasions*, **4**:431–439.

Shurin, J. B., J. E. Havel, M. A. Leibold, and B. Pinel-Alloul, 2000, Local and regional zooplankton species richness: a scale independent test for saturation, *Ecology* **81**:3062–3073.

Simard, Y., and D. L. Mackas, 1989, Mesoscale aggregations of euphausiid sound scattering layers on the continental Shelf of Vancouver Island, *Can. J. Fish. Aquat. Sci.* **46**:1238–1249.

Simard, Y., P. Legendre, G. Lavoie, and D. Marcotte, 1992, Mapping, estimating biomass, and optimizing sampling programs for spatially autocorrelated data: Case study of the Northern Shrimp (*Pandalus borealis*), *Can. J. Fish. Aquat. Sci.* **49**:32–45.

Smith, V. H., 2003, Eutrophication of freshwater and coastal marine ecosystems: a global problem, *Environ. Sci. Pollut. Res.* **10**:126–139.

Smith, E., and Y. T. Prairie, 2004, Bacterial metabolism and growth efficiency in lakes: the importance of phosphorus availability, *Limnol. Oceanogr.* **49**:137–141.

Smith, S. L., and M. Madhupratap, 2005, Mesozooplankton of the Arabian Sea: Patterns influenced by seasons, upwelling, and oxygen concentrations, *Prog. Oceanogr.* **65**:214–239.

Smith, V. H., S. B. Joye, and R. W. Howarth, 2006, Eutrophication of freshwaters and marine ecosystems, *Limnol. Oceanogr.* **51**:351–355.

Souissi, S., L. Seuront, F. G. Schmitt, and V. Ginot, 2005, Describing space-time patterns in aquatic ecology using IBMs and scaling and multi-scaling approaches, *Nonlinear Anal. – Real* **6**:705–730.

Sparrow, A. D., 1999, A heterogeneity of heterogeneities, *Tree* **14**:422–423.

Sprules, W. G., and M. Munawar, 1986, Plankton size spectra in relation to ecosystem productivity, *Can. J. Fish. Aquat. Sci.* **43**:1789–1794.

Sprules, W. G., S. B. Brandt, D. J. Stewart, M. Munawar, E. H. Jin, and J. Love, 1991, Biomass size spectrum of the Lake Michigan pelagic food web, *Can. J. Fish. Aquat. Sci.* **48**:105–115.

Sprules, W. G., B. Bergstrom, H. Cyr, B. R. Hargreaves, S. S. Kilham, H. J. MacIsaac, K. Matshushita, R. Stemberger, and R. Williams, 1992, Non-video optical instruments for studying zooplankton distribution and abundance, *Arch. Hydrobiol. Beih. Ergebn. Limnol.* **36**:45–58.

Sprules, W. G., E. H. Jin, A. W. Herman, and J. D. Stockwell, 1998, Calibration of an optical counter for use in fresh water, *Limnol. Oceanogr.* **43**:726–733.

Steele, J. H., and E. W. Henderson, 1998, Vertical migration of copepods, *J. Plankton Res.* **20**: 787–799.

Stemberger, R. S., and J. M. Lazorchak, 1994, Zooplankton assemblage responses to disturbance gradients, *Can. J. Fish. Aquat. Sci.* **51**:2453–2447.

Stemberger, R. S., A. T. Herlihy, D. L. Kugler, and S. G. Paulsen, 1996, Climate forcing on zooplankton richness in lakes of the Northeastern United States, *Limnol. Oceanogr.* **41**: 1093 1101.

Stirling, D. G., D. J. McQueen, and M. R. S. Johannes, 1990, Vertical migration *in Daphnia galeata mendotae* (Brooks): demographic responses to changes in planktivore abundance, *Can. J. Fish. Aquat. Sci.* **47**:395–400.

Stockwell, J. D., and W. G. Sprules, 1995, Spatial and temporal patterns of zooplankton biomass in Lake Erie, *ICES J. Mar. Sci.* **52**:557–564.

Strickland, J. D. H., 1968, Continuous measurement of *in vivo* chlorophyll; a precautionary note, *Deep Sea Res.* **15**:225–227.

Strutton, P. G., J. G. Michell, and J. S. Parslow, 1997, Using non-linear analysis to compare the spatial structure of chlorophyll with passive tracers, *J. Plankton Res.* **19**:1553–1564.

Sutor, M. M., T. J. Cowles, W. T. Peterson, and S. D. Pierce, 2005, Acoustic observations of finescale zooplankton distributions in the Oregon upwelling region, *Deep Sea Res. II* **52**: 109–121.

Suzuki, M., C. M. Preston, F. P. Chavez, and E. F. de Long, 2001, Quantitative mapping of bacterioplankton populations in seawater: field tests across an upwelling plume in Monterey bay, *Aquat. Microb. Ecol.* **24**:117–127.

Swadling, K. M., R. Pienitz, and T. Nogrady, 2000, Zooplankton community composition of lakes in the Yukon and Northwest Territories (Canada): relationship to physical and chemical limnology, *Hydrobiologia* **431**:211–224.

Tada, K., M. Morishita, K. Hamada, and S. Mondtani, 2001, Standing stock and production of phytoplankton and a red tide outbreak in a heavily eutrophic embayment, Dokai Bay, Japan, *Mar. Pollut. Bull.* **42**:1177–1186.

Tang, D. L., D. R. Kester, I. H. Ni, Y. Z. Qi, and H. Kawamura, 2003, *In situ* and satellite observations of a harmful algal bloom and water condition at the Pearl River Estuary in late autumn 1998, *Harmful Algae* **2**:89–99.

Tarling, G. A., F. Buchholz, and J. B. L. Matthews, 1999, The effect of a lunar eclipse on the vertical migration behaviour of *Meganyctiphanes norvegica* (Crustacea, Euphausiacae) in the Ligurian Sea, *J. Plankton Res.* **21**:1475–1488.

ter Braak, C. J. F., 1986, Canonical correspondence analysis: a new eigenvector technique for multivariate direct gradient analysis, *Ecology* **67**:1167–1179.

ter Braak, C. J. F., 1988, Partial Canonical Analysis, in: *Classification and Related Methods of Data Analysis*, H. H. Bock, ed., Elsevier Science Publishers, Amsterdam, pp. 551–558.

Teubner, K., M. Tolotti, S. Greisberger, H. Morscheid, M. T. Dokulil, and H. Morscheid, 2003, Steady state phytoplankton in a deep pre-alpine lake: species and pigments of epilimnetic versus metalimnetic assemblages, *Hydrobiologia* **502**:49–64.

Thackeray, S. J., D. Glen George, R. J. Jones, and I. J. Winfield, 2004, Quantitative analysis of the importance of wind-induced circulation for the spatial structuring of planktonic populations, *Freshw. Biol.* **49**:1091–1102.

Thioulouse, J., D Chessel, and S. Champely, 1995, Multivariate analysis of spatial patterns: a unified approach to local and global structures, *Environ. Ecol. Stat.* **2**:1–14.

Thorisson, K., 2006, How are the vertical migrations of copepods controlled? *J. Exp. Mar. Biol. Ecol.* **329**:86–100.

Thorpe, A. P., and J. R. Jones, 2005, Bacterial abundance in Missouri (USA) reservoirs in relation to trophic state and global patterns, *Verh. Internat. Verein. Limnol.* **29**:239–245.

Tiselius, P., 1992, Behavior of *Acartia tonsa* in patchy food environments, *Limnol. Oceanogr.* **37**:1640–1651.

Tiselius, P., 1998, An *in situ* video camera for plankton studies: design and preliminary observations, *Mar. Ecol. Prog. Ser.* **164**:293–299.

Tomlinson, M. C., R. P. Stumpf, V. Ransibrahmanakul, E. W. Truby, G. J. Kirkpatrick, B. A. Pederson, G. A. Vargo, and C. A. Heil, 2004, Evaluation of the use of SeaWiFS imagery for detecting *Karenia brevis* harmful algal blooms in the eastern Gulf of Mexico, *Remote Sens. Environ.* **91**:293–303.

Townsend, D. W., S. L. Bennett, and M. A. Thomas, 2005, Diel vertical distributions of the red tide dinoflagellates *Alexandrium fundyense* in the Gulf of Maine, *Deep Sea Res. II* **52**:2593–2602.

Van de Meutter, F., R. Stoks, and L. De Meester, 2004, Behavioral linkage of pelagic prey and littoral predators: microhabitat selection by *Daphnia* induced by damselfly larvae, *Oikos* **107**:265–272.

Van Gool, E., and J. Ringelberg, 1996, Daphnids responds to algal associated odours, *J. Plankton Res.* **18**:197–202.

Vezzulli, L., and P. C. Reid, 2003, The CPR survey (1948–1997): a gridded database browser of plankton abundance in the North Sea, *Prog. Oceanogr.* **58**:327–336.

Voss, S., and H. Mumm, 1999, Where to stay by night and day: size-specific and seasonal differences in horizontal and vertical distribution of *Chaoborus flavicans* larvae, *Freshw. Biol.* **42**:201–213.

Wackernagel, H., 1995, *Multivariate Geostatistics, An Introduction with Applications*. Springer, Heidelgerg, Berlin.

Wærvågen, S. B., N. A. Rukke, and D. O. Hessen, 2002, Calcium content of crustacean zooplankton and its potential role in species distribution, *Freshw. Biol.* **47**:1866–1878.

Walks, D. J., and H. Cyr, 2004, Movement of plankton through lake-stream systems, *Freshw. Biol.* **49**:745–759.

Walsby, A. E., A. Avery, and F. Schanz, 1998, The critical pressures of gas vesicles in *Planktothrix rubescens* in relation to the depth of winter mixing in Lake Zürich, Switzerland, *J. Plankton Res.* **20**:1357–1375.

Warner, A. J., and G. C. Hays, 1994, Sampling by the Continuous Plankton Recorder survey, *Prog. Oceanogr.* **34**:237–256.

Waters, R. L., and J. G. Mitchell, 2002, Centimeter-scale spatial structure of estuarine *in vivo* fluorescence profiles, *Mar. Ecol. Prog. Ser.* **237**:51–63.

Waters, R. L., J. G. Mitchell, and J. R. Seymour, 2003, Geostatistical characterization of centimetre-scale spatial structure of *in vivo* fluorescence, *Mar. Ecol. Prog. Ser.* **251**:49–58.

Weber, A., and A. Van Noordwijk, 2002, Swimming behaviour of *Daphnia* clones: differentiation through predator infochemicals, *J. Plankton Res.* **24**:1335–1348.
Webster, I. T., 1990, Effect of wind on the distribution of phytoplankton cell in lakes, *Limnol. Oceanogr.* **35**:989–1001.
Weisse, T., and E. MacIsaac, 2000, Significance and fate of bacterial production in oligotrophic lakes in British Columbia, *Can. J. Fish. Aquat. Sci.* **57**:96–105.
Weithoff, G., 2004, Vertical niche separation of two consumers (Rotatoria) in an extreme habitat, *Oecologia* **139**:594–603.
Werdell, P. J., and S. W. Bailey, 2005, An improved *in situ* bio-optical data set for ocean color algorithm development and satellite data product validation, *Remote Sens. Environ.* **98**:122–140.
Wetzel, R. G., 2001, *Limnology: Lake and River Ecosystems*. Academic Press, San Diego, pp. 1006.
Wiafe, G., and C. L. J. Frid, 1996, Short-term temporal variation in coastal zooplankton communities: the relative importance of physical and biological mechanisms, *J. Plankton Res.* **18**:1485–1501.
Wieland, K., D. Petersen, and D. Schnack, 1997, Estimates of zooplankton abundance and size distribution with the Optical Plankton Counter (OPC), *Arch. Fish. Mar. Res.* **45**:271–280.
Wiens, J. A., 1997, Metapopulation dynamics and landscape ecology, in: *Metapopulation Biology: Ecology, Genetics and Evolution*, I. Hanski, M. E. Gilpin, eds., Academic Press, San Diego, pp. 43–62.
Wiens, J. A., 1989. Spatial scaling in ecology, *Funct. Ecol.* **3**:385–397.
Williamson, C. E., R. W. Sanders, R. E. Moeller, and P. L. Stutzman, 1996, Utilization of subsurface food resources for zooplankton reproduction: implications for diel vertical migration theory, *Limnol. Oceanogr.* **41**:224–233.
Winder, M., P. Spaak, and W. M. Mooij, 2004, Trade-offs in *Daphnia* habitat selection, *Ecology* **85**:2027–2036.
Wissel, B., and C. W. Ramacharan, 2003, Plasticity of vertical distribution of crustacean zooplankton in lakes with varying levels of water colour, *J. Plankton Res.* **25**:1047–1057.
Wissel, B., A. Gaçe, and B. Fry, 2005, Tracing river influences on phytoplankton dynamics in two Louisiana estuaries, *Ecology* **86**:2751–2762.
White, P. A., J. Kalff, J. B. Rasmussen, and J. M. Gasol, 1991, The effect of temperature and algal biomass on bacterial production and specific growth rates in freshwater and marine habitats, *Microb. Ecol.* **21**:99–118.
White, J. R., Z. Xinsheng, L. A. Welling, M. R. Roman, and H. G. Dam, 1995, Latitudinal gradients in zooplankton biomass in the tropical Pacific at 140°W during the JGOFS EqPac Study: effects of El Niño, *Deep Sea Res. II* **42**:715–733.
Wojciechowska, W., M. Poniewozik, and A. Pasztaleniec, 2004, Vertical distribution of dominant cyanobacteria species in three lakes – evidence of tolerance to different turbulence and oxygen conditions, *Pol. J. Ecol.* **52**:347–351.
Wojtal, A., P. Frankiewick, K. Izydorczyk, and M. Zaliewski, 2003, Horizontal migration of zooplankton in a littoral zone of the lowland Sulejow reservoir (Central Poland), *Hydrobiologia* **506/509**:339–346.
Wolk, F., L. Seuront, and H. Yamazaki, 2001, Spatial resolution of a new micro-optical probe for chlorophyll and turbidity, *J. Tokyo Univ. Fish.* **87**:13–21.
Wolk, F., H. Yamazaki, L. Seuront, and R. G. Lueck, 2002, A new free-fall profiler for measuring biological microstructure, *J. Atmos. Oceanic Technol.* **19**:780–793.

Wolk, F., L. Seuront, H. Yamasaki, and S. Leterme, 2003, Comparison of biological scale resolution from CTD and microstructure measurements, Chapter 1, in: *Handbook of Scaling Methods in Aquatic Ecology. Measurement, Analysis, Simulation,* eds., L. Seuront, and P. G. Strutton, CRC Press, Boca Raton, FL, pp. 3–15.

Woodd-Walker, R. S., C. P. Gallienne, and D. B. Robins, 2000, A test model for optical plankton counter (OPC) coincidence and a comparison of OPC-derived and conventional measures of plankton abundance, *J. Plankon Res.* **22**:473–483.

Woodd-Walker, R. S., P. Ward, and A. Clarke, 2002, Large-scale patterns in diversity and community structure of surface water copepods from the Atlantic Ocean. *Mar. Ecol. Prog. Ser.* **236**:189–203.

Wurtsbaugh, W. A., H. P. Gross, P. Budy, and C. Luecke, 2001, Effects of epilimnetic versus metalimnetic fertilization on the phytoplankton and periphyton of a mountain lake with a deep chlorophyll maxima, *Can. Fish. Aquat. Sci.* **58**:2156–2166.

Xu, X., and L. V. Madden, 2003, Considerations for the use of SADIE statistics to quantify spatial patterns, *Ecography* **26**:821–830.

Yahel, R., G. Yahel, T. Berman, J. S. Jaffe, and A. Genin, 2005, Diel pattern with abrupt crepuscular changes of zooplankton over a coral reef, *Limnol. Ocenogr.* **50**:930–944.

Yamaguchi, A., Y. Watanabe, H. Isgida, T. Harimoto, K. Furusawa, S. Suzuki, J. Ishizaka, T. Ikeda, and M. Mac Takahashi, 2004, Latitudinal differences in the planktonic biomass and community structure down to the greater depths in the western North Pacific, *J. Oceanogr.* **60**:773–787.

Yan, N. D., 1986, Empirical prediction of crustacean zooplankton biomass in nutrient-poor Canadian Shield lakes, *Can. J. Fish. Aquat. Sci.* **43**:788–796.

Yannarell, A. C., and E. W. Triplett, 2004, Within and between-lake variability in the composition of bacterioplankton communities: investigations using multiple spatial scales, *Appl. Environ. Microbiol.* **70**:214–223.

Yannarell, A. C., and E. W. Triplett, 2005, Geographic and environmental sources of variation in lake bacterial community composition, *Appl. Environ. Microbiol.* **71**:227–239.

Zeichen, M. M., and I. S. Robinson, 2004, Detection and monitoring of algal blooms using SeaWiFS imagery, *Int. J. Remote Sens.* **25**:1389–1395.

Zhang, X., M. Roman, A. Sanford, H. Adolf, C. Lascara, and R. Burgett, 2000, Can an optical plankton counter produce reasonable estimates of zooplankton abundance and biovolume in water with high detritus? *J. Plankton Res.* **22**:137–150.

Zhao, B., and Q. Cai, 2004, Geostatistical analysis of Chlorophyll *a* in a freshwater ecosystem, *J. Freshw. Ecol.* **19**:613–621.

Chapter 9

THE INTERRELATIONSHIP BETWEEN THE SPATIAL DISTRIBUTION OF MICROORGANISMS AND VEGETATION IN FOREST SOILS

Sherry J. Morris[1] and William J. Dress[2]
[1]*Biology Department, Bradley University, Peoria, Illinois 61625, USA;* [2]*Science Department, Robert Morris University, Moon Township, PA 15108, USA*

Abstract: Recent advances in techniques for investigating soil organisms and evaluating spatial structure have improved our understanding of the spatial dynamics of the soil microbial community. Identifying the scale at which microbial community function and interact in forest soils is essential to designing sampling schemes that will allow us to adequately evaluate the complex relationships between the microbial community and vegetation. Geostatistical tools useful for evaluating these relationships include tools that allow researchers to identify the extent to which the data are spatially structured and allow for the creation of maps for linking organisms and ecosystem characteristics that might exist at different scales. Research on the microbial community in forest soils using these and other scaling techniques has demonstrated that microbial communities both are patterned by and influence the spatial dynamics of the vegetation in their environment at scales that range from centimeter to stand size. Microbes are key to nutrient cycling and microbial community dynamics respond to the vegetation in their immediate vicinity in ways that reflect both the specific identity of the microbe and plant and the spatially patterning of the processes. The mechanisms that underlie these tight relationships of pattern and function reflect the dependence of autotrophs on decomposers and mutualists for nutrient acquisition and the long evolutionary history of these organisms. Improved understanding of the complex spatial relationships between the microbial community and vegetation will improve our ability to provide management guidelines that will allow managers to protect our forest resources.

Keywords: forest soils, bacteria, fungi, microorganism, ecosystem function, community structure

1. INTRODUCTION

The distribution and abundance of organisms is a fundamental question addressed in the field of ecology (Clements, 1936; Gleason, 1926). Soil ecologists have become increasingly interested in evaluating the spatial distribution of organisms and properties in soils as evaluating impact of soil communities on ecosystem processes requires an understanding of small-scale mechanisms and dynamics. Soil organisms represent a highly diverse community (Lawton et al., 1996) that can have tremendous control on ecosystem function and structure (Wall and Moore, 1999). These organisms are of great concern because they number in the hundreds of billions in soil, are exceedingly difficult to culture or separate from soils for study, and the majority have not been taxonomically classified. Determining the spatial organization of soil organisms and the influence of this organization on ecosystem properties is essential for hypothesis testing and evaluating the large-scale impacts of disturbances to ecosystems.

Aside from a few early publications (Zinke, 1962; Zinke and Crocker, 1962; Mitchell, 1978; Lussenhop and Wicklow, 1984), the vast majority of research into the spatial structure of forest-soil organisms has occurred in the last 10–15 years. Two main factors have lead to the increased interest in the spatial distribution of soil organisms. First, there has been tremendous advancement in techniques for sampling, quantifying, and characterizing microbial communities, including automated microscopy, biochemical characterization (i.e., PLFA, FAME), and DNA analysis. Second, geostatistical techniques (semi-variance analysis, kriging, mantel's analysis) have allowed a more precise analysis of spatial structure and description of the spatial distribution of soil biota (Jackson and Caldwell, 1993; Goovaerts, 1998; Rossi et al., 1992). As techniques have improved there has been increased awareness of the importance of small-scale processes on ecosystem dynamics.

It is well documented that soil organisms play a central role in controlling many ecosystem processes, including organic matter decomposition, detrital accumulation, nutrient mineralization, immobilization, and development of soil structure (Coleman et al., 2004). In respect to concerns over global climate change, there is great interest in resolving models of carbon (C) exchange into the atmosphere. Forest soils contain 80% of aboveground and 40% of belowground terrestrial carbon (Dixon et al., 1994) and afforestation is receiving increasing attention as a strategy to increase terrestrial C sequestration. Clearly, understanding the spatial structure of soil microbes will greatly enhance our understanding of natural and disturbed systems, and enhance any models of C dynamics.

Determining the appropriate scale at which to examine microbial community structure or function is difficulty. It is likely that the impact of any species or community is nested at several scales (Franklin and Mills, 2003). Typically, researchers will analyze the spatial distribution of a particular organism or community in relation to some process or factor of interest. For example, if examining the spatial distribution of organisms in relation to landscape topography, the scale of interest may be very large: 25–100 m. However, understanding spatial structure soil microbial communities in relation to soil structure (i.e., micro- and macroaggregates) might investigate a scale of <1 cm. Indeed, Nunan et al. (2002) found bacterial samples in the top layer of soil from an agricultural field to be structured at the micrometer–millimeter scale. Small-scale processes impact landscape scale dynamics and mechanistic understanding of microbial function can only be truly appreciated if evaluated from multiple scales.

This review will examine the spatial distribution of soil biota, particularly soil microbial communities, in temperate forest ecosystems. We will primarily investigate the relationship at the vegetation scale, assessing the relationship between soil biota and vegetation. We will examine:

- Methodology important for evaluating small-scale dynamics
- The degree to which the patterning and spatial structure of microbial communities is a consequence of vegetation
- Whether fine-scale spatial patterns in microbial communities influences relatively larger-scale patterns in vegetation
- Mechanisms that may explain the observed relationships between vegetation and the microbial community

2. METHODOLOGY

Spatial statistics have become a valuable tool for investigating questions of scale. There are a number of useful tools for evaluating the scale at which organisms or properties are spatially structured. Geostatistics is one of the more common tools for examining the spatial distribution of materials in soils. The field of geostatistics was developed in the mining industry in response to the need to develop unbiased estimation of resource reserves but was then adopted by the fields of soil science and ecology in the 1980s as the need for information of the spatial distribution of organisms and materials arose. Matheron (1963) provided the basis to the approach to geostatistics. His observation was that data collected from short intervals is more alike than data collected at longer distances.

There are many tools available within geostatistics for examining the extent to which data collected are spatially structured. Autocorrelation indices such as Morans I and Geary's C are useful to determine whether values at one point are independent of values at another location. If this is true, the data set being examined is not autocorrelated, or is not autocorrelated at the scale being examined. Autocorrelation indices are often used together. For example Moran's *I* (Moran, 1948, 1950) is used to measure correlation between neighboring sites, where Geary's *c* (Geary, 1954) is used to determine the difference between values or, more specifically, the degree of dissimilarity between two values. Autocorrelation measurements are also useful to determine the landcover distribution or overall landscape patchiness (e.g., Legendre and Fortin, 1989; Qi and Wu, 1996; Wu and Jelinski, 1995).

Another of the most important and basic statistics in geostatistics is the variogram. The variogram is a summary of the relationship between measured values and the physical distance between two points. The variogram is useful to determine the distance (range) at which the samples taken are related in a random fashion (i.e., the distance at which the variance level off, the sill). It is possible that the variogram, which is plotted on an x, y grid will not pass through the origin. This suggests that the measurements taken do not explain all of the spatial relationships within the item of interest and there is spatial structure within the variable measured at a scale smaller than that measured in the study.

Kriging is also a valuable geostatistical tool. Conceptually, data are plotted on a grid and a complete map of the surface of the area from which the data were collected by producing estimates of the areas not sampled can be produced using kriging variance or the minimized error variance through a least-squares regression procedure that allows weights to be calculated which minimize standard error. The patterns of distribution of the variable of interest can then be observed. Cokriging can also be used to examine the relationship of multiple variables across a site based on the data points and maps produced with the kriging procedure.

Fractal geometry is also making an appearance in studies of the spatial distribution of organisms and soil dynamics. Although, only really used since the early 1990s, the application of fractal analysis has provided insight to the spatial or temporal variability to biological and geological variables. The fractal dimension, D, (Mandelbrot, 1977) allows one to examine shapes and the degree to which shapes repeat across a landscape. This allows one to evaluate the extent to which patterning is similar across scales. Fractal analyis is exceedingly useful as it allows distributions across multiple scales to be quantitatively characterized. Monofractal approaches, which were used early on by soil scientists, assumed that soil spatial distribution could be described using a single fractal dimension (Kravchenko et al., 1999). Since then there has been a push towards using multifractal analyses that view the data set as a multifractal spectrum instead of a single value. The ability to view soils as a more complex set of interwoven, covarying entities that differ across multiple scales is invaluable for evaluating spatial dynamics in soils.

Tools such as those described above have allowed us to develop a more comprehensive understanding of soils as landscapes of great complexity that are spatially structured from the microscale to the kilometer scale. While all of the tools above are all useful and important for examining spatial scale one must be careful that the data collected meet the criteria of the tool used for analysis. These tools have been used successfully to evaluate small-scale structure allowing us to better link structure and function in communities and allowing us to design sampling schemes that are better suited to our needs as ecologists. Without the ability to acquire information about phenomenons that occur at exceedingly small scales one must oversample to decrease the variation relative to the mean to detect differences across treatments or sites. A sampling design that is constructed with an inherent understanding of the scale of the processes involved will allow more samples to be collected that are directed specifically at evaluating hypotheses rather than getting statistical power to detect differences.

Obviously, we have not included all tools available for examining the degree to which soil properties or entities are spatially structured. As in other fields of science, the need for new tools drives their development, and of course, this is the case for spatial analysis. The introduction of scale as an important concept in ecology has also resulted in a new expansion into statistical tools for evaluating scale.

3. PATTERNING AND SPATIAL STRUCTURE OF MICROBIAL COMMUNITIES AS A CONSEQUENCE OF VEGETATION

Vegetation has wide-reaching effects on ecosystem attributes (see reviews by Hobbie, 1992; Binkley, 1995; and Chapin, 2003). In temperate forest ecosystems, in particular, the influence of individual tree species is very strong (i.e., Zinke, 1962; Finzi et al., 1998a). Trees can influence the soils at their base through the processes of stemflow and throughfall where nutrients from the tree are transported to the base of the tree in nonrandom patterns as a result altered rainfall distribution through interception by leaves, stems, and bole. The type and amounts of materials translocated to the base of the tree differs in terms of overall nutrient content. Trees also influence soils and soil nutrient dynamics, thorough differences in root structure and chemical contributions to surrounding soils. Smaller herbaceous species also influence the environment in which they live. The amount of influence is often dependent on specific plant characteristics (i.e., loss of toxins such as glucosinolates from roots) or plant density (i.e., grassroot density has great impact on soil structure). Traits exhibited by various species can have long-term impacts on ecosystem functions.

The most widely studied relationship has been between plant species and nitrogen (N) cycling. Boerner and Koslowsky (1989) documented differences among dominant tree species (including *Acer saccharum, Fraxinus americana*, and *Fagus grandifolia*) by comparing soils collected beneath tree trunks of each species and between canopy areas. Samples collected near the bases of trees had significantly greater N-mineralization rates, and this difference was greatest for samples collected near the base of *A. saccharum*. The authors concluded that estimates of ecosystem N mineralization would be biased by 8–20% if the influence of individual trees was not explicitly included.

Recently, numerous studies have documented the influence of individual tree species on N cycling in temperate forest ecosystems, (Binkley and Valentine, 1991; Finzi et al., 1998a; Lovett et al., 2004; Dijkstra, 2003; Templer et al., 2005; Washburn and Arthur, 2003; Peterjohn et al., 1999; Priha and Smolander, 1999; Boerner and Brinkman, 2005). In particular, Lovett and Mitchell (2004) provided evidence that sugar maple (*A. saccharum*) has a significant effect on nitrogen cycling in eastern North American forests. There was a significant correlation between the relative abundance of *A. saccharum* and nitrification and N leaching from forested ecosystems.

In addition to N cycling, individual tree species have been shown to alter many other ecosystem properties, including pH and cations (Boerner and Koslowsky, 1989; Finzi et al., 1998b), soil C (Boerner and Koslowsky, 1989; Finzi et al., 1998a) Ca mineralization (Djikstra, 2003), organic acid production (Dijkstra et al., 2001), P availability (Boerner and Koslowsky, 1989), and nutrient uptake (Brandtberg et al., 2004; Templer and Dawson, 2004).

Differences among tree species also influence the composition and spatial distribution of microbial communities. Several authors have demonstrated differences in microbial community structure associated with individual tree species (Turner and Franz, 1985; Grayston and Campbell, 1996; Grayston et al., 1996; Grayston and Prescott, 2005; Bauhus et al., 1998; Templer et al., 2003; Gallardo et al., 2000; Leckie et al., 2004). In a series of experiments, researches in Finland established robust differences in ecosystem properties and microbial communities under different tree species. Field surveys (Priha and Smolander, 1997, 1999; Priha et al., 2001) of Scots pine, Norway spruce and silver birch established at low- and high-fertility sites showed different microbial community characteristics (i.e., microbial biomass N, PLFA profiles, etc.) in soils collected form adjacent stands of each species. In addition, there were differences among litter, organic soil, and mineral soil samples.

Priha et al. (1999) also examined the influence of individual tree species in pot studies comparing microbial community characteristics with different tree seedlings grown in a common soil. There were significant differences in microbial biomass C and N, and microbial community structure analyzed by PLFA among birch, spruce, and pine seedlings grown in an organic soil, but not a mineral soil after 1 year. Both the bulk soil in which the seedlings were grown and the rhizosphere soil associated with roots were significantly different from control soil (with no seedling). In addition, microbial parameters were significantly different among the individual seedlings (Priha et al., 1999).

The spatial structure of microbial communities has also been described and analyzed using geostatistics (see Table 9-1). Studies of the spatial distribution of soil microbial communities show that these are also influenced by individual tree species. Bacterial biomass, fungal/bacterial ratio and the relative concentration of PLFA 16:1ω5 had a range of spatial autocorrelation between 3–4 m (Pennanen et al., 1999). The authors attributed this range to the distribution of trees (Norway spruce).

Similarly, Saetre et al. (1999) found similar results studying microbial biomass and activity in a mixed spruce–birch (*P. abies–Beutula pubescens*) stand. Microbial respiration was significantly greater under birch trees compared to spruce trees and intercanopy areas. The range of autocorrelation for microbial respiration was approximately 4–5 m, corresponding to the zone of influence of *B pubescens* (Saetre et al., 1999). Microbial biomass (measured as the sum of PLFA) was signficanlty affected by both *P. abies* and *B. pubescens*; with significantly lower microbial biomass under *P. abies* and signifycantly greater microbial biomass under *B. pubescens*. The range of spatial autocorrelation for soil microbial biomass was 7–8 m, which the author attributed to the distance between *P. abies* and *B. pubescens* trees (Saetre et al., 1999).

Dress and Frey (2003) found similar patterns in an oak–hickory forest. The range of autocorrelation for mite abundance, collembola abundance, and fungal biomass was 2–8 m, with the majority of estimates from 3–5 m. The authors attributed the spatial distribution to the presence of dominant trees and major coarse-woody debris.

In addition to influencing the broad-scale patterns of microbial communities between dominant trees, vegetation can influence spatial patterns at smaller scales. Decker et al. (1999) compared microbial enzyme activity 1 m above and below dominant red oak trees in southern Ohio oak–hickory forests. Organic matter and acid phosphatase activity was significantly greater 1 m downslope while β-glucosidase activity was significantly greater 1 m upslope, with no difference in the activity of chitinase or phenol oxidase.

Table 9-1. Summary of studies utilizing geostatistics to characterize microbial communities in forested ecosystems.

Citation	Ecosystem Type	Soil/microbial property	Sampling range	Range of spatial autocorrelation
Pennanen et al. (1999)	Coniferous forest	Bacterial biomass, F:B ratio, PLFA 16:1ω5	10×30 m sampling plot	3–4 m
Pennanen et al. (1999)	Coniferous forest	Total microbial biomass, fungal biomass	10×30 m sampling plot	Up to 1 m
Saetre et al. (1999)	Mixed coniferous forest (spruce, birch)	Soil microbial biomass	14×22 grid; sampling plot	7–8 m
Saetre et al. (1999)	Mixed coniferous forest (spruce, birch)	Microbial respiration, ground vegetation	14×22 grid sampling plot	4–5 m
Stoyan et al. (2000)	Poplar (9-year-old saplings)	Soil respiration	1×2 m plot	6–11 cm
Morris (1999)	Oak-Hickory hardwood	Fungal biomass	0.5–2.0 m macroplots	0.36–0.76 m
Morris (1999)	Oak–hickory hardwood	Bacterial biomass	0.5–2.0 m macroplots	1.09–1.96 m
Morris (1999)	Oak–hickory hardwood	Fungal biomass	10×10 cm microplots	2.9–3.6 cm
Morris (1999)	Oak-hickory hardwood	Bacterial biomass	10×10 cm microplots	3.6–7.5 cm
Dress and Frey (2003)	Oak–hickory hardwood	Mite and collembola abundance; fungal biomass	10×5 m plot	2–8 m
Saetre and Bååth (2000)	Norway spruce–birch	PLFA	14×22 grid; 2×2 m samples and 1.2×1.2 m nested grid with 0.2×0.2 m samples	1–11 m

Similarly, Morris (1999) quantified the spatial variation of microbial biomass in 2.0×0.5 m macroplots and 10×10 cm nested microplots upslope and downslope of a single red oak tree. Overall, there was significantly greater OC downslope and significantly greater fungal biomass, soil pH, and soil moisture upslope of a single red oak tree but no difference in bacterial biomass. Semi-variance analysis showed significant microbial spatial structure at both the macro- and microplot scale. Spatial structure accounted for 33–56% of the total variance in fungal and bacterial biomass at the macroplot scale and 28–100% of the total variance of fungal and bacterial biomass at the microplot scale. The maximum distance at which fungal biomass, the fungal:bacterial ratio and organic carbon were spatially autocorrelated was approximately 25–35 cm upslope of the tree and 70–80 cm downslope of the tree. The range of spatial autocorrelation for bacterial biomass was larger, from 1–2 m, and did not show the same pattern with the slope. In these steeply sloping sites, the base of trees represents a barrier to the downslope movement of water and materials, potentially compressing the range at which biological and biochemical processes operate.

In one of the few other studies to examine microbial properties at this microscale, Stoyan et al. (2000) examined the spatial distribution of soil respiration within a 9-year-old poplar stand by sampling a 1×2 m plot. The range of spatial autocorrelation was 6.0 and 11.7 cm at two sampling dates. At this smaller scale, the primary mechanisms controlling the spatial structure of soil respiration was hypothesized to be root growth and mortality, and rhizodeposition patterns.

Although it is clear that single trees have a significant influence on the distribution of microbes, the mechanisms that cause variability associated with single trees has received less attention. In general, "traits" of individual trees has been hypothesized to control the effect individual trees have on ecosystem properties. Detritus quality (foliar root litter) has been hypothesized to control N availability through a positive feedback loop, where higher N availability leads to litter production with higher N concentration, lower defensive chemical concentration (i.e., tannins), faster decomposition, and finally greater N availability (Hendricks et al., 2000). Although foliar litter is subject to redistribution (Boerner and Kooser, 1989) that would limit the local differentiation, root litter is known to proliferate in nutrient rich patches (Jackson et al., 1990), has lower nutrient resorption (Nambiar, 1987), and thus could certainly cause differences in N dynamics among tree species. Rhizosphere deposition may also influence the may also explain the observed patterns of microbial communities and individual tree species (Hobbie, 1992).

Stemflow, rainwater that is intercepted by trees and moves along branches and the stem to be deposited at the base of a tree, is another potentially important mechanism to explain differences among trees. Stemflow that moves along trees becomes enriched with nutrients that either leach for leaves and stems, or is residual dry deposition (Gesper and Holowaychuk, 1970; Crozier and Boerner, 1986). Different tree species have been shown to have different amounts of stemflow based on bark and canopy characteristics (Gesper and Holowaychuk, 1970) and different nutrient concentrations in stemflow. These differences in stemflow characteristics have been correlated with differences in soil characteristics at the base of different trees (Gesper and Holowaychuk, 1970; Crozier and Boerner, 1986).

It is likely that none of these potential mechanisms is mutually exclusive. Likely, multiple mechanisms could work in concert to produce the observed patterns in the field. In addition, mechanisms may vary among different tree species, as proposed by Eviner (2004).

4. IMPACT OF FINE-SCALE PATTERNING IN MICROBIAL COMMUNITIES ON LARGER-SCALE PATTERNS IN VEGETATION

While there is certainly less data to support the contention that spatial patterning of microbial communities impacts plant diversity and growth in forest soils, evidence is accumulating that microbes have impact on the types and number of species in a given area and process rates in ecosystems (Bever, 1994; Bever, 2003; Fitter, 1977; Van der Heijden et al., 1998; Wardle et al., 2004; Klironomos, 2002). Host plant growth is altered in a host specific fashion by association with pathogens or mycorrhizal fungi thus a mechanism exists for belowground diversity to alter plant community dynamics. A greater understanding of the spatial dynamics of the microbial communities relationship to plant structures such as roots, leaves, and material as they hit the surface, will allow us to create the necessary bridges to develop mechanisms for microbial impact on plant growth and the degree to which microbial spatial patterns impact this growth.

Our greatest challenge in evaluating the impact of spatial patterning of microbes on plant dynamics is not the analysis of the data, as geostatistics have provided ample analytical tools for analyzing data sets collected. Our greatest challenge today is actually acquiring the samples necessary at a scale that is relevant to microbial growth and function. Scaling from microbial growth to ecosystem function is a challenge that must be successfully met before we can provide more broad evidence for the impact of microbes on structural and functional diversity in plant communities. Solid evidence for this link requires a mechanistic understanding at a scale that is smaller than has been available to researchers almost until the advent of molecular techniques. Analysis of characteristics of microbial communities has been hampered by the lack of information available from culture data, which in the past has been the most common microbial tool. The poor capture with this technique (<7% of the organisms in soil) allows the majority of organisms to avoid capture. Molecular techniques now allow for the identification of the genomes of the organisms in a given area and the genetic makeup of the organisms in a given soil.

However great the challenge, there is subtle evidence currently available that the spatial structure of the microbial community does alter community structure and function at higher scales. Some of the stronger evidence of these impacts comes from the mycorrhizal literature (Klironomos, 2002, 2003). Yet, there are additional pieces of evidence in the bacterial and saprophytic fungal literature that contribute to an evolving picture of the importance of microbes in ecosystem dynamics. While there are many wonderful studies that have focused on the impact of belowground systems on plant dynamics, we will limit our study to forested ecosystems.

Mycorrhizae, the mutualistic symbiosis between fungi and plants, have long been associated with the success of plant communities. There is a great deal of evidence that the development of this relationship allowed aquatic plants to transition to land plants (Brundrett, 2002). Water was not likely the challenge for plant transitioning to terrestrial systems, the access to phosphorus, which is immobile in soil systems, was the challenge. Access to phosphorus through a fungal partner allowed plants to emerge from aquatic systems. As such early plant communities were influenced highly by the availability of fungal partners.

Recovery of systems following disturbance has always provided insight into the mechanisms important to system productivity and stability. Research on areas disturbed by the 1980 eruption of Mt. St. Helens has provided information on the importance of mycorrhizae in plant establishment. Ten years following the eruption, the Pumice Plain, formerly a conifer forest and the area most disturbed by the volcano, had very sparse recovery by conifers. Allen et al. (1992) found poorly developed ectomycorrhizae (ECM) and few fruiting bodies. Hence, lack of ECM fungal recovery in this highly disturbed area retarded the recovery of vegetation. Across the Pumice Plain, ECM and arbuscular mycorrhizal (AM) fungi were returned to the site by animals (Allen et al., 2005). As animals graze patches of vegetation, fungi were distributed in ways that increased the patchiness of the landscape and promoted the development of resource patches. These resource patches have greater organic matter content than if fungi were randomly distributed improving the probabilities of survival for fungi and host.

Boerner et al. (1996) examined six vegetation patterns of different ages (stripmine, soybean field, 5-year-old field, 10-year-old field, prairie, woods) across a single landscape to determine ECM and AM mycorrhizal infectivity across land-use types. They found reduced AM and ECM infectivity in agricultural fields compared to areas with native vegetation. While AM fungi had lower infectivity in disturbed sites compared to native, ECM infectivity was low- or nonexistent and extremely patchy across the disturbed site. The high heterogeneity in ECM across site suggests recovery is a consequence of discrete return events possibly establishment by animal transport. As successional age increases, the ECM infectivity increased such that approaching 30 years, an ECM-dependent seed dropped anywhere in the matrix would encounter ECM. However AM species were more likely to be able to establish anywhere in the matrix and ECM species limited to patches and margins, possibly altering spatial patterns of trees early in succession.

There have been a large number of studies that have provided evidence that mycorrhizal species have different impacts on plant growth. One recent study in a broadleaf temperate forest demonstrated the impact of mycorrhizal inoculum on tree growth (Lovelock and Miller, 2002). While there were no detectable differences in inoculum potential across sites within a single forest, inoculum taken from different trees in that forest had different impacts on tree growth when seedlings were established with the inoculum in a greenhouse. Thus, differences in effectiveness resulting from spatial distribution of trees will likely alter success of seedlings establishing in specific areas.

There are also recent reports that introduced mycorrhizae may have detrimental impacts on ecosystem dynamics. In a study on pine plantations and soil carbon Chapela et al. (2001) examined ectomycorrhizal fungi in pine plantations. Their research detected poor diversity in pine plantations, having three mycorrhizal fungi present compared to 100 in native stands. They also detected soil carbon loss and exceedingly old carbon being used for fruitbody formation. Their conclusions were that introduced fungi may be altering carbon dynamics in forests leading to the decrease in soil carbon being detected in pine stands across the country. The degree to which introduced fungi remain in or spread to native forests requires investigation.

The spatial distribution of bacteria and saprophytic fungi in soil also has great potential to alter the diversity and distribution of plants. Negative feedback between plant and soil communities has been detected in a number of studies (Packer and Clay, 2000; Bever, 2003). Poor seedling recruitment was detected within 10 m of maternal tree for *Prunus serotina* (black cherry)

(Packer and Clay, 2000). Presence of the fungal pathogen, identified as a *Pythium* spp. in the soil associated with the maternal plant resulted in death of seedlings of the related black cherry seedlings but not other tree species. This pattern was detected only in the native range of the black cherry where the *Pythium* spp. was located but not when black cherry was found in European forests, where it was an introduced invasive species (Reinhart et al., 2003). These studies suggest that at the local scale the soil microbial community is spatially structured in response to the distribution of the aboveground biomass and the negative feedback of the soil microbial community in turn, impacts the spatial distribution of plant species.

There are a number of studies have begun to detect beneficial bacteria, root zone bacteria that impact plant growth (Cooper, 1959; Brown, 1974; Schippers et al., 1995; Domenech et al., 2004; Lucas Garcia et al., 2004). Beneficial bacteria have been reported to alter plant growth by increasing stem length, neck diameter, and shoot weight of the plants for which they are associated (Lucas Garcia et al., 2004). As the bacteria were primarily detected in the rhizosphere the degree to which they are spatially structured at other scales and the specific roles played in positive or negative feedback have not been fully evaluated. It is likely that they are spatially structured at small scales in forests and differences in the degree to which they are beneficial or host specific could have large impacts at the level of community dynamics. These bacteria are being used as inoculants in horticultural practices to improve establishment of nursery transplants so are introduced species across large areas. Interest in these organisms in the near future will likely increase with improvement in our ability to use molecular techniques to track them.

5. MECHANISMS

One of the basic tenets of the field of ecology is that organisms respond to their environments either on ecological or evolutionary time frames. Changes in the ecological or short time frames often reflect the genetic plasticity that is inherent to variation in trait expression. The evolutionary time frame represents long-term changes to genetic materials that are a consequence of adaptations to new environments. The short generation time of microbes allows very rapid changes in response to environments. The mechanism by which microbes and plants become spatially structured is a consequence of environmental conditions as discussed above but over longer evolutionary times likely results in evolutionary changes to both organisms. There is some evidence on this ancient historical signature in the literature.

Theoretical studies suggest the local adaptation of microbial communities at exceedingly small spatial scales. In a reciprocal transplant study, bacterial colonies isolated from an old-growth forest had greater fitness when grown in soil collected from their home site compared to soils collected at a distance (Belotte et al., 2003). Bacterial isolates grown in soil media collected 10 m from the home site showed only 60% of the maximum growth of the home site. Overall, bacterial isolates grown on its home soil had approximately 50% greater fitness (Belotte et al., 2003). The spatial extent of local adaptation was estimated to be 6.1 m (Belotte et al., 2003), a range that is similar to many estimates of spatial autocorrelation of soil microbes in forested ecosystems.

In another example of potential local adaptation, Hansen (1999) exposed litter bags of individual species (yellow birch, sugar maple, and red oak) and mixed litter types in forested ecosystems in North Carolina, USA. The species composition of the oribatid mite assemblage differed among each individual litter type (Hansen, 1999), and litter bags with red oak promoted an assemblage of endophagous mites that increased litter decay. This pattern is consistent with local adaptation of decomposer organisms for a particular type of detritus, similar to local adaptation.

The study of Packer and Clay (2000) found survival of black cherry seedlings under maternal plants to be quite low for related seedlings but survival for unrelated black cherry seedlings and species such as dogwood and tulip poplar to be unaffected.

These data shows a potential evolutionary mechanism to explain the patterns observed in forested ecosystems. Dominant trees typically have a life span of at least 200 years, with some species with typical life spans of 0much longer. Clearly, the life span of soil biota is much shorter compared to the life span of dominant trees. Even relatively long-lived soil biota (such as oribatid mites), have at least 1–2 generations per year, allowing a minimum of 200 generations to respond to the selective influence of the dominant trees. Local adaptation is a potential mechanism that could explain the observed patterns of spatial structure.

More important than simple demonstrations of local adaptation, there is growing sentiment that the evolution of mutualisms was dependent on microbes being spatially structured in soil (Denison et al., 2003). The problem of cheating or being a free rider, taking from the host without contributing to host maintenance, is often discussed as a reason that mutualisms should be inherently unstable. Spatial patterning in soil microbes is considered a consequence of poor dispersal skills that increases the likelihood that related organisms will be found close together. This spatial patterning may provide a mechanism, primarily kin selection, for the long-term stability and wide distribution of mutualisms across organisms.

Spatial patterning results in repeated contact of would-be mutualists that increases the probability of forming relationships.

In this way, spatial structure of soil microbial communities may also have been a driver in the development of symbiotic nitrogen fixation. Bever and Simms (2000) found the only way to explain the development of symbiotic nitrogen fixation was through benefits to relatives through kin selection. For some rhizobial strains to fix N, the organisms must be in a terminal non-reproductive state. As such, the evolutionary benefit of fixation cannot be passed on to offspring. The benefit therefore must be to free-living N fixers distributed in proximity to the root that are capable of benefiting from increased root exudation. In their study, mathematical models were used to examine the importance of spatial structure in the evolution of N fixation and determined that spatial structure was necessary for the evolution of nitrogen fixation.

6. CONCLUSIONS

Ultimately, the failure to understand microbial dynamics at the smallest scale will limit our ability to fully evaluate the impact of microbes on plant dynamics. The most obvious component of our story is that trees alter the spatial distribution of soil microorganisms.

However, the more difficult challenge but ultimately the most important scientifically, will be elucidating the impact of the spatial distribution of soil organisms on aboveground structure and function. Our goal should be to integrate from a single bacterium in the forest floor to its impact on global nutrient cycles. While some would argue that individual microbes impact global nutrient cycles in a redundant fashion that make knowledge of identity and specific role unnecessary, recent studies on the impact of microbes on aboveground community dynamics show this to be a weak argument that will keep us from fully understanding and protecting natural systems. Microbes are spatially distributed in forest soils. Continued investigation into the cause and effects of the patterns that develop will improve our ability to protect and manage forest systems.

REFERENCES

Allen, M. F., C. Crisafulli, C. F. Friese, and S. L. Jeakins, 1992, Re-formation of mycorrhizal symbioses on Mount St Helens, 1980–1990 – interactions of rodents and mycorrhizal fungi, *Mycol. Res.* **96**:447–453.

Bauhus, J., D. Pare, and L. Cote, 1998, Effects of tree species, stand age and soil type on soil microbial biomass and its activity in a southern boreal forest, *Soil Biol. Biochem.* **30**: 1077–1089.

Belotte, D., J. B. Curien, R. C. Maclean, and G. Bell, 2003, An experimental test of local adaptation in soil bacteria, *Evolution* **57**:27–36.

Bever, J. D., 1994, Feedback between plants and their soil communities in an old field community, *Ecology* **75**:1965–1977.

Bever, J. D., 2003, Soil community feedback and the coexistence of competitors: conceptual frameworks and empirical tests, *New Phytol.* **157**:465–473.

Bever, J. D., and E. L. Simms, 2000, Evolution of nitrogen fixation in spatially structured populations of *Rhizobium*, *Heredity* **85**:366–372.

Binkley, D., 1995, The influence of tree species on forest soils: processes and patterns, in: *Proceedings of the Trees and Soils Workshop*, 28 February–2 March, Canterbury, N. Z., D. J. Mead, and I. S. Cornforth, eds., Agronomy Society of New Zealand, Canterbury, New Zealand, Special Publication 10, pp. 1–34.

Binkley, D., and D. Valentine, 1991, Fifty-year biogeochemical effects of green ash, white pine, and Norway spruce in a replicated experiment, *For. Ecol. Manage.* **40**:13–25.

Boerner, R. E. J., and J. G. Kooser, 1989, Leaf litter redistribution among forest patches within an Allegheny Plateau watershed, *Landscape Ecol.* **2**:81–92.

Boerner, R. E. J., and S. D. Koslowsky, 1989, Microscale variations in nitrogen mineralization and nitrification in a beech-maple forest, *Soil Biol. Biochem.* **21**:795–801.

Boerner, R. E. J., B. G. DeMars, and P. N. Leicht, 1996, Spatial patterns of mycorrhizal infectiveness of soils long a successional chronosequence, *Mycorrhiza* **6**:79–90.

Boerner, R. E. J., J. A. Brinkman, and A. Smith, 2005, Seasonal variations in enzyme activity and organic carbon in soil of a burned and unburned hardwood forest, *Soil Biol. Biochem.* **37**:1419–1426.

Brandtberg, P. O., J. Bengtsson, and H. Lundkvist, 2004, Distributions of the capacity to take up nutrients by *Betula* spp. and *Picea abies* in mixed stands, *For. Ecol. Manage.* **198**:193–208.

Brown, M. E., 1974, Seed and root bacterization, *Annu. Rev. Phytopathol.* **12**:181–197.

Brundrett, M. C., 2002, Coevolution of roots and mycorrhizas of land plants, *New Phytol.* **154**:275–304.

Chapela, I. H., L. J. Osher, T. R. Horton, and M. R. Henn, 2001, Ectomycorrhizal fungi introduced with exotic pine plantations induce soil carbon depletion, *Soil Biol. Biochem.* **33**: 1733–1740.

Chapin, F. S., 2003, Effects of plant traits on ecosystem and regional processes: a conceptual framework for predicting the consequences of global change, *Ann. Bot.* **91**:455–463.

Clements, F. E., 1936, Nature and structure of the climax, *J. Ecol.* **24**:252–284.

Coleman, D. C., D. A. Crossley, and P. F. Hendrix, 2004, *Fundamentals of Soil Ecology*. Elsevier Academic Press, New York, NY.

Cooper, R., 1959, Bacterial fertilizers in the Soviet Union, *Soil Fertil.* **22**:327–333.

Crozier, C. R., and R. E. J. Boerner, 1986, Stemflow induced forest floor heterogeneity in a mixed mesophytic forest, *Bartonia* **52**:1–8.

Decker, K. L. M., R. E. J. Boerner, and S. J. Morris, 1999, Scale-dependent patterns of soil enzyme activity in a forested landscape, *Can. J. For. Res.* **29**:232–241.

Denison R .F., C. Bledsoe, M. Kahn, F. O'Gara, E. L. Simms, and L. S. Thomashow, 2003, Cooperation in the rhizosphere and the "free rider" problem, *Ecology* **84**:838–845.

Dijkstra, F. A., 2003, Calcium mineralization in the forest floor and surface soil beneath different tree species in the northeastern US, *For. Ecol. Manage.* **175**:185–194.

Dijkstra, F. A., C. Geibe, S. Holmstrom, U. S. Lundstrom, and N. van Breeman, 2001, The effect of organic acids on base cation leaching from the forest floor under six North American tree species, *Eur. J. Soil Sci.* **52**:205–214.

Dixon, R. K., S. Brown, R. A. Houghton, M. A. Solomon, M. C. Trexler, and J. Wisniewski, 1994, Carbon pools and flux of global forest ecosystems, *Science* **263**:185–190.

Domenech, J., B. Ramos-Solano, A. Probanza, J. A. Lucas-García, J. J. Colón, and F. J. Gutiérrez-Mañero, 2004, *Bacillus* spp. and *Pisolithus tinctorius* effects on *Quercus ilex* ssp. ballota: a study on tree growth, rhizosphere community structure and mycorrhizal infection, *For. Ecol. Manage.* **194**:293–303.

Eviner, V. T., 2004, Plant traits that influence ecosystem processes vary independently among species, *Ecology* **85**:2215–2229.

Finzi, A. C., N. van Breeman, and C. D. Canham, 1998a, Canopy tree-soil interactions within temperate forests: species effects on soil carbon and nitrogen, *Ecol. Appl.* **8**:440–446.

Finzi, A. C., C. D. Canham, and N. van Breeman, 1998b, Canopy tree-soil interactions within temperate forests: species effects on soil pH and cations, *Ecol. Appl.* **8**:447–454.

Fitter, A. H., 1977, Influence of mycorrhizal infection on competition for phosphorus and potassium by two grasses, *New Phytol.* **79**:119–125.

Franklin, R. B., and A. L. Mills, 2003, Multi-scale variation in spatial heterogeneity for microbial community structure in an eastern Virginia agricultural field, *FEMS Microbiol. Ecol.* **44**: 335–346.

Gallardo, A., J. J. Rodriguez-Saucedo, F. Covelo, and R. Fernandez-Ales, 2000, Soil nitrogen heterogeneity in a Dehesa ecosystem, *Plant Soil* **222**:71–82.

Geary, R. C., 1954, The contiguity ratio and statistical mapping, *Inc. Ststcian.* **5**:115–145.

Gesper, P. L., and N. Holowaychuk, 1971, Some affects of stem flow from forest canopy trees on chemical properties of soils, *Ecology* **52**:691–702.

Gleason, H. A., 1926, The individualistic concept of the plant association, *Bull. Torrey Bot. Club.* **53**:7–26.

Goovaerts, P., 1998, Geostatistical tools for characterizing the spatial variability of microbiological and physico-chemical soil properties, *Biol. Fertil. Soils.* **27**:315–334.

Grayston, S. J., and C. D. Campbell, 1996, Functional biodiversity of microbial communities in the rhizosphere of hybrid larch (*Larix eurolepis*) and Sitka spruce (*Picea stichensis*), *Tree Physiol.* **16**:1031–1038.

Grayston, S. J., and C. E. Prescott, 2005, Microbial communities in forest floors under four tree species in coastal British Columbia, *Soil Biol. Biochem.* **37**:1157–1167.

Grayston, S. J., D. Vaughan, and D. Jones, 1996, Rhizosphere carbon flow in trees, in compareson with annual plants: the importance of root exudation and its impact on microbial activity and nutrient availability, *Appl. Soil Ecol.* **5**:29–56.

Hansen, R. A., 1999, Red oak litter promotes a microarthropod functional group that accelerates its decomposition, *Plant Soil* **209**:37–45.

Hendricks, J. J., J. D. Aber, K. J. Nadelhoffer, and R. D. Hallett, 2000, Nitrogen controls on fine root substrate quality in temperate forest ecosystems, *Ecosystems* **3**:57–69.

Hobbie, S. E., 1992, Effects of plant species on nutrient cycling, *Trends Ecol. Evol.* **7**:336–339.

Jackson, R. B., and M. M. Caldwell, 1993, The scale of nutrient heterogeneity around individual plants and its quantification with geostatistics, *Ecology* **74**:612–614.

Jackson, R. B., J. H. Manwaring, and M. M. Caldwell, 1990, Rapid physiological adjustment of roots to localized soil enrichment, *Nature* **344**:58–60.

Klironomos, J. N., 2002, Feedback with soil biota contributes to plant rarity and invasiveness in communities, *Nature* **417**:67–70.

Klironomos, J. N., 2003, Variation in plant response to native and exotic arbuscular mycorrhizal fungi, *Ecology* **84**:2292–2301.

Kravchenko, A. N., C. W. Boast, and D. G. Bullock, 1999, Multifractal analysis of soil spatial variability, *Agron. J.* **91**:1033–1041.
Lawton J. H., D. E. Bignell, G. F. Bloemers, P. Eggleton, and M. E. Hodda, 1996, Carbon flux and diversity of nematodes and termites in Cameroon forest soils, *Biodivers. Conserv.* **5**:261–273.
Leckie, S. E., C. E. Prescott, and S. J. Grayston, 2004, Forest floor microbial community responce to tree species and fertilization of regenerating coniferous forests, *Can. J. For. Res.* **34**:1426–1435.
Legendre, P., and M. J. Fortin, 1989, Spatial pattern and ecological analysis, *Vegetatio*, **80**: 107–138.
Lovett, G. M., and M. J. Mitchell, 2004, Sugar maple and nitrogen cycling in the forests of eastern North America, *Front. Ecol. Env.* **2**:81–88.
Lovelock C. E., and R. Miller, 2002, Heterogeneity in inoculum potential and effectiveness of arbuscular mycorrhizal fungi, *Ecology*, **83**:823–832.
Lucas García, J. A., J. Domenech, C. Santamaría, M. Camacho, A. Daza, and F. J. Gutierrez Mañero, 2004, Growth of forest plants (pine and holm-oak) inoculated with rhizobacteria: relationship with microbial community structure and biological activity of its rhizosphere, *Environ. Exp. Bot.* **52**:239–251.
Lussenhop, J., and D. T. Wicklow, 1984, Changes in spatial distribution of fungal propagules associated with invertebrate activity in soil, *Soil Biol. Biochem.* **16**:601–604.
Mandelbrot, B. B., 1977, *Fractals: Form, Chance and Dimension.* Freeman Press, San Francisco, CA.
Matherton, G., 1963, Principles of geostatistics, *Econ. Geol.* **58**:1246–1266.
Mitchell, M. J., 1978, Vertical and horizontal distributions of oribatid mites (Acari: Cryptostigmata) in an aspen woodland soil, *Ecology* **59**:516–525.
Moran, P. A. P., 1948, The interpretation of statistical maps, *J. Roy. Statist. Soc. Ser. B* **10**: 243–251.
Moran, P. A. P., 1950, Notes on continuous stochastic phenomena, *Biometrika* **37**:17–23.
Morris, S. J., 1999, Spatial distribution of fungal and bacterial biomass in southern Ohio hardwood forest soils: fine scale variability and microscale patterns, *Soil Biol. Biochem.* **31**:1375–1386.
Morris, S. J., and R. E. J. Boerner, 1999, Spatial distribution of fungal and bacterial biomass in southern Ohio hardwood forest soils: scale dependency and landscape patterns, *Soil Biol. Biochem.* **31**:887–902.
Nambiar, E. K. S., 1987, Do nutrients retranslocate from fine roots? *Can. J. For. Res.* **17**:913–918.
Nunan, N., K. Wu, I. M. Young, J. W. Crawford, and K. Ritz, 2002, *In situ* spatial patterns of soil bacterial populations, mapped at multiple scales, in an arable soil, *Microb. Ecol.* **44**: 296–305.
Packer, A., and K. Clay, 2000, Soil pathogens and spatial patterns of seedling mortality in a temperate tree, *Nature* **404**:278–281.
Pennanen, T., J. Liski, E. Bååth, V. Kitunin, J. Uotila, C. J. Westman, and J. Fritze, 1999, Structure of the microbial communities in coniferous forest soils in relation to site fertility and stand development stage, *Microb. Ecol.* **38**:168–179.
Peterjohn, W. T., C. J. Foster, M. J. Christ, and M. B. Adams, 1999, Patterns of nitrogen availability within a forested watershed exhibiting symptoms of nitrogen saturation, *For. Ecol. Manage.* **119**:247–257.
Priha, O., and A. Smolander, 1997, Microbial biomass and activity in soil and litter under *Pinus sylvestris*, *Picea abies* and *Betula pendula* at originally similar field afforestation sites, *Biol. Fertil. Soils* **24**:45–51.

Priha, O., and A. Smolander, 1999, Nitrogen transformations in soil under *Pinus sylvestris*, *Picea abies* and *Betula pendula* at two forest sites, *Soil Biol. Biochem.* **31**:965–977.

Priha, O., S. J. Grayston, R. Hiukka, T. Pennanen, and A. Smolander, 2001, Microbial community structure and characteristics of the organic matter in soil under *Pinus sylvestris*, *Picea abies* and *Betula pendula* at two forest sites, *Biol. Fertil. Soils* **33**:17–24.

Priha, O., S. J. Grayston, T. Pennanen, and A. Smolander, 1999, Microbial activities related to C and N cycling and microbial community structure in the rhizospheres of *Pinus sylvestris*, *Picea abies* and *Betula pendula* seedlings in an organic and mineral soil, *FEMS Microbiol. Ecol.* **30**:187–199.

Qi, Y., and J. Wu, 1996, Effects of changing spatial scales on analysis of landscape patterns using spatial autocorrelation indices, *Landscape Ecol.* **11**:39–49.

Reinhart, K. O., A. Packer, W. H. Van der Putten, and K. Clay, 2003, Plant-soil biota interactions and spatial distribution of black cherry in its native invasive ranges, *Ecol. Lett.* **6**: 1046–1050.

Rossi, R. E., D. J. Mulla, A. G. Journel, and E. H. Franz, 1992, Geostatistical tools for modeling and interpreting ecological spatial dependence, *Ecol. Monogr.* **62**:277–314.

Saetre, P., P. O. Brandtberg, H. Lundkvist, and J. Bengtsson, 1999, Soil organisms and carbon, nitrogen and phosphorus mineralization in Norway spruce and mixed Norway spruce-Birch stands, *Biol. Fertil. Soils* **28**:382–388.

Schippers, B., R. J. Scheffer, B. J. J. Lugtenberg, and P. J. Weisbeck, 1995, Biocoating of seeds with plant growth promoting rhizobacteria to improve plant establishment, *Outlook Agric.* **24**:179–185.

Stoyan, H., H. De-Polli, S. Bohm, G. P. Robertson, and E. A. Paul, 2000, Spatial heterogeneity of soil respiration and related properties at the plant scale, *Plant Soil* **222**:203–214.

Templer, P., S. Findlay, and G. Lovett, 2003, Soil microbial biomass and nitrogen transformations among five tree species of the Catskill Mountains, New York, USA, *Soil Biol. Biochem.* **35**:607–613.

Templer, P. H., and T. E. Dawson, 2004, Nitrogen uptake by four tree species of the Catskill Mountains, New York: implications for forest N dynamics, *Plant Soil* **262**:251–261.

Templer, P. H., G. M. Lovett, K. C. Weathers, S. E. Findlay, and T. E. Dawson, 2005, Influence of tree species on forest nitrogen retention in the Catskill Mountains, New York, USA, *Ecosystems* **8**:1–16.

Turner, D. P., and E. H. Franz, 1985, The influence of western hemlock and western red cedar on microbial numbers, nitrogen mineralization, and nitrification, *Plant Soil* **88**:259–267.

Van Der Heijden, M. G. A., J. N. Klironomos, M. Ursic, P. Moutoglis, R. Streitwolf-Engel, T. Boller, A. Wiemken, and I. R. Sanders, 1998, Mycorrhizal fungal diversity determines plant biodiversity, ecosystem variability and productivity, *Nature* **396**:69–72.

Wall, D. H., and J. C. Moore, 1999, Interactions underground – soil biodiversity, mutualisms and ecosystem processes, *Bioscience* **49**:109–117.

Wardle, D. A., R. D. Bardgett, J. N. Klironomos, H. Setälä, W. H. van der Putten, and D. H. Wall, 2004, Ecological linkages between aboveground and belowground biota, *Science*, **304**:1629–1633.

Washburn, C. S. M., and M .A. Arthur, 2003, Spatial variability in soil nutrient availability in an oak-pine forest: potential effects of tree species, *Can. J. For. Res.* **33**:2321–2330.

Wu, J., and D. E. Jelinski. 1995, Pattern and scale in ecology: the modifiable area unit problem, in: *Lectures in Modern Ecology*, B. Li, ed., Science Press, Beijing.

Zinke, P. J., 1962, The pattern of influence of individual forest trees on soil properties, *Ecology* **43**:130–133.

Zinke, P. J., and R. L. Crocker, 1962, The influence of giant Sequoia on soil properties, *For. Sci.* **8**:2–11.

Index

aquifer 16, 135, 136, 139–143, 145, 146, 149–154, 156, 158, 160–165, 169, 171
autocorrelation 20, 31–37, 43, 44, 46–48, 50, 51, 55, 95, 109, 119–121, 125, 192, 273, 314, 317–319, 324

bacteria 1, 2, 4, 13, 15, 17, 21–23, 39, 40, 51, 61, 64–68, 70–72, 74, 75, 78, 87–97, 99, 101–103, 109, 111–115, 118, 120, 122, 124, 125, 127, 136, 138, 140, 141, 143, 144, 148, 150, 151, 153–158, 168, 169, 188, 196, 204, 205, 207, 211, 214, 216–218, 220, 222–224, 226, 234, 240–243, 245, 247, 249–251, 253–255, 257, 264–266, 270, 272–274, 282, 283, 311, 322, 323
bacterial diversity 87, 89, 100–102
bacterial spatial distribution 61, 69, 88, 90
biological control 109–112
biological thin sections 61, 66, 67, 70, 71, 78
biotic interactions 179, 222, 254

community analysis 1, 11
community structure 10, 16, 17, 20, 23, 47, 48, 55, 63, 67, 74, 75, 77, 90, 138, 160, 165, 169, 192, 194, 197, 207, 222, 229, 235, 311, 313, 317, 321

distribution 1–4, 8, 11–21, 23, 32, 35, 39–42, 48, 50–52, 54, 61–75, 77, 78, 87–91, 93–98, 100, 102, 103, 109, 111, 113, 114, 120, 124, 135, 137, 139, 141, 142, 146, 149, 159–162, 164, 165, 169, 179, 181, 186–190, 192, 194–196, 203–207, 210–214, 222–225, 227, 229–239, 242, 244–262, 266–283, 311–317, 319, 321–325

ecosystem function 2, 72, 178, 188, 311, 312, 316, 320
environmental spatial heterogeneity 177

forest soils 16, 138, 186, 188, 311, 312, 320, 325
fungi 13, 23, 51, 52, 54, 64–66, 70, 109, 111, 112, 114, 115, 125, 126, 179, 199, 211, 320–322

geostatistics 31, 40–42, 49, 52, 91, 95, 109, 110, 121–123, 313, 314, 317, 318, 320

332 *Index*

interaction scale 1

kriging 31, 32, 37, 42, 49, 50, 272, 274–276, 312, 314

marine and freshwater ecosystems 201, 207, 209, 211, 240, 249, 253–259, 262, 264–266, 272
microbial communities 2–4, 8–11, 13, 14, 16–18, 20–23, 31, 32, 41, 42, 48, 54–56, 62, 63, 67, 75–77, 89, 101, 110, 113, 116, 118, 155, 160, 161, 163–165, 168, 171, 179, 203–208, 210, 211, 224, 228–234, 252, 253, 266, 273, 282, 311–313, 315–321, 323–330, 332–334
microbial ecology 1, 2, 4, 9–12, 21, 31–33, 36, 39, 40, 48, 50–52, 54, 56, 75, 87, 88, 91, 99, 102, 110, 112
microhabitat 2, 13, 15, 61, 62, 64, 74–76, 78, 87–96, 99–101, 204, 212, 244
microorganisms 1–4, 8, 9, 12–15, 18, 20, 23, 32, 33, 40, 54, 62, 65, 73, 75, 77, 89, 91, 93, 113, 116, 135, 136, 145, 146, 148, 149, 154, 155, 157–159, 163, 165–167, 170, 203–208, 210, 222, 230, 232, 251–253, 257–260, 265, 270–272, 274, 283, 311, 325
micro-sampling 61, 64, 68
microscale 1, 4, 12–15, 21, 39, 40, 61–75, 67–69, 74–77, 79, 87–90, 94, 96, 99–101, 215, 250, 255, 256, 266, 315, 319
multi-scale 1, 12, 17, 120, 211, 278

nutrient cycling 2, 19, 21, 179, 190, 219, 311, 325
nutrient transport 179, 197

plankton communities 203, 206–208, 225, 278, 279

rhizosphere 65, 72, 109–111, 113, 114, 118–120, 125, 317, 319, 323

scale 1–4, 7, 8, 12, 13, 15–23, 33, 34, 36, 38, 39, 41, 44, 45, 48, 54, 61–65, 68, 70, 71, 74–76, 78, 87–89, 91, 93, 95, 100, 101, 103, 110, 113, 119, 120, 122, 146–148, 153, 155, 156, 166–168, 179–183, 187, 191, 192, 194–197, 203–215, 217, 218, 220, 222–230, 232–242, 244, 245, 247, 249–253, 256–260, 262–266, 268, 270, 271, 273, 274, 278, 281, 282, 313–315, 317, 319–321, 323, 325
soil 8, 10, 13, 15, 17, 18, 20, 39, 53–55, 61–68, 71–78, 87–103, 109–111, 113–119, 126, 127, 136, 138, 140, 141, 154, 179, 183–197, 204, 311–325
soilspatial organization 87
spatial 1–2, 4, 8, 10–23, 31–55, 61–66, 71–73, 76–91, 93–103, 109–113, 115–128, 135, 136, 146, 148, 159, 160, 164, 166, 168, 179, 183, 186–189, 191–198, 203–220, 222–230, 236–239, 246, 247, 249, 251–266, 268–283, 311, 325
spatial autocorrelation 20, 31–37, 43, 44, 46, 47, 50, 95, 109, 192, 273, 317–319, 324
spatial distribution 1, 2, 4, 8, 12, 13, 15, 17, 18, 20, 21, 32, 40, 48, 52, 61, 62, 64, 65, 67–69, 71–73, 76, 78, 87–91, 93–95, 102, 103, 109, 111, 117, 119, 121, 135, 136, 159, 164, 166, 179, 189, 192, 196, 204–207, 211, 212, 225, 227, 231, 234, 235, 239, 246, 247, 251, 255, 259, 261, 262, 270, 272–274, 276, 278, 283, 311–317, 319, 322, 323, 325
spatial heterogeneity 11, 13, 14, 16, 21, 54, 62, 109, 112, 119, 179, 188, 189, 203–209, 211, 212, 222, 228, 248, 251, 257, 266, 270, 272, 281, 282
spatial statistics 35, 109–111, 117, 118, 121, 123, 124, 126, 128, 273, 313

Index 333

spatial structure 14, 15, 17–19, 31–38, 40, 41, 43, 46, 49–55, 62, 76, 77, 99–101, 120–122, 124, 126, 127, 181, 192–194, 197, 210, 211, 220, 222, 239, 247, 253, 254, 272, 273, 275, 277, 279–281, 311–315, 317, 319, 321, 324, 334
statistical analysis 20, 31, 33, 35, 50, 54, 109, 111, 117–119, 121, 122, 124–128, 156, 158, 160, 165, 168, 192, 273, 276, 281
subsurface 135–149, 155, 158–160, 162–170, 225, 226, 240, 242, 245, 246, 261, 264

vadose 13, 139, 142–146, 154, 160, 162, 166, 168